Halogenated Fire Suppressants

Richard G. Gann, EDITOR

Naval Research Laboratory

A symposium hosted by the

Southwest Research Institute,

San Antonio, Texas,

April 23-24, 1975.

ACS SYMPOSIUM SERIES **16**

AMERICAN CHEMICAL SOCIETY

WASHINGTON, D. C. 1975

Library of Congress CIP Data

Halogenated fire suppressants.
(ACS symposium series; 16 ISSN 0097-6156)

Includes bibliographical references and index.

1. Fire extinguishing agents—Congresses. 2. Organohalogen compounds—Congresses. I. Gann, Richard G., 1944- II. Series: American Chemical Society. ACS symposium series; 16.

TP266.H34 628.9'254 75-25638
ISBN 0-8412-0297-4 ACSMC8 16 1-453 (1975)

PRINTED IN THE UNITED STATES OF AMERICA

ACS Symposium Series

Robert F. Gould, *Series Editor*

FOREWORD

The ACS SYMPOSIUM SERIES was founded in 1974 to provide a medium for publishing symposia quickly in book form. The format of the SERIES parallels that of its predecessor, ADVANCES IN CHEMISTRY SERIES, except that in order to save time the papers are not typeset but are reproduced as they are submitted by the authors in camera-ready form. As a further means of saving time, the papers are not edited or reviewed except by the symposium chairman, who becomes editor of the book. Papers published in the ACS SYMPOSIUM SERIES are original contributions not published elsewhere in whole or major part and include reports of research as well as reviews since symposia may embrace both types of presentation.

CONTENTS

PREFACE

"In many areas of the fire problem, proposed solutions rest on limited experience, shaky assumptions, and guesswork." (*1*)

"More research is needed on extinguishing agents . . . , to improve the effectiveness of existing agents and *to investigate the chemical and physical mechanisms of new agents.*" (*1*)

Any of us who have been near a roaring fire or who have seen gutted buildings or who have known burn victims are aware of the need for rapid, effective fire suppression. In many cases, especially for large buildings, the use of fire trucks or sprinkler systems is unavoidable. Despite the foreknowledge that the water damage could match or exceed the direct fire damage, water is still the most deliverable, often the only available suppressant for a fully developed conflagration. However, in certain situations, particularly indoors, something less massive is desired. Perhaps people need to remain in the area to secure valuables, perhaps expensive equipment is present, or perhaps the fire is too small to warrant calling the fire department. Whatever the specific reason, the need exists for a conveniently local, non-destructive and physiologically tolerable fire suppressant.

An early alternative candidate to water was carbon tetrachloride, CCl_4. Although it is effective for flame quenching, we now know that CCl_4 is too toxic to use indoors. However, the idea of halogenated agents was sufficiently attractive that copious testing was performed on a wide variety of these compounds, commonly referred to as halons (*2*). Many were found effective; many others posed patently unacceptable health hazards. The three major halon agents in commercial use today evolved from this previous work. These are CF_3Br (halon 1301), CF_2ClBr (halon 1211), and $C_2F_4Br_2$ (halon 2402).

In 1972, in response to controversy concerning the safety and effectiveness of these compounds, the National Academy of Science's Committee on Fire Research and the National Research Council's Committee on Toxicology held a symposium to appraise the toxicology and usage of at least the three aforementioned principal halons (*3*). They concluded that the halons displayed variable degrees of effectiveness in fire suppression and that some potential physiological hazard exists in the agents themselves as well as in their fire-generated breakdown products.

Although thoroughly covering the two principal goals, the 1972 appraisal, by its definition, stopped short of considering what else was needed to better the existing situation. There are several complementary approaches to improving our knowledge in this direction. The one chosen for this symposium was to look at the basic processes occurring in inhibited flames and to deduce mechanistic elements for suppressing fires by halons. Fortified with this knowledge, we can perhaps better discuss their usage conditions, screen suggested compounds, and even provide guidelines for developing better agents. Thus, the authors of these papers, representing a wide range of disciplines, assembled to examine each other's work in flame inhibition, to look for mutual assistance, and to ascertain how far along we are in understanding the processes involved. The papers review prior classical and contemporary bulk flame inhibition studies, describe various studies of chemically and dynamically different flame–halon systems, review the current state of the pertinent kinetics, and attempt to gather, model, and interpret the previously presented data. Much of the oral discussion has been transcribed and included, as has the general discussion which took place at the conclusion of the symposium. Hopefully, this collaborative effort will isolate and highlight the significant information we have already evolved and point out research recommendations for each other and for new workers in the field.

Acknowledgments

In the process of organizing this symposium I have received help from several sources, and these deserve a large measure of credit for the success and the very existence of the conference. My thanks go to E. I. du Pont de Nemours & Co., Inc., and the National Bureau of Standards, especially to Charles Ford and Andrej Macek, respectively, for financial assistance. The officers and program chairmen of the Central States and Western States Section and the national office of the Combustion Institute helped to publicize and coordinate the symposium. In particular, I acknowledge the efforts of the Southwest Research Institute, especially David Black, William McLain, and W. D. Weatherford for invaluable assistance with the local arrangements and with the techniques of running a meeting. I am also grateful to the Naval Research Laboratory for allowing me the time to prepare both the conference and this volume and to Homer W. Carhart for his active encouragement throughout. In addition, I thank Bunny Hampton for her able help in preparing these proceedings for publication.

Literature Cited

1. "America Burning: The Report of the National Commission on Fire Prevention and Control," U. S. Government Printing Office, Washington, D. C. 20402 (1973).
2. Halon is an abbreviation devised by the Army Corps of Engineers to designate halogenated hydrocarbon. If the compound under discussion has the formula $C_aF_bCl_cBr_dI_e$, it is designated as halon abcde. Terminal zeros are dropped, allowing halon "names" to consist of fewer than five digits.
3. "An Appraisal of Halogenated Fire Extinguishing Agents," National Academy of Sciences, Washington, D. C., 1972.

RICHARD G. GANN

Naval Research Laboratory
Washington, D. C. 20375
June 1975

1

An Overview of Halon 1301 Systems

CHARLES L. FORD

E. I. du Pont de Nemours & Co., Wilmington, Del. 19898

A symposium to focus upon the mechanism by which halogenated extinguishants operate is commendable, and I am honored to be a participant. This paper does not speak directly to the mechanism of halogenated agents, or even of Halon 1301, but instead speaks around it. In the thirty-odd years since Halon 1301 was first synthesized, the Du Pont Company has assumed a major role in developing practical application data on this agent for the fire protection industry. Although virtually none of this effort has been aimed directly at mechanisms, there are nevertheless many relationships which have been observed and which must be explained by a valid mechanism. Further, perhaps an awareness of these relationships in light of the papers which follow will help to form the nucleus of a viable mechanism. These relationships include differences in extinguishing requirements between agents, between fuels, between methods of application, between conditions of application, and other variables which exist in a fire situation. This paper will therefore be of a historical nature, examining 28 series of tests conducted in the past. Because most of the work with which Du Pont has been associated has dealt with Halon 1301, most of the studies included here also deal with Halon 1301. However, I have also included studies on Halon 1211 and a few points of comparison with other agents. As the title implies, the discussion will be heavy on systems applications, but again not solely so. Three studies on portable extinguishers are reported. All in all, I believe this paper represents one of the most comprehensive compilations of prior fire test work on the "modern" halogenated agents which has been published to date. I offer it with the hope that it will indeed contribute to the formulation of an acceptable mechanism for these agents.

History. Halogenated compounds have been employed as fire suppressants throughout the 20th Century, although until recently their use was rather limited. It was recognized in the late 1800's that carbon tetrachloride was an effective fire extinguishant. In the early 1900's, when the electrolytic process for the production of caustic soda from brine provided large quantities of cheap chlorine, hand pump extinguishers containing carbon tetrachloride were introduced commercially(1). These extinguishers became very popular and continued in service up until the mid-1960's when their listings were discontinued by the testing laboratories. In the 1920's, carbon tetrachloride was applied experimentally in small aircraft systems to extinguish engine fires. These systems were later adopted in Europe, but were rejected by the U. S. Army Air Corps in favor of carbon dioxide systems in 1931.

In the late 1920's, methyl bromide was considered as a fire extinguishant and was found to be more effective than carbon tetrachloride, but also was considerably more toxic. Its use in the United States was rejected, largely based on its high toxicity, but in England about 1938, it became the standard agent for aircraft extinguishing systems. Some of these systems remain in use today. Methyl bromide was also widely used by the Germans in World War II, both for aircraft and for marine fire protection systems. A number of casualties in the German Navy were attributed to methyl bromide poisoning(2).

Beginning in the late 1930's, the Germans developed a third fire extinguishing agent, chlorobromomethane, somewhat less toxic than methyl bromide but still a highly effective extinguishing agent. It quickly replaced methyl bromide in both the Luftwaffe and the Kriegsmarine. Chlorobromomethane was evaluated by the U. S. Civil Aeronautics Administration in 1939 to 1941, and by the C-O-Two Company in 1948. Chlorobromomethane was later adopted by the U. S. Air Force for on-board fire extinguishing systems for aircraft power plants, for use in portable extinguishers aboard aircraft, and for airport ramp patrol vehicles(3). Many of these units are still in service.

About 1948, the U. S. Army initiated a research program to develop an extinguishing agent which had the high effectiveness of chlorobromomethane or methyl bromide but without the attendant high toxicity of these agents. Approximately 60 candidate agents were collected and evaluated both for extinguishing effectiveness and toxicity. Most of these compounds were halogenated hydrocarbons. Because the names of these compounds are rather long, the Army devised a numbering system known as the "halon" (contraction of halogenated hydrocarbon) nomenclature system. This system identifies the empirical formula of the compound with a five-digit number which represents, in order, the number of carbon, fluorine, chlorine, bromine, and iodine atoms in the molecule. Terminal zeroes are dropped so that the resulting number may have only two, three, or four digits. In the remainder of the paper, halogenated extinguishants are identified by their Halon numbers.

As a result of the Army studies, four agents were selected for further evaluations. These were: Halon 1301, bromotrifluoromethane, Halon 1211, bromochlorodifluoromethane, Halon 1202, dibromodifluoromethane, Halon 2402, dibromotetrafluoroethane. These further evaluations resulted eventually in the selection of Halon 1301 by the Army for a small 2-3/4 pound portable extinguisher for battle tanks and electronic vans, and by the Federal Aviation Administration for protecting aircraft engines in commercial transport aircraft. Halon 1211 was selected by the British for use in both military and commercial aircraft. Halon 1202 was selected by the U. S. Air Force in military aircraft protection. Recently, Halon 2402 has seen some limited use in southern and eastern Europe.

On both continents, halogenated agent systems were applied to protect the engines of many types of military land and water craft such as battle tanks, personnel carriers, trucks, hydrofoils, and hovercraft. Gradually, the systems were expanded to cover other areas of these vehicles.

In the United States, the use of Halon 1301 in commercial total flooding systems began in the early 1960's. Many of these early systems utilized carbon dioxide hardware and technology. In 1965, a project was begun with Factory Mutual Research Corporation to define the requirements for Halon 1301 in total flooding systems. In 1966, the National Fire Protection Association organized a technical committee on halogenated fire extinguishing systems to develop standards for systems using Halon 1301 and other halogenated agents. A tentative Standard NFPA 12A-T was adopted in 1968 and made a permanent Standard NFPA No. 12A in 1970(4). At the present time, there are perhaps 10,000 Halon 1301 systems in service in the world. Typical applications include computer rooms, magnetic tape storage vaults, electronic control rooms, storage areas for art work, books, and stamps, aerosol filling rooms, machinery spaces in ships, cargo areas in large transport aircraft, processing and storage areas for paints, solvents, and other flammable liquids, instrument trailers, and compressor stations.

In Europe, development of Halon 1211 closely paralled that of Halon 1301 in the States. However, there has been somewhat greater emphasis on the use of Halon 1211 in portable fire extinguishers than in the U. S. Halon 1211 total flooding systems, using suitable safeguards such as time delays, are employed for similar applications as previously mentioned for Halon 1301. NFPA Standard No. 12B on Halon 1211 systems was adopted by NFPA tentatively in 1971, and permanently in 1972(5). Some work has been initiated in Europe towards applying Halon 1211 to local application systems. Halon 1211 portables have recently been introduced into the United States and currently several models from several manufacturers have been listed or approved by the testing laboratories.

Applications for halogenated fire suppressants can be categorized as follows:

Portable extinguishers, in which a supply of agent is directed onto the burning surface by a person;

Total flooding systems, in which a supply of agent is arranged to discharge into an enclosure and fill it uniformly with a concentration of agent sufficient to extinguish the fire. Subclassifications of total flooding systems include:

Flame extinguishment, in which the agent is applied to an existing fire, usually a diffusion flame;

Inerting, in which a suppressing concentration of agent is developed within an enclosure to prevent ignition by an ignition source;

Explosion suppression, in which agent is rapidly injected into a developing flame front to suppress it before it has reached damaging proportions;

Local application systems, in which a supply of agent is arranged to discharge directly onto a burning surface.

Special systems designed to protect specific hazards such as aircraft engines, race cars, restaurant hood and ducts, etc.

Conceivably, all the halogenated agents could be used in any of the above applications, and most of them have been evaluated in several of them. We may, therefore, utilize these past evaluations to provide comparisons between agents, between method of application, and between various fuels to perhaps shed some light on mechanisms of these agents. The primary objective of this paper is to examine some of these relationships. We shall begin by examining the evaluations on portable extinguishers.

Portable Extinguisher Evaluations

Some factors which are known to influence the performance of portable extinguishers are:

1. Extinguishing agent
2. Fuel and fuel temperature
3. Fuel configuration
4. Discharge rate
5. Discharge pattern
6. Nozzle motion

Fuel configuration includes the shape of the fire pan, the amount of freeboard, and obstructions which are present. The discharge pattern is largely governed by the type of extinguishing agent and the design of the discharge nozzle. Nozzle motion is imparted by the operator and can be quite influential on the performance of the extinguisher.

Three evaluation programs for portable extinguishers are reviewed here:

U. S. Army Corps of Engineers - 1954

Ansul Company Portable Evaluations - 1955

Du Pont Portable Extinguisher Design - 1958-1973

U. S. Army Corps of Engineers(2). The project undertaken by the U. S. Army Corps of Engineers in 1947 had as its primary goal the development of a small portable fire extinguisher. Following an initial screening of agents by Purdue Research Foundation using an explosion burette technique(3), the Corps of Engineers conducted a series of practical extinguishment tests in which 12 halogenated agents were applied to two standardized fires. One fire was a flammable liquid contained in a 2 ft diameter steel tub filled with water to within 10-1/2 in of the rim. Two quarts of gasoline were poured on top of the water and a 20 sec preburn was allowed before the fire was attacked. A second fire was composed of 8 pounds of cotton waste spread over a 2 x 4 ft area and wetted with two quarts of gasoline. A 10 sec preburn time was allowed before attacking the fire. A 2-1/2 pound carbon dioxide extinguisher was used to apply the agents. Tests were run with the extinguishers pressurized with nitrogen to both 800 psi and to 400 psi total pressure. For each agent/fire/pressure level combination, 10 tests were made and the results averaged. From the average weight of each agent required to extinguish each fire, a weight effectiveness value was calculated based on Halon 1001 equal to 100. Results of these tests are shown in Table I for Halons 1201, 1301, 2402, and for carbon dioxide.

TABLE I
PORTABLE EXTINGUISHER EVALUATIONS
Army Corps of Engineers R&D Laboratories
1951

Agent	24 in Tub Fire: Weight to Extinguish oz	24 in Tub Fire: Weight Effectiveness %	2x4 ft Cotton Waste Fire: Weight to Extinguish oz	2x4 ft Cotton Waste Fire: Weight Effectiveness %
Halon 1001	8.0	100	20	100
Halon 1202	6.6	120	20	100
Halon 1301	7.6	105	15.6	126
Halon 2402	10.8	74	19.8	N.E.
CO_2	9.1	88	32	N.E.

Of the three halogenated candidates, Halon 1202 was most effective on the tub fire and Halon 1301 most effective on the cotton waste fire. Halon 1211 was not evaluated in this series of tests, having been eliminated as a result of the Purdue study (see below). Based upon this test program and upon toxicology studies conducted by the Army Chemical Center which showed Halon 1301 to be the least toxic of these candidates, Halon 1301 was

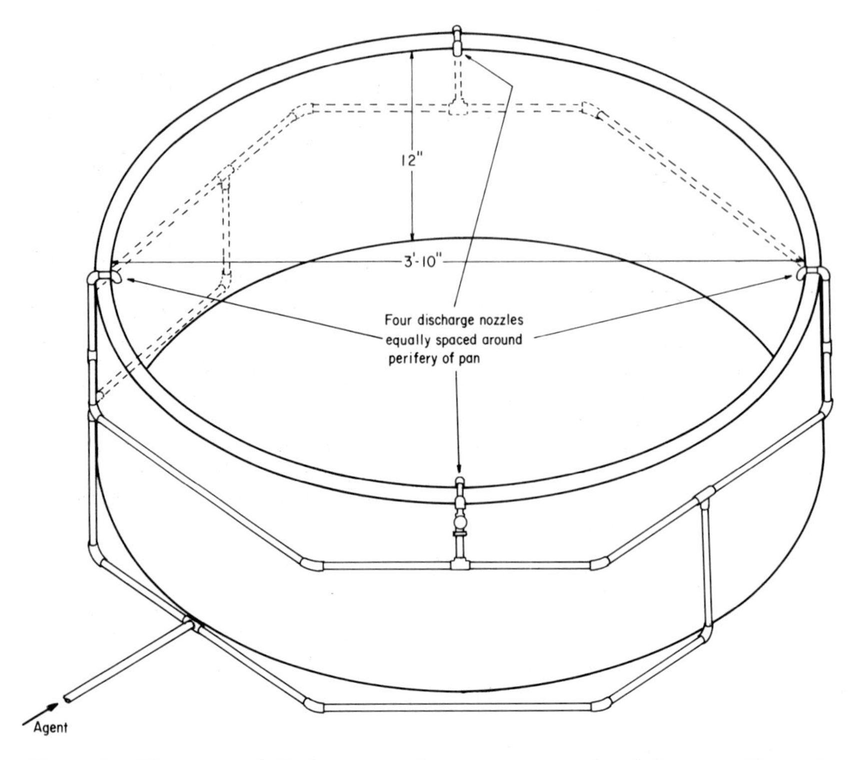

Figure 1. Fire pan and discharge nozzle arrangement. Ansul Co. portable evaluations—1955.

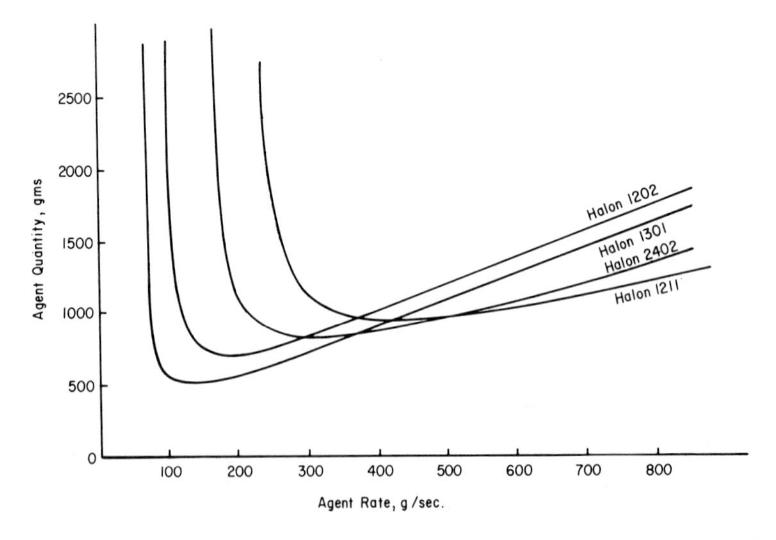

Figure 2. Performance curves for Halons 1301, 1211, 1202, and 2402. Ansul Co. portable evaluations—1955. Nozzle AX-11, P = *550 psi.*

selected by the Army as its sole halogenated fire extinguishing agent.

Ansul Company Portable Evaluations (6). In 1955, the Ansul Company and Du Pont jointly conducted a second series of agent evaluations aimed at portable extinguishers. For this series of tests, the human operator factor was eliminated by utilizing fixed nozzles. A sketch of the test arrangement is shown in Figure 1. Four fan spray nozzles were installed along the inner periphery of a circular tank of approximately 4 ft diameter. The nozzles were connected through a manifold to the agent supply cylinder. The tank was filled with water and a 2 in layer of unleaded gasoline to leave a 12 in freeboard in the tank. Eight halogenated agents and a sodium bicarbonate-base dry chemical agent were evaluated in these tests.

The evaluation was conducted in a manner described by Guise (7). Each agent was applied to the fire at various discharge rates, observing in each case the quantity of agent required for extinguishment. For each agent, several different discharge nozzles were used, and the nozzle giving optimum performance was selected. The data for each agent was then plotted to show quantity of agent for extinguishment vs discharge rate. Figure 2 shows curves obtained in this manner for the four primary halogenated agents, Halons 1301, 1211, 1202, and 2402. The minimum in each curve represents the minimum quantity of agent Q_m required for extinguishant and the optimum discharge rate R_o. These values are tabulated in Table II for the eight halogenated agents and for the sodium-based dry chemical. From these tests also, Halon 1301 was shown to be more effective than the other halogenated candidates and somewhat more effective than sodium bicarbonate-base dry chemical. Also, compounds containing bromine were all more effective than any of the agents containing no bromine. This data was utilized to reinforce the Army's previous selection of Halon 1301.

TABLE II
ANSUL PORTABLE EVALUATIONS
1955

Agent	Q_m g	R_o g/sec
Halon 1301	509	125
Halon 1202	670	183
Halon 2402	836	300
Halon 1011	884	232
Halon 1211	952	417
Halon 113 (R-11)	988	350
Halon 104	1064	458
Halon 122	1103	472

Sodium Bicarbonate Dry Chemical	813	218

Accompanying the main development thrust of halogenated agents on portable extinguishers in these early years, a number of tests were made by the Du Pont Company to determine the requirements of Halon 1301 in portable extinguishers having standardized Underwriters' Laboratories ratings. The UL rating system is based on a square fire pan 12 in deep containing 4 in of water and 2 in of normal heptane given a 1 minute preburn time. The number of the rating refers to the area in square feet of fuel surface which can be extinguished by an untrained operator. At the testing laboratories, the extinguisher must be shown capable of extinguishing a fire 2.5 times greater in surface area than the numerical UL size.

Du Pont Portable Extinguisher Design (8,9,10,11,12,13). Du Pont between the years 1958 and 1973 has determined Halon 1301 requirements for portable extinguishers ranging from size 1B (2.5 ft^2) to 40B (100 ft^2). These requirements are summarized in Table III.

TABLE III
DU PONT DATA - PORTABLE EXTINGUISHERS

UL Fire Size	Surface Area ft^2	Minimum Discharge Rate g/sec	Minimum Quantity for 10 sec Discharge lb
1B	2.5	25	0.61
2B	5	50	1.25
4B	10	100	2.5
(5B)	(12.5)	(125)	(3.2)
6B	15	150	3.75
20B	50	589	14.5
30B	75	760	18.5
40B	100	(1000)	(25)

For each rate shown, an optimum nozzle design is required. This design represents a balance between agent penetration and nozzle coverage of the fire area. In the case of Halon 1301, it must also exhibit a minimum amount of nozzle "drool" (liquid agent which is separated from the discharging stream within the nozzle and which drips from the end of the nozzle, thus representing loss of agent). The data show that over a 40-fold range of fuel surface area, a constant discharge rate per unit surface area of 10 g/sec/ft^2 (based on actual fuel area) is required.

Subsequent testing by others has shown that Halons 1211 and

2402, when used in optimized equipment, are much more effective than the foregoing analyses indicate. In addition, these agents have some limited effectiveness on standard Class A wood crib fires which Halon 1301 does not possess. Therefore, most of the current activity in halogenated agents for portable extinguishers is being conducted with these two agents, rather than with Halon 1301. Their lower volatility also gives them the capability to be projected for a considerable distance to the fire. Their discharge streams are also more visible, and hence more easily directed than that of Halon 1301.

Total Flooding Systems

The basic components of total flooding systems are shown in Figure 3, Schematic Diagram of Total Flooding Fire Suppression Systems. A supply of agent contained under pressure (or pressurized at the time of discharge) is connected by piping to one or more discharge nozzles located within the enclosure. Upon detection of a fire or a hazardous condition, a control valve on the agent supply container releases agent through the piping to one or more discharge nozzles located within the enclosure. The nozzles disperse the agent in such a manner as to form a uniform extinguishing (or inerting) concentration throughout the enclosure. Total flooding systems may be of two types: flame extinguishment and inerting (defined on page 4). For the purpose of this discussion, we will also define a third type of total flooding suppression system, i.e., for Class A fires.

Halogenated agents are particularly attractive for use in total flooding systems because of their cleanliness (lack of particulate residue), low toxicity, lack of vision obscuration, compactness and cost. Halon 1301, having the lowest toxicity of the halogenated agents currently in use, has found its greatest application in this area. Most Halon 1301 systems are for flame extinguishment or for Class A protection, although a few have been installed for inerting.

Flame Extinguishment Evaluations. Two test methods have been employed to determine the requirements of halogenated agents to achieve flame extinguishment in total flooding systems. The first is a "dynamic" test in which a flame, usually a diffusion flame, is surrounded by a flowing mixture of air and agent. Combustion and decomposition gases are not recirculated into the fire zone. "Static" test methods are also employed, particularly at large scale, wherein pure agent is dispersed into an enclosure containing the fire. This method is often desirable for demonstration since it closely approximates the conditions under which the system operates in practice. A number of investigations have been conducted using both test methods. In this discussion the results of six investigations of dynamic tests and four of static tests are examined:

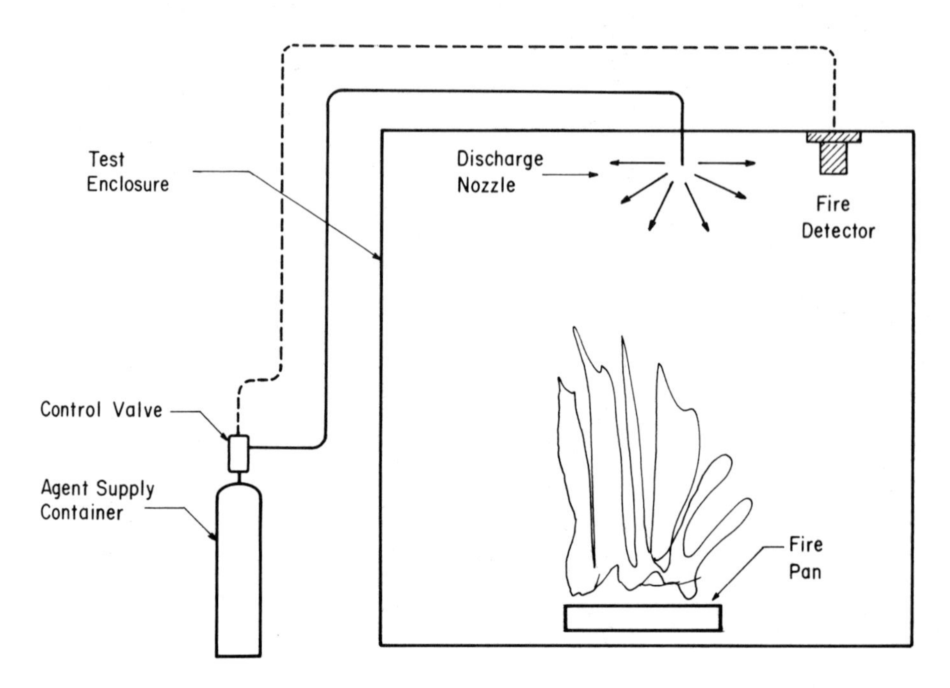

Figure 3. Schematic of total flooding fire suppression system

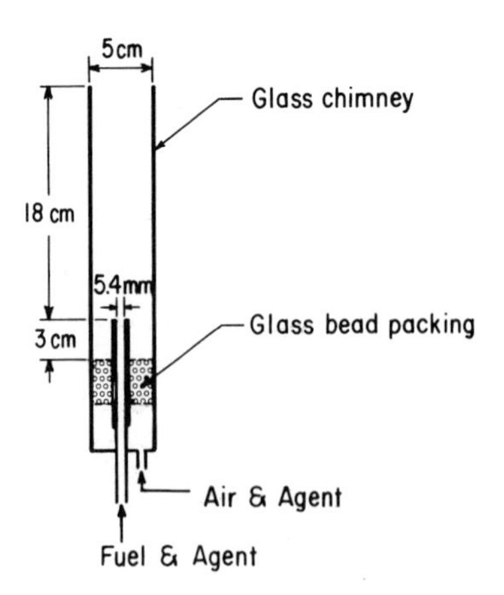

Figure 4. Creitz' diffusion flame extinction apparatus (1961)

Dynamic Tests
- Creitz - 1961
- ICI Hemispherical Burner - 1970
- ICI Cup Burner - 1973, 1974
- Factory Mutual Research Corporation Cup Burner - 1973
- Factory Mutual Research Corporation Vapor Burner - 1974
- Cardox Intermediate-Scale - 1974

Static Tests
- Factory Mutual Research Corporation Small-Scale - 1967
- Factory Mutual Research Corporation Intermediate and Large Scale - 1967 to 1968
- ICI Large-Scale - 1970
- Ansul Large-Scale - 1972

Dynamic Tests.

Creitz(14). In 1961, Creitz studied the effects on diffusion flames of four gaseous hydrocarbon fuels plus carbon monoxide and hydrogen, of adding Halons 1001 and 1301 and nitrogen to the air side and to the fuel side of the flame. His apparatus, shown in Figure 4, consisted of a 5.4 mm i.d. tube upon which a diffusion flame of the fuel was supported. The flame was surrounded by a 5 cm i.d. glass chimney through which a mixture of air plus extinguishing agent flowed upwards past the flame. He observed the minimum concentration of agent in either the air or in the fuel to produce extinguishment of the flame. In some cases, his point of extinguishment was considered to be the point at which the flame lifted off the 5.4 mm tube. His results are shown in Table IV.

TABLE IV
CREITZ FLAME EXTINCTION
1961

Flame supported on 5.4 mm i.d. glass tube
Surrounded by 5 cm i.d. glass chimney
Air velocity past flame = 16 cm/sec
Fuel, agent, and air at room temperature

Agent Concentration in Air or Fuel

	Added to Air			Added to Fuel		
	N_2	Halon 1001	Halon 1301	N_2	Halon 1001	Halon 1301
H_2	94.1	11.7	17.7	52.4	52.6	58.1
CH_4	83.1	2.5	1.5	51.0	22.9	28.1
C_2H_6	85.6	4.0	3.0	57.3	35.1	36.6
C_3H_8	83.7	3.1	2.7	58.3	37.6	34.0

C_4H_{10}	83.7	2.8	2.4	56.8	37.9	40.0
CO	90	7.2	0.8	42.8	-	19.9

The results illustrate several points. First, extinction concentrations of both Halon 1001 and Halon 1301, in terms of additional concentrations added to the system, are considerably lower than for nitrogen. The nitrogen values are somewhat deceptive. For example, to increase the concentration of nitrogen in air from 80% to 85.6%, requires the addition of 39% additional nitrogen by volume. This compares to 4% additional Halon 1011 or 3% Halon 1301 to air to extinguish ethane. Second, when added to the air side, Halon 1301 is more effective than Halon 1011 on the four hydrocarbon fuels and on carbon monoxide. Halon 1011 is more effective than Halon 1301, however, on the hydrogen fire. When added to fuel, the relationship appears to be reversed. Here Halon 1011 is generally more effective than Halon 1301. For the hydrogen fire, nitrogen appears to be the most effective extinguishant when added to the fuel. Third, the concentration of extinguishants in the fuel are considerably higher than for the agents added to the air side. Again, however, this is somewhat deceptive. Considering the relative ratios of fuel to air, a smaller quantity of agent would be required when added to the fuel than when added to the air. This would be especially true when air-to-fuel ratios greatly exceed stoichiometric. A fourth observation is that the lower concentration of Halon 1301 required to extinguish carbon monoxide is considerably lower than for the alkanes; with Halon 1011, a considerably greater concentration is required for CO than for the hydrocarbon fuels.

ICI Hemispherical Burner (15). Imperial Chemical Industries reported in 1970 a series of flame extinction tests with Halon 1211 on 18 liquid and gaseous fuels which were supported on a 28.5 mm diameter porous metal hemisphere. The apparatus is illustrated in Figure 5. Fuels were fed through jacketed tubing onto the hemisphere until the lower surface of the hemisphere was wetted with fuel. The fuel was ignited. For gaseous fuels, the fuel rate was adjusted to give the flame approximately 6 to 9 in long before addition of Halon 1211. Once the flame was established, Halon 1211 was gradually added to the air stream until extinction was achieved. The concentration of Halon 1211 at the extinguishment point was calculated from the respective air and Halon 1211 addition rates from calibrated rotameters. For these tests, the air flow rate was held constant at 45 l/min, which produced a velocity past the burner of approximately 13 cm/sec. Extinction concentrations obtained are shown in Table V.

The data show that for the alkanes butane through hexane, the Halon 1211 requirement for extinction is relatively constant at 3.5 to 3.7% by volume. Avtur (aviation turbine fuel) and

TABLE V
ICI HEMISPHERICAL BURNER
1970

1 1/8 in dia Sintered Stainless Steel Hemisphere
96 mm i.d. glass chimney
Air velocity past burner = 13.4 cm/sec
Fuel at room temperature

	Halon 1211 Concentration at Extinction % by Vol.
Methane	3.91
Ethane	5.02
Propane	4.37
n-Butane	3.7
i-Butane	3.5
n-Hexane	3.49
Cyclohexane	3.45
n-Heptane	3.45
Avtur	2.50
Kerosene	2.60
Ethylene	5.87
Methanol	7.70
Ethanol	4.50
Acetone	3.54
Carbon Disulfide	2.40
Hydrogen*	20.5

*Air velocity past hemisphere = 7.6 cm/sec

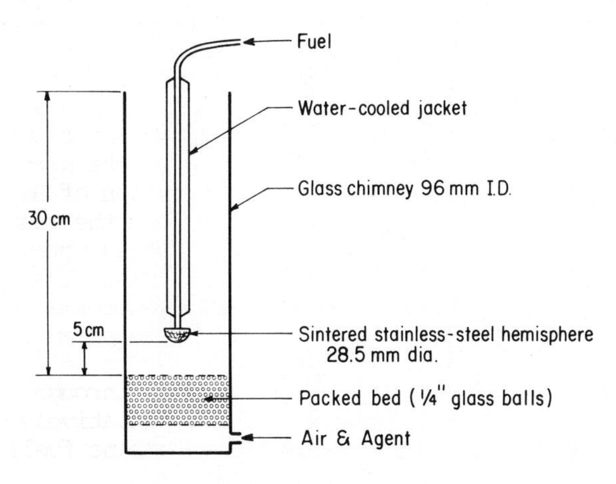

Figure 5. ICI hemispherical burner (1970)

TABLE VI
ICI CUP BURNER (1974)

28 mm o.d. Glass Cup in 85 mm i.d. Glass Chimney
Air Velocity Past Cup: 13 cm/sec

	Agent Conc. in Air at Extinction, %				
	Fuel at 20°C		Fuel at t°C		
Fuel	Halon 1211	Halon 1301	t°C	Halon 1211	Halon 1301
Methane	1.53	1.33			
Ethane	3.35	2.85			
Propane	3.5	3.1			
n-Butane	3.0	2.7			
i-Butane	3.0	2.5			
n-Pentane	3.7	3.3			
n-Hexane	3.7	3.3	60	4.0	
n-Heptane	3.8	3.5	90	4.4	4.3
2,2,5 Trimethyl Hexane	3.3	3.0			
Avtag	4.0	3.8			
Avgas	3.5	3.3	80	3.8	3.6
Avtur	3.7	3.4	170	4.0	
Avcat	3.5				
Gasoline (98 octane)	3.9	3.5	45	3.8	3.4
Diesel Oil			150	3.4	
Transformer Oil			200	2.8	
Cyclohexane	3.9	3.5	75	4.0	
Benzene	2.9	2.9	70	2.7	
Toluene	2.2	1.9	100	2.7	
Ethyl Benzene	3.1	3.0	130	3.7	
Mixed Xylenes	2.5	2.3	100	2.5	2.4
Ethylene	5.3	5.1			
Propylene	4.1	3.9			
1-Butene	3.8	3.7			
1,3-Butadiene	4.4	4.5	170	4.0	
Decalin	2.9	2.6			
Methanol	8.2	7.3	55	8.5	8.0
Ethanol	4.5	3.9	75	4.5	4.3
n-Propanol	4.3	3.8	90	4.3	3.7
i-Propanol	3.8	3.4	75	3.9	
n-Butanol	4.4	3.9	100	4.7	
i-Butanol	4.3	4.0	100	4.6	
sec-Butanol	3.8	3.4	95	4.0	
Amyl Alcohol	4.2	3.7	130	4.2	
n-Hexanol	4.5	3.9	150	4.7	
Benzyl Alcohol	2.9	3.1	190	4.3	

Ethylene Glycol	3.0	2.6	100	3.1	
			125	3.5	
			150	4.0	
			165	5.0	
			180	5.2	
Acetone	3.8	3.5	50	3.7	3.6
Acetyl Acetone	4.1	3.7	130	4.0	
Methyl Ethyl Ketone	3.9	3.6			
Ethyl Acetate	3.3	2.9	70	3.3	
Ethyl Aceto-acetate	3.6	3.2	170	3.9	
Methyl Acetate	3.3	2.7	50	3.4	2.9
Diethyl Ether	4.4	3.9			
Petroleum Ether	3.7	3.3			
Ethylene Gylcol	3.0	2.6	180	5.2	
Chlorobenzene	0.9	0.9	120	1.4	
Acetonitrile	3.0	2.6	75	30	
Acrylonitrile	4.7	4.7			
Epichlorohydrin	5.5	5.3	100	5.5	
Nitromethane	4.9	4.1	85	8.0	7.0
NN-Dimethyl Formamide	3.6	3.1	140	3.8	
Carbon Disulfide	1.6	2.6			
Isopropyl Nitrate	7.5		95	7.7	
Hydrogen	15.0				

kerosene flames required somewhat less agent for extinction. Ethylene, methanol, and hydrogen all required considerably higher concentrations. Because some back-diffusion of agent decomposition products onto the surface of the hemisphere occurred, and because the burner proved difficult to clean, the use of this apparatus was discontinued in favor of the ICI cup burner described next.

ICI Cup Burner(16). The hemispherical burner described above was modified by replacing the hemisphere with a streamlined 28.5 mm o.d. glass cup, as shown in Figure 6. The inner rim of the cup was ground to a sharp 45° edge so that the liquid level of fuel could be maintained at the very outside edge of the cup. In 1973, a series of tests were reported in which 58 gaseous and liquid fuels were examined with Halon 1211. In 1974, the cup was further modified to include a heating coil within the structure of the cup itself. For gaseous fuels the cup was replaced with an 8 mm o.d. glass tube having the same length as the cup. Air rates of 10 l/min to 50 l/min were examined, with a rate of 40 l/min becoming standard. Air velocity past the cup was approximately 13 cm/sec. For gaseous fuels, the linear velocity of the gas stream was matched to that of the air plus extinguishing agent. Many liquid fuels were examined both at room temperature (20°C) and at elevated temperatures near the boiling point of each fuel. Several fuels were evaluated with Halon 1301 as well as with Halon 1211. Results from these tests are shown in Table VI.

Generally speaking, Halon 1211 requirements determined in these tests are somewhat greater than for Halon 1301. For gaseous fuels with both agents, the alkanes of ethane through butane show relatively constant requirements of 3 to 3-1/2% Halon 1211 and 2-1/2 to 3% Halon 1301. Olefins showed higher requirements, as did most oxygenated fuels. The highest requirement was observed with hydrogen, requiring 15% Halon 1211 for extinction. Some fuels showed a higher requirement for both agents when heated than when the fuel was at room temperature; for others the difference was only marginal. This suggests that some types of fuels exhibit some temperature sensitivity in their extinction requirements, while others do not.

FMRC Cup Burner(17). Factory Mutual Research Corporation in 1973 attempted to reproduce the ICI cup burner apparatus with a series of measurements on 10 liquid fuels with Halons 1211 and 1301. A diagram of their apparatus is shown in Figure 7. An internal heater was placed in the cup so that the fuels could be heated. The range of volumetric air flows examined was similar to that examined by ICI, but due to a larger diameter chimney, the velocity of air past the cup was only about 60% of that obtained in the ICI apparatus at an equivalent volumetric rate. The results are shown in Table VII.

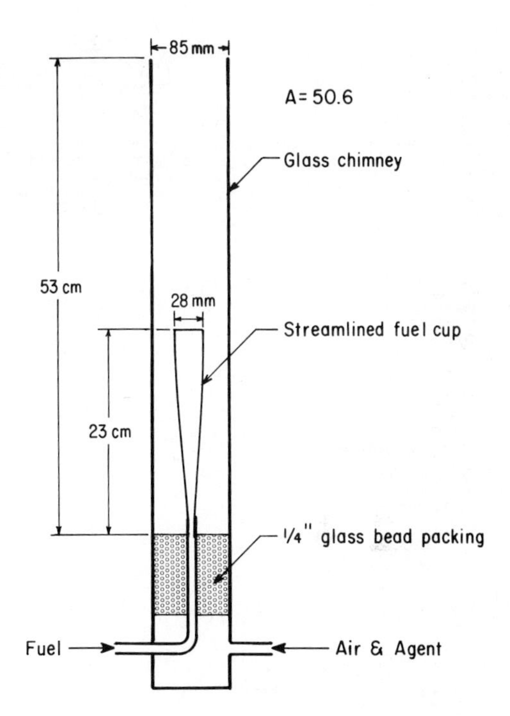

Figure 6. ICI cup burner (1973, 1974)

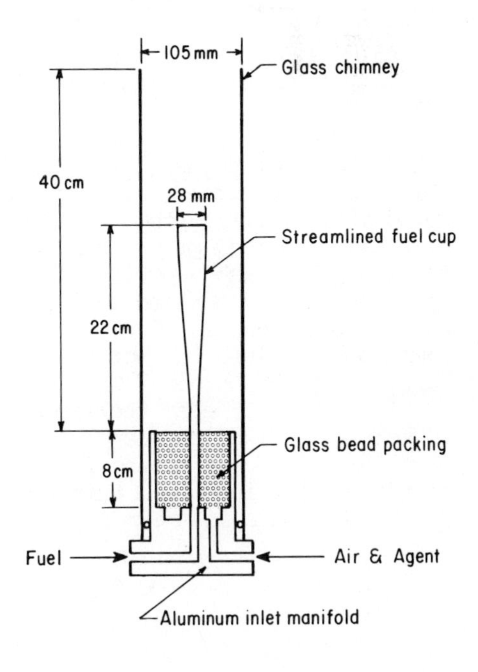

Figure 7. FMRC cup burner (1973)

TABLE VII
FMRC CUP BURNER
1973

	Peak Agent Concentration in Air at Extinction, %				
	Fuel at Room Temp. (24°C)		Fuel at Elevated Temp.		
Fuel	Halon 1211	Halon 1301	t,°C	Halon 1211	Halon 1301
Methanol	7.85	7.06			
Ethanol	4.16	3.66	75	3.43	
Acetone	3.62	3.29			
Diethyl Ether	4.17	3.67			
Ethylene Glycol			150	5.45	4.66
Nitromethane	4.74	4.11	85		7.40
n-Heptane	3.55	3.29			
Kerosene			60	3.63	3.48
Navy Distillate			90	3.75	3.38
			250	2.91	2.93
Xylene	1.96	2.08	100	2.29	

The FMRC room temperature data shows the same relationships as observed by ICI, but the extinction concentrations seem to be uniformly lower than ICI's by approximately 0.5% (absolute). In this series, ethanol exhibited a lower extinction concentration with Halon 1211 at elevated temperature than at room temperature. With Navy distillate, increasing the fuel temperature from 90°C to 250°C reduced the extinction concentrations of both Halon 1211 and Halon 1301. For nitromethane, a substantially higher concentration of Halon 1301 was required at 85°C than at room temperature.

FMRC Vapor Burner (18). Following the series of tests with the liquid cup burner reported above, Factory Mutual Research Corporation modified the apparatus to permit liquid fuels to be flash-vaporized prior to being fed to the burner. Fuels were fed into the flash vaporizer from a syringe driven by an Orion injector. The vaporized fuel then passed up the 38.5 mm burner tube and was ignited at the top. Air containing extinguishing agent was fed into the bottom of the 105 mm i.d. glass chimney surrounding the burner. The apparatus is shown diagrammatically in Figure 8. For each fuel/agent combination, the concentration to achieve extinction was plotted against the ratio of the air plus agent velocity to the fuel vapor velocity. The peak concentration obtained from this plot was then reported for each fuel. This data is shown in Table VIII below.

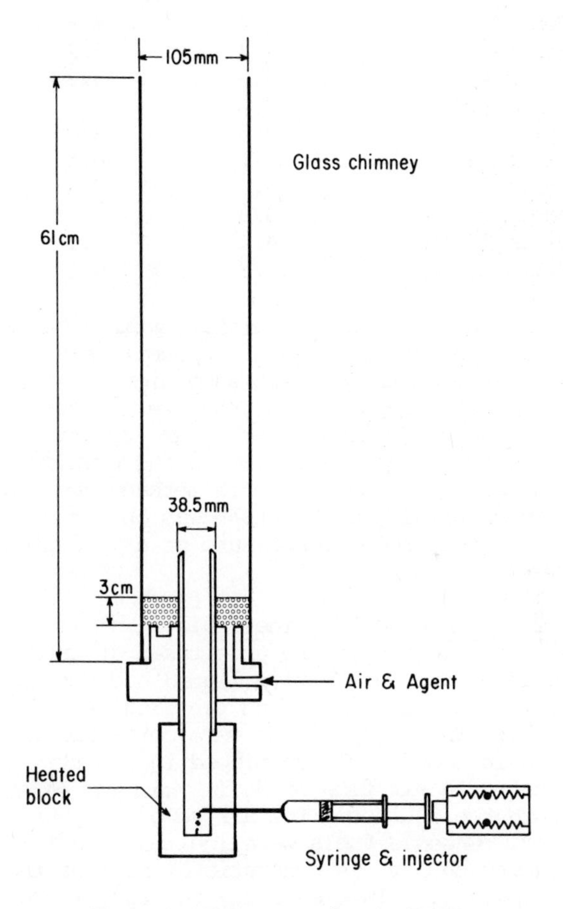

Figure 8. FMRC vapor burner (1974)

TABLE VIII
FMRC VAPOR BURNER
1974

	Peak Agent Concentration at Extinction, % (Vapor Temp. Approx. 3-4°C Above Boiling Point)	
	Halon 1301	Halon 1211
Pentane	4.3	4.6
Hexane	4.4	5.1
Heptane	4.0	5.2
Kerosene	2.7	-
Xylene	2.5	2.6
Methanol	9.1	8.6
Ethanol	4.6	5.2
Acetone	3.9	4.8
Diethyl Ether	4.4	5.1
Nitromethane	12.3	10.1

This data generally shows a higher extinction requirement for Halon 1211 than for Halon 1301. Exceptions are methanol and nitromethane. Many of these fuels show higher extinguishing concentrations in the vapor burner than were observed in the cup burner. In general, however, the same relationships between fuels was observed in this series as in the previous ones. The extinguishing concentrations for both methanol and for nitromethane are substantially higher than was obtained from cup burner data at either room temperature or at elevated temperatures.

Cardox 21.5 Cubic Foot Cabinet(19). A series of intermediate scale dynamic tests were undertaken at Cardox R&D Laboratories in 1974. Thirty-five liquid and ten gaseous fuels were evaluated with Halon 1301, and ten liquids and two gases with Halon 1211. The liquid fuels were contained in a 5 in diameter x 3/8 in deep steel pan placed in a 2 ft x 2 ft x 5.4 ft high cabinet. An air flow of 50 ft^3/min was established upwards through the cabinet. The air velocity past the fuel pan was 6.5 cm/sec. Gaseous fuels were ignited in a 1-1/4 in diameter Fisher laboratory burner. A schematic diagram of the apparatus is shown in Figure 9. In each test, the fuel was allowed to burn until an equilibrium temperature was established. Then the concentration of extinguishant in air was slowly increased until extinction was obtained. Data for 18 fuels is presented in Table IX.

In almost every case, a slightly lower extinguishing concentration was required for Halon 1301 than for Halon 1211. In the case of xylene, the two agents displayed equal effectiveness. For alkanes, alcohols, ketones, and ethers, the data shows good agreement with that from the laboratory cup burner with the fuels

TABLE IX
CARDOX INTERMEDIATE-SCALE TESTS (1974)

21.5 ft^3 Cabinet
50 ft^3/min Air Flow Rate

Fuel	Agent Concentration at Extinction, % Halon 1301	Halon 1211
Methane	1.3	1.5
Ethane	2.8	
Propane	2.9	3.2
n-Butane	2.3	
i-Pentane	3.0	
neo-Pentane	2.9	
n-Heptane	3.8	4.0
Petroleum Naphtha	3.4	
JP-4	3.3	
JP-5	3.5	
Diesel Fuel	3.8	
Navy Distillate	3.3	
Kerosene	3.7	
Stoddard Solvent	3.5	
Transformer Oil	3.5	
Cyclopentane	3.3	
Cyclohexane	3.5	4.1
Benzene	3.4	3.4
Toluene	2.7	
Xylene (mixed)	2.9	2.9
Acetylene	8.8	
Ethylene	4.6	
Propylene	3.9	
Methanol	7.3	
Ethanol	3.9	4.7
i-Propanol	3.3	4.2
Dimethyl Ether	4.6	
Diethyl Ether	3.7	4.2
Acetone	3.2	3.7
Methyl Ethyl Ketone	3.7	4.0
Ethyl Acetate	3.0	
Amyl Acetate	3.8	
Vinyl Acetate	4.3	
Methyl Methacrylate	5.6	
Ethylene Oxide	7.9	
Dioxane	6.0	
Acrylic Acid	4.4	
Ethylene Glycol	3.9	4.5
i-Propyl Amine	2.6	
Acetonitrile	2.8	
Acrylonitrile	4.6	

Tetrahydrofuran	4.2
Carbon Disulfide	2.8 (>10)*
Hydrogen	18

*With wire screen obstruction in fire pan.

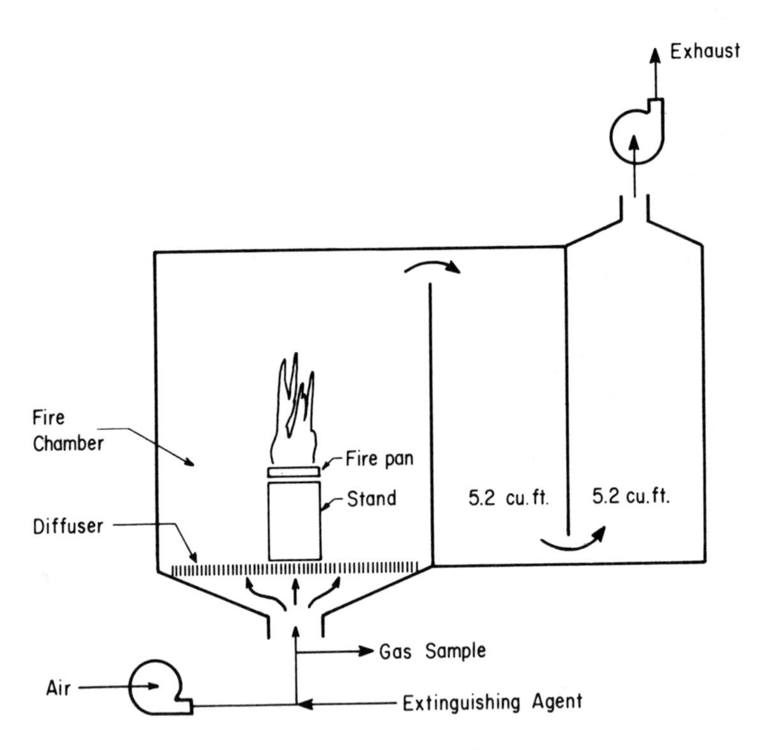

Figure 9. Schematic of Cardox apparatus

at room temperature. For nitromethane, reasonably good agreement is obtained with the laboratory cup burner with the fuel at elevated temperature. For Halon 1301, the extinction concentrations range from 1.3% for methane up to about 18% for hydrogen. Most of the alkanes require concentrations between 2.3 and 3.8% by volume. Oxygenated fuels and unsaturated hydrocarbon fuels require higher concentrations of Halon 1301 for extinction.

Static Flame Extinguishment Tests

FMRC Small-Scale Tests(20). In 1967, Factory Mutual Research Corporation conducted a series of tests using a 180 ft^3 glove box as a test enclosure. Fuels examined were CDA-19 (commercial denatured alcohol) and n-heptane using both Halon 1301 and carbon dioxide as extinguishing agents. The enclosure was equipped with recirculating fans as well as an exhaust fan which could maintain approximately 75 ft^3/min air flow through the cabinet. Four sizes of fire pans were utilized:

1. 2-1/8 in diameter x 1-1/8 in high copper
2. 3-1/8 in diameter x 1-1/2 in high copper
3. 6 x 6 x 2 in high steel
4. 12 x 12 x 2 in high steel

Agent concentrations of both Halon 1301 and carbon dioxide were monitored using a Ranarex specific gravity meter. A Bailey oxygen meter recorded oxygen concentrations before and during extinction. In one series of tests, the extinguishing agent was injected directly to the cabinet with the ventilating fan turned off. In a second series of tests, the ventilating fan remained in operation and the extinguishing agent was slowly added to the inlet air stream until extinction of the fire was obtained. The results from this series of tests are shown in Table X. A schematic diagram of the apparatus employed is shown in Figure 10.

TABLE X
FMRC TOTAL FLOODING TESTS
1967

180 ft^3 Glovebox
Fire Pan Sizes: 2-1/8 in diameter
3-1/8 in diameter
12 in x 12 in x 2 in

	Min. Halon 1301 Concentration for Extinction, %			
	Direct Injection		Added to Vent. Inlet	
	1301	CO_2	1301	CO_2
CDA-19	3.6	23	3.6	23
n-Heptane			3.3	19

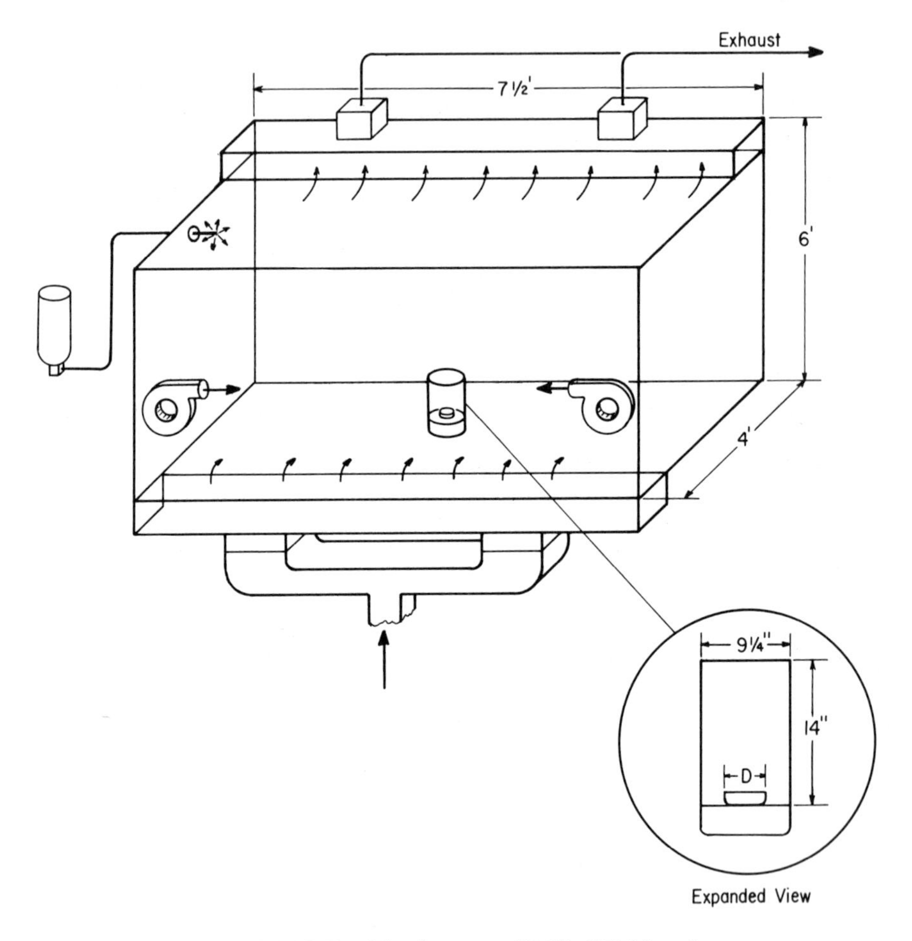

Figure 10. FMRC Total flooding tests (1967). 180-ft³ enclosure.

With the exception of the 12 in x 12 in fire pan, the extinguishing concentration was found to be dependent upon the agent employed and the fuel. Essentially no difference was observed between the direct injection tests and the agent slowly added to the air inlet. A somewhat lower concentration was observed for both Halon 1301 and carbon dioxide on n-heptane than on CDA-19.

FMRC Intermediate and Large Scale Tests (21). To demonstrate the validity of the determinations made in the glovebox tests above, Factory Mutual Research Corporation undertook in 1968 a series of intermediate- and large-scale tests with Halon 1301 on two fuels, CDA-19 and n-heptane. The tests were conducted in three stages. First, using a 1,000 ft^3 enclosure (10 x 10 x 10 ft) and two small fire pans, 2-1/8 in diameter and 3-1/8 in diameter copper pan, the fuel was ignited and the agent slowly bled into the enclosure over a 6 to 18 min time period. At extinction, the agent flow was shut off and the agent concentration in the enclosure was determined by calculation from the net weight of agent discharged. In a second series of tests in an 893 ft^3 enclosure using the same two fire pans plus a 12 x 12 in steel pan, the extinguishing agent was discharged rapidly into the enclosure within a 1-min period. The minimum weight of agent to extinguish the fire was determined and the agent concentration calculated from this weight. Finally, two tests were conducted in a 9,282 cu ft enclosure using a 38 in square fire pan (10 ft^2 surface), again injecting the agent into the enclosure rapidly. In the large tests, only n-heptane was used as fuel. Results from this series is shown in Table XI below.

TABLE XI
FRMC INTERMEDIATE AND LARGE SCALE FLAME EXTINGUISHMENT TESTS
1968

Minimum Halon 1301 Concentration in Enclosure at Extinction, %

Enclosure Size, ft^3	1000	893	9242
Fire Size	2-1/8 in dia 3-1/8 in dia	2-1/8 in dia 3-1/8 in dia 12 in x 12 in	38 in x 38 in
Agent Discharge Time:	6-18 min	<1 min	<1 min
CDA-19	2.35	3.3	
n-heptane	2.50	3.5	3.4

Tests in which the agent was slowly bled into the room required a lesser amount of agent than ones in which the agent was discharged quickly. This difference is attributed to oxygen depletion during the extinguishment period. With the rapid discharge tests, excellent agreement was obtained with the smaller scale work. In these tests, however, a somewhat lower extinguishing concentration was found for CDA-19 than for n-heptane, an observation which conflicts somewhat with the laboratory-scale tests reported previously. In general, however, the Halon 1301 requirements agreed well with the smaller-scale work.

ICI Total Flooding Tests (22). Imperial Chemical Industries also attempted to verify their laboratory-scale data with a series of large-scale total flooding tests. This series was conducted in a 2,500 ft^3 room (15 x 15 x 11 ft high) with a CENTRI 8B (approximately 30 in diameter) fire pan elevated 30 in above the floor. The minimum extinction concentrations of Halon 1211 for the five fuels was determined. Data from this series are shown in Table XII below.

TABLE XII
ICI LARGE-SCALE TESTS
1970

2,500 ft^3 Room
5 ft^2 Circular Fire Pan

	Halon 1211 Concentration to Produce Extinguishment % by Vol.
n-Heptane	3.3-3.4
Crude Naphtha	3.5-3.7
Methylated Spirit	3.7-3.8
Ethylene Glycol	3.3
Methanol	7.5-7.7

N-heptane, crude neptha, and methylated spirit, which were all similar fuels, exhibited similar Halon 1211 concentrations for extinction. Methanol showed a considerably higher concentration, with ethylene glycol slightly lower. Since only a minimum preburn time was permitted (5 sec), the bulk fuel temperature in all of these tests was probably near ambient. Nevertheless, the data agreed closely with that obtained from the laboratory cup burner with fuels at room temperature.

Ansul Large-Scale Tests (23). In 1972, the Ansul Company undertook a series of large-scale tests in a 9,856 ft^3 test enclosure with Halon 1301 to verify prior laboratory tests with this agent. Tests were run with seven liquid fuels in a 38 in square (10 ft^2) fire pan with fuels heated to near their boiling point prior to ignition. For the two gaseous fuels, methane and ethylene, an open 2 in jet burner was placed within a wood crib arrangement to act as a flame holder. The gas rate was adjusted to produce a flame height approximately equal to that of the flammable liquid fires (4 to 6 ft high). Halon 1301 was injected from standard Ansul Halon 1301 commercial hardware with the discharge time being approximately 10 sec for each test. For each fuel, a number of determinations were made with varying quantities of Halon 1301 injected into the enclosure. Samples of the atmosphere were withdrawn, analyzed by vapor-phase chromatography, and averaged to obtain the mean agent concentration in each test. For each fuel, two values of extinguishing concentration were reported: a lower limit, which was equal to the maximum concentration which did not extinguish the fire; and an upper limit, representing the minimum concentration of agent which did produce extinction. The results from this study are shown in Table XIII.

TABLE XIII
ANSUL LARGE-SCALE TESTS
1972

9,856 ft^3 Test Enclosure
38 in x 38 in x 11 in Deep Heated Fire Test Pan

	Limits of Halon 1301 Extinction Conc., %		
Fuel	Fuel Temp. °F	Lower	Upper
Acetone	132	-	3.00
Diethyl Ether	94	3.34	3.95
Ethylene Glycol	332	3.72	3.96
Nitromethane	176	5.00	5.82
Navy Distillate	440	-	3.28
Methyl Alcohol	138	6.33	6.36
Xylene	258	1.60	1.81
Methane	70	0.62	0.93
Ethylene	70	3.63	3.83

Almost all the extinguishing concentrations for these fuels either agree with, or are somewhat lower than, those obtained in the small-scale laboratory cup burner or in the Cardox intermediate-scale tests. Again, a similar relationship between fuels appears to hold. The gaseous fuels appear to require significantly lower concentrations of Halon 1301 for extinguishment in these tests, however, than was observed in smaller-scale work.

Considering all the flame extinguishment tests above, both static and dynamic, a rough comparison can be made to show the effect of scale. This is shown in Table XIV.

For the fuels included in this comparison, the ones examined at large-scale generally show lower Halon 1301 extinguishment requirements than for those at intermediate- or small-

TABLE XIV
EFFECT OF SCALE

	Minimum Halon 1301 Concentration for Extinction, %				
	Dynamic Tests		Static Tests		
	FMRC Cup Burner	Cardox 21.5 ft^3	FMRC 1000 ft^3	FMRC 10,000 ft^3	Ansul 10,000 ft^3
Methane		1.3			0.62-0.93
n-Heptane	3.30	3.8	3.5	3.4	
Navy Distillate	3.38	3.3			-3.28
Ethylene		4.6			3.63-3.83
Acetone	3.29	3.2			-3.00
Diethyl Ether	3.66	3.7			3.34-3.95
Ethylene Glycol	4.65	3.9			3.72-3.96
Methanol	7.05	7.3			6.33-6.36
Ethanol	3.65	3.9	3.3		
Nitro-methane	4.13	7.3			5.00-5.82
Xylene	2.08	2.9			1.60-1.81

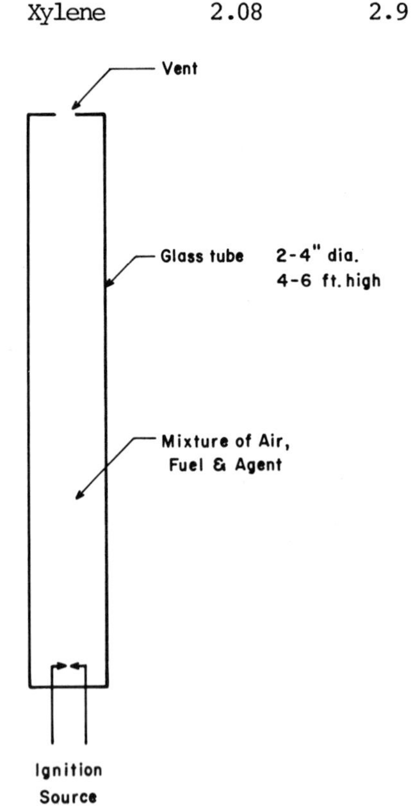

Figure 11. Schematic of explosion burette

scale. The tests made under dynamic conditions appear to give somewhat higher extinction concentrations than those made under static conditions. Presumably this is due to the lack of oxygen depletion in the dynamic test procedure. Thus, it appears that a laboratory-scale test in which various fuel temperatures are examined, or an intermediate-scale test of the Cardox type wherein the fuel temperature is allowed to reach equilibrium through the combustion process, provide adequate design concentrations for larger-scale application. A dynamic test procedure appears preferable to a static test procedure in this regard.

Inerting Evaluations. Inerting tests and applications have one common objective: to prevent the ignition of a combustible gas mixture through the addition of a suppressing agent. The process of testing a suppressant on a given fuel mixture is basically simple. A typical apparatus for inerting determinations is shown in Figure 11. It consists of a glass tube, often called an "explosion burette", to contain the test mixture, and a means of producing three-component test mixtures accurately and conveniently. For a laboratory apparatus, calibrated rotameters are commonly used for this purpose. Once the test mixture has been established in the tube, an ignition source, such as a high-voltage spark, is activated near the bottom of the tube. The amount of resulting flame propagation within the tube is observed, and the mixture is declared "flammable" or "nonflammable". A "flammable" mixture is one which permits flame propagation through the full length of the tube. However, most laboratories introduce a safety factor into the test by specifying a lesser amount of propagation, such as halfway up the tube, as the break point between flammable and nonflammable.

After a number of gas mixtures have been tested, the results are presented in the form of a "flammability peak" diagram, as shown in Figure 12. The curve is drawn to include within the envelope all the experimental points which were flammable. The Y-axis intercepts (no suppressant in the mixture) represent the upper and lower flammable limits of the fuel. As increasing concentrations of suppressant are added, the limits are narrowed. At the "flammability peak", or nose of the curve, all proportions of fuel and air are nonflammable. The flammability peak concentration of agent is the minimum suppressant concentration in the system which will render all proportions of fuel and air nonflammable.

Six series of inerting tests are reported here:

- Purdue Research Foundation - 1959
- Du Pont Mason Jar - 1971
- Du Pont Intermediate-Scale - 1971
- Du Pont Explosion Burette - 1972
- ICI Explosion Burette - 1973-1974
- Factory Mutual Research Corporation - 1973-1974
- Fenwal, Inc. Large-Scale - 1973-1974

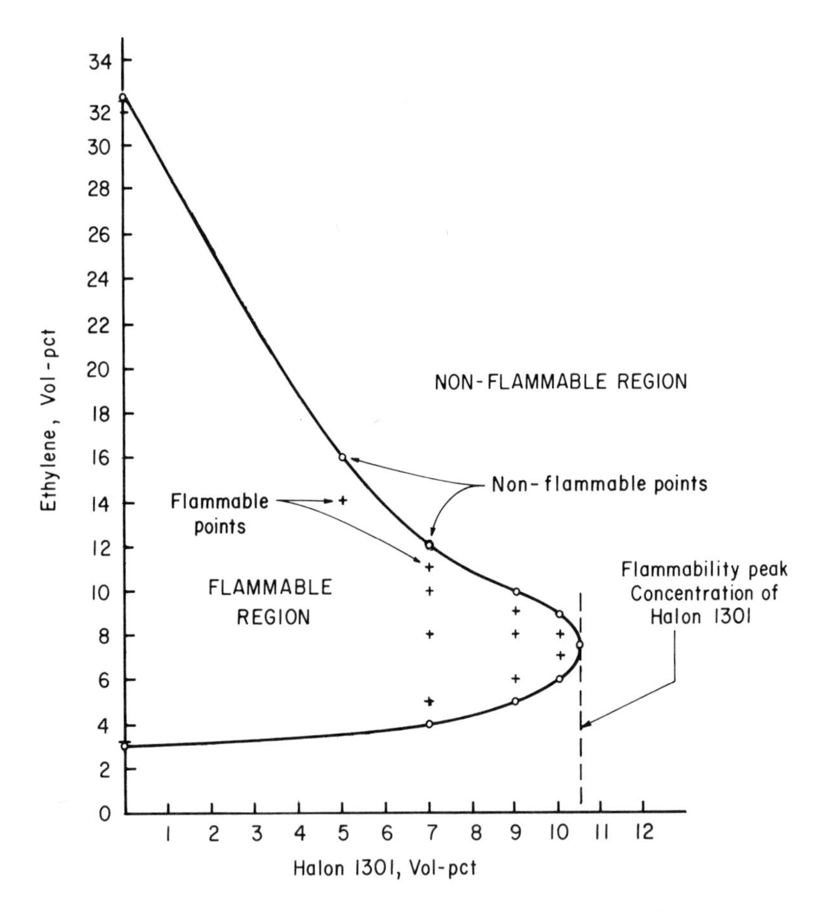

Figure 12. Typical flammability peak diagram. FMRC Explosion Burette (1973). Halon 1301/ethylene/air system. Top venting: 25°C and 1 atm.

TABLE XV
PURDUE RESEARCH FOUNDATION (1950)
FIRE SUPPRESSANT EVALUATIONS

51 mm i.d. x 120 cm Long Burette Spark Ignition
27°C at 400 mm Hg Total Pressure n-heptane Fuel

	Observed Flammability Peak Conc. of Agent % (vol)	Agent Weight Effectiveness %
Halon 1001	9.7	100
Halon 2001	6.2	136
Hydrogen Bromide	9.3	122
Halon 1202	4.2	107
Halon 3301	4.9	106
Halon 10001	6.1	106
Halon 20001	5.6	105
Halon 1002*	5.2	102
Halon 1301	6.1	101
Halon 2202	4.3	95
Halon 1011	7.6	94
Halon 1103	4.3	81
Halon 2501	6.1	76
Halon 2312	4.6	73
Halon 2402	4.9	72
Halon 131	12.3	71
Carbon dioxide	29.5	71
Boron trifluoride	20.5	66
Halon 1211	9.3	60
Silicon tetrachloride	9.9	55
Halon 104	11.5	52
Halon 122	14.9	51
Halon 14	26	40
Perfluorobutane	9.8	39
Perfluoromethylcyclohexane	7.5	35
Perfluoroheptane	7.5	32
Sulfur hexafluoride	20.5	31
Phosphorus trichloride	22.5	30

*Methylene bromide

Purdue Research Foundation (24). The U. S. Army Corps of Engineers employed Purdue Research Foundation in 1948 to scout approximately 60 candidate fire extinguishants for effectiveness using the Bureau of Mines explosion burette procedure. The tube employed was 51 mm i.d. x 120 cm high, with venting at the lower end of the tube. The spark source was a Model T Ford ignition coil of unspecified energy. Four test variables were explored: effect of agent, effect of fuel, effect of temperature, and effect of pressure. Data from these tests are shown in Tables XV, XVI, XVII, and XVIII, respectively.

TABLE XVI
PURDUE RESEARCH FOUNDATION
1950

Effect of Fuel

Combustion tube 51 mm i.d. x 120 cm high
400 mm Hg total pressure, room temperature
Spark ignition

	Concentration of Agent at Flammability Peak, % by volume		
	Halon 1301	Halon 1001	Halon 14
n-Heptane	6.1	9.7	26
Isopentane	6.3	8.4	20.4
Benzene	4.3	8.4	23.6
Ethyl alcohol	3.7	6.2	19.8
Diethyl ether	6.3	7.2	22.4
Acetone	5.3	7.3	18.7
Ethyl acetate	4.6	6.8	21.4

TABLE XVII
PURDUE RESEARCH FOUNDATION
1950

Effect of Temperature

Isobutane
400 mm Hg Total Pressure

	Concentration of Agent at Peak, %		
	-78°C	+27°C	+145°C
Halon 1301	3.25	4.7	7.3
Halon 1001	3.75	6.75	8.3
Halon 14	18.25	23.75	21.4

TABLE XVIII
PURDUE RESEARCH FOUNDATION
1950

Effect of Pressure

n-Heptane Fuel
27°C

Agent	Total System Pressure mm Hg	Agent Peak Flammability Concentration %
Halon 1001	200	5.8
	300	6.6
	400	9.7
	500	7.2
Halon 1301	200	4.3
	300	6.1
	400	6.1
	500	6.3
Halon 122	200	13.4
	300	14.8
	400	14.9
	500	13.4

Table XV shows the relative effectiveness of various halogenated extinguishants on suppressing n-heptane. From the observed flammability peak concentrations, the weight effectiveness of each agent was determined by multiplying by the molecular weight and expressing the result as a percent of the requirement for Halon 1001, which was taken to be 100. Table XV lists 28 candidate agents in order of decreasing weight effectiveness. From these tests, Halon 1202 was the most effective fluorinated agent, with Halon 1301 following closely. Active halogen atoms were ranked in increasing effectiveness as F<Cl<Br. While one atom of bromine gave a marked increase in effectiveness, a second atom of bromine gave only a marginal additional increase. Iodine was found approximately equal to bromine in effectiveness.

The effect of fuels is shown in Table XVI for three halogenated agents, Halons 1301, 1001, and 14. Although the same relationship appears to hold between agents for all fuels, there is no clear relationship between fuels. N-heptane carries the highest requirement for Halons 1001 and 14, while i-pentane and diethyl ether have the highest requirement for Halon 1301. Ethyl alcohol has the lowest requirements for Halons 1301 and 1001, whereas acetone has the lowest requirement for Halon 14. It would appear that a value determined with an aliphatic hydrocarbon, such as n-heptane or i-pentane, would be adequate to suppress the other fuels. However, this conclusion would be improper in light of later testing.

Table XVII shows the effect of temperature on three halogenated agents, Halons 1301, 1001, and 14, using isobutane fuel. Increasing the temperature from -78°C to 145°C increased the requirements of Halons 1301 and 1001 by 125%, while the requirement of Halon 14 was not substantially increased. In fact, the flammability peak concentration of Halon 14 was lower at 145°C than at room temperature. From these data, however, it seems fair to conclude that increasing temperature produces an increase in the flammability peak concentration of suppressant.

The effect of pressure is shown in Table XVIII for three halogenated suppressants, Halons 1001, 1301, and 122, using n-heptane as fuel. The total pressure was varied from 200 mm Hg to 500 mm Hg (0.26 to 0.66 atm). For all three agents, the lowest agent requirement was observed at the lowest pressure. At the three higher pressures, however, the results do not show a consistent trend. Halon 1001 shows a sharp maximum requirement at 400 mm Hg; Halon 122 also shows a maximum at 400 mm Hg, but a considerably less-pronounced one. For Halon 1301, the requirement seems to increase slightly as the pressure is increased from 300 to 500 mm Hg. Perhaps it would be fair to conclude from this series that decreasing the pressure below 0.4 atm tends to reduce the agent requirement, but pressure changes within the range of 0.4 to 0.66 atm do not greatly affect flammability peak concentrations of halogenated agents.

The Purdue data has often been critized for the large number of discrepancies which it contains. Wall-quenching effects caused by deposits from previous tests, inaccuracies in making gas mixtures, inadequate gas mixing, and venting procedure, have been found in subsequent test programs to be of critical importance. Yet, little recognition is given to these variables in the Purdue report. Finally, the almost ironclad interpretation of data points in drawing of flammability peak diagrams (see Figure 13) causes some skepticism. As a result, the absolute flammability peak concentrations reported by Purdue are often cursorily dismissed by current investigators. The relationships which this study determined are, however, useful.

Du Pont Mason Jar Test (25). To eliminate wall-quenching effects, Du Pont in 1966 replaced the long slender burette in the Bureau of Mines test with a standard one-quart Mason Jar. A kitchen match head was used for the ignition source, determined by calorimetry to produce a total energy of 176 J. The criterion for nonflammability was that no visible flame extension was permitted beyond that produced by the match head in air. A diagram of the test apparatus is shown in Figure 14. Test mixtures were made by evacuating the jar and admitting suppressant and fuel by partial pressure. The contents were mixed by swirling a chain within the jar for about 30 sec. The cap was

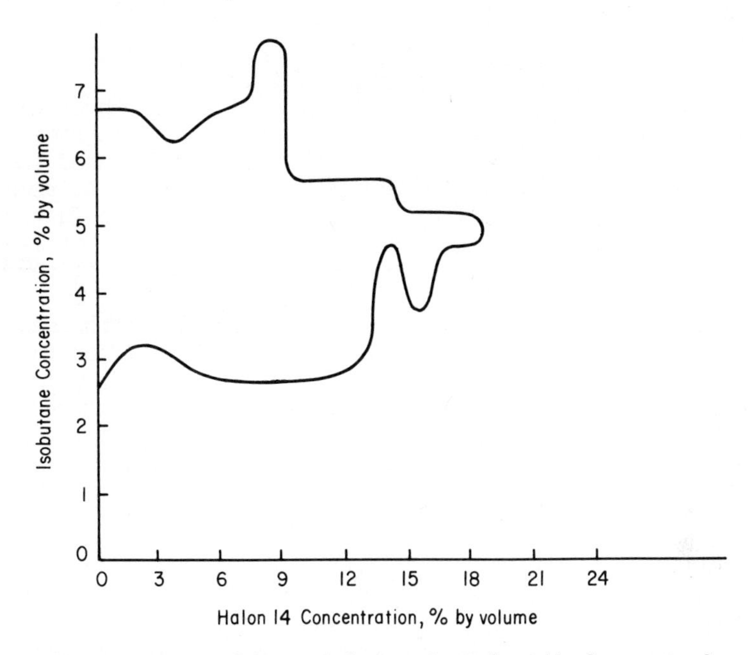

Figure 13. Flammability peak diagram for Halon 14/isobutane. Purdue Research Foundation. −78°C at 400 mm Hg.

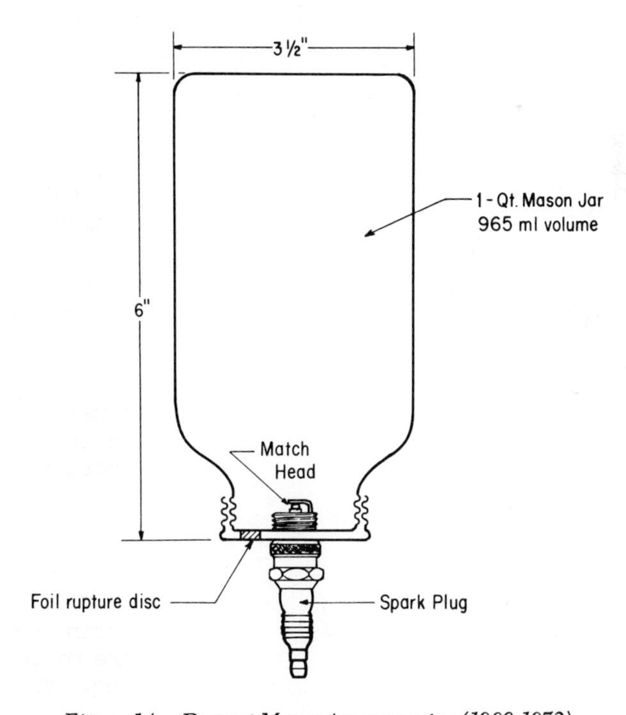

Figure 14. Dupont Mason jar apparatus (1966-1972)

TABLE XIX
DU PONT MASON JAR - 1971

1-Quart Mason jar
Match head ignition source (176 J)
Zero flame propagation
1 atm at 75°F

Fuel	Flammability Peak Concentration of Halon 1301 % by Volume
Methane	9
Ethane	9.5
Propane	9
n-Butane	9
i-Butane	9
n-Pentane	9
i-Pentane	8.5
neo-Pentane	8.5
Cyclopentane	8
n-Heptane	8
Gasoline (unleaded)	8
JP-4	8
Benzene	6.5
Acetylene	15
Ethylene	11
Propylene	9.5
Methanol	35
Ethanol (absolute)	9
Dimethyl Ether	30
Diethyl Ether	25
Ethyl Acetate	8
Acetone	8
Methyl Ethyl Ketone	8
Carbon Monoxide	6
Ethylene Oxide	37.5
Ammonia (anhydrons)	5.5
Methyl Amine	17.5
Acetonitrile	8.5
Acrylonitrile	8.5
Tetrahydrofuran	9.5
Ethyl Chloride	7.5
1,1 Difluoroethane	6
Hydrogen	30

then removed and replaced with one containing the match head. The jar was inverted, the match head activated, and the degree of flame propagation observed. For each determination, a new jar was employed, thus eliminating possible effects of incomplete cleaning between tests. The data were plotted into a flammability peak diagram, and the flammability peak concentration reported. Data from the Mason jar test is shown in Table XIX.

The flammability peak concentrations determined by the Mason jar method are considerably higher than those reported above by Purdue. Alkane fuels all show rather uniform requirements, with olefins, particularly acetylene, showing higher peak concentrations. The oxygenated fuels generally showed exceptionally high peak concentrations, as did hydrogen. The highest value was for ethylene oxide, requiring 37.5% Halon 1301 at the peak.

The Du Pont Mason jar data is criticized from three aspects:

a. The severe criterion of "zero" permissible flame propagation.
b. The high-energy ignition source for the relatively small volume of the jar.
c. Exposure of the test mixture to air when switching from the mixing cap to the igniting cap.

All of these procedures would tend to increase the agent requirement above the explosion burette type test, producing results which are too conservative. The test data do show a high degree of internal consistency, however, and many of the relationships exhibited by the data are considered valid.

Du Pont Intermediate-Scale Tests (26). To determine the validity of the Mason jar test method for establishing inerting requirements for Halon 1301, a series of larger-scale tests were made in a 5-gal pail (0.71 ft^3) and a 55-gal drum (8.23 ft^3) using n-heptane, i-butane, and methane as fuels. The ignition source and criterion for nonflammability was the same used in the Mason jar tests described above. Finally, a demonstration was conducted in a 1,695 ft^3 (12.5 x 15.5 x 8.75 ft high) enclosure in which i-butane and Halon 1301 concentrations were developed corresponding to the intermediate-scale tests. A match head was placed in the center of the enclosure and ignited. The mixture proved nonflammable. Data from these three test series are presented in Table XX.

TABLE XX
DU PONT INTERMEDIATE-SCALE
1971

70°F, 1 atm
No flame extension beyond ignition source

Enclosure Volume ft3	Fuel	Minimum Halon 1301 Concentration for Inerting, %
0.71	n-Heptane	7
8.23	n-Heptane	8
8.23	i-Butane	8
(1695	i-Butane	8)
8.23	methane	9.0

These intermediate-scale tests showed no scaling effects when the ignition source and criterion for flammability was held constant. It thus appears that the flammability peak concentration of Halon 1301 determined for a fuel by the Mason jar technique can be applied to large-scale applications with a high degree of confidence.

Du Pont Explosion Burette(27). A series of inerting studies were conducted in 1972 in which the effects of ignition energy and flame extension were investigated. The studies employed both the Mason jar and a 4 in i.d. x 48 in long vertical burette vented at the top. Ignition energies of 176 J (match head), 27 J/sec (A-C spark), 1300 J/sec (A-C spark) and 165 J (D-C spark) were examined. In the 4-in burette, the effects of permitting flame propagation to 50% of the tube height was compared to essentially zero propagation. Five fuels (methane, i-butane, ethylene oxide, dimethyl ether, and hydrogen) were used with Halon 1301 as the suppressant. The data from these tests are presented in Tables XXI and XXII.

TABLE XXI
DU PONT MASON JAR & EXPLOSION BURETTE
1972

Effect of Ignition Energy

70°F at 1 atm
No flame extension beyond ignition source
Flammability Peak Concentration of Halon 1301, % by vol

Fuel	Ignition Chamber	Ignition Source			
		Match Head 176 J	A-C Spark 27 J/sec	A-C Spark 1300 J/sec	D-C Spark 165 J
Methane	Mason Jar	9.0	4.3	40.0	-
i-Butane	"	9.0	6.5	31.0	-
Ethylene Oxide	"	37.5	-	-	-
Dimethyl ether	"	30.0	8.0	36.0	-
Hydrogen	"	30.0	-	-	-

Methane	4 in dia x 48 in high	6.75	4.25	10.0	-
i-Butane	"	6.5	6.25	11.0	7.0
Ethylene Oxide	"	34.0	21.0		
Dimethyl Ether	"	27.0	4.0		
Hydrogen	"	24.0	17.0		

TABLE XXII
DU PONT EXPLOSION BURETTE
1972

Effect of Flame Extension and Ignition Energy
4 in i.d. x 48 in high tube
70°F at 1 atm

Fuel	Flame Extension %	Match Head 176 J	A-C Spark 27 J/sec	A-C Spark 1300 J/sec	D-C Spark 165 J
Methane	0	6.75	4.25	10.0	
	50	5.5	4.25	5.25	
i-Butane	0	6.5	6.25	11.0	7.0
	50	6.25	6.25	6.5	6.0
Ethylene Oxide	0	34.0	21.0		
	50	33.0	21.0		
Dimethyl Ether	0	27.0	4.0		
	50	26.0	4.0		
Hydrogen	0	24.0	17.0		
	50	24.0	17.0		

From Table XXI, the effect of the various levels of ignition energy and a comparison between the Mason jar and explosion burette may be observed. In the Mason jar, an accompanying increase in peak concentration with increasing energy level is observed. At equal ignition energies, peak concentrations obtained in the 4-in burette are lower than those obtained in the Mason jar. This is probably due to the greater dissipation of ignition energy in the burette because of its larger volume (10 l vs 1 l for the Mason jar). At the 27 J/sec ignition energy, the difference between the jar and burette are smaller than at the higher energy levels. This is probably because excessive ignition energy increases the temperature of reactants in advance of the propagating flame front beyond that which the mixture per se is capable of generating. Lower ignition energies therefore would reflect more accurately the properties of the fuel/agent system.

Table XXII compares flammability peak concentrations based on zero flame propagation to ones determined using 50% tube length propagation as a criterion. At low ignition energy levels, the difference between using zero flame propagation and using 50% propagation is slight. At the high energy level of 1300 J/sec, the difference becomes significant. The Halon 1301 concentration is nearly doubled if zero flame propagation is used as a criterion over the value taken from 50% propagation. Again, this can perhaps be explained by considering the possible alteration of the fuel by the higher energy source. Some flame propagation will be produced in the vicinity of a very high energy source, even in the presence of a high suppressant concentration. The further the flame is required to travel from the source, the effect caused by the ignition source would be expected to diminish, and again more accurately reflect a property of the fuel/agent system.

Table XXII illustrates a fuel-dependent ignition energy effect. Compared to the 27 J/sec spark, the 176 J match head produced a slightly higher peak concentration for methane and isobutane, moderately higher values for ethylene oxide and hydrogen, and a grossly higher value for dimethyl ether. This suggests that some fuels are more sensitive to the ignition energy "override" discussed earlier.

ICI Explosion Burette (28). Imperial Chemical Industries has evaluated 24 gaseous and volatile liquid fuels using a 56 mm i.d. x 149 cm long burette. The primary purpose of the tests has been to evaluate the inerting requirements of Halon 1211. For comparison, four fuels have been evaluated with Halon 1301. The ignition source used in all tests was an A-C spark coil providing 20 to 30 J/sec energy at approximately 40 Hz. The tube was vented at the bottom and the criterion of half-tube propagation was applied. With the exception of n-hexane and methyl isobutyl ketone, the evaluations were performed at 25, 50 or 65°C. For n-hexane, additional determinations have been made with Halon 1211 at temperatures up to 200°C. ICI also has evaluated a larger burette, 105 mm i.d. x 210 cm long, for three fuels: methane, iso-butane, and ammonia. Results from these studies are shown in Table XXIII.

The alkane series from ethane through n-hexane, which were run at 25°C, all show uniform peak concentrations of Halon 1211 between 5.5 and 5.9%. Methane shows a somewhat lower requirement. N-heptane shows a higher requirement than the other alkanes, probably due to the higher temperature at which it was evaluated. The olefins, ethylene and proplyene, show somewhat higher requirements also. Requirements for methanol, ethylene oxide, and hydrogen are all very high. The few points obtained with Halon 1301 uniformly exhibit a somewhat lower requirement than for Halon 1211.

The series of tests with n-hexane at elevated temperature show a dramatic effect. As the temperature is increased from 25°C to 140°C, the Halon 1211 requirement is increased by about

TABLE XXIII
ICI EXPLOSION BURETTE (1973-1974)

56 mm i.d. x 149 cm long burette
A-C spark ignition 20-30 J/sec 40 Hz
50% tube length flame propagation
Bottom venting

Fuel	Temp. °C	Agent Concentration at Flamm. Peak, % Halon 1211	Halon 1301
Methane	25	4.1	
Methane*	25	4.1	
Ethane	25	5.8	
Propane	25	5.9	
n-Butane	25	5.9	
i-Butane	25	5.7	
i-Butane*	25	5.5	
n-Hexane	25	5.8	4.8
n-Hexane	120	7.6	
n-Hexane	140	7.8	
n-Hexane	160	15.6	
n-Hexane	200	20.2	
Cyclohexane	25	5.7	
n-Heptane	50	6.6	6.1
Benzene	25	4.0-4.8	3.8
Toluene	50	3.3	
Ethylene	25	9.6	
Propylene	25	6.2	
Methanol	50	24.7	
Ethanol	50	6.2	4.9
i-Propanol	50	5.6	
Diisopropyl Ether	50	6.7	
Acetone	25	4.9	
Methyl Ethyl Ketone	50	5.8	
Methyl Isobutyl Ketone	65	6.1	
Ethylene Oxide	25	23.5	
Vinyl Chloride	25	5.5	
Ethylene Dichloride	50	2.7	
Ammonia	25	0.34	
Ammonia*	25	0.34	
Hydrogen	25	27.0	

*Measurements made in 105 mm i.d. x 210 cm long burette

35%. At 160°, however, the requirement is nearly triple that at 25°C, and at 200° it is almost four times higher than the room temperature value. In the flammability peak diagram (not shown) other differences are also exhibited, in that a second, higher, peak is formed. The magnitude of this shift is similar to the one observed in Table XXII (Du Pont 4-in tube data) for dimethyl ether in going from the 27 J/sec spark to the 176 J match head ignition source. This investigation is continuing and will eventually be published.

ICI considers the technique of venting the bottom of the burette prior to ignition to be critical. In their tests, a cone joint containing the electrodes is slowly and carefully lowered away from the burette just prior to ignition, leaving an anular vent area of about 1.9 cm^2. This technique has been perfected by placing a 5% HCl/air mixture inside the tube, and ammonia vapor outside the lower firing bung. The operator then practices lowering the bung until no clouding occurs within the tube. The meticulous procedures which ICI has developed with this apparatus are evident from the very high degree of internal consistency exhibited by their data.

Tests using a 4 in (105 mm) i.d. tube produced results in excellent agreement with the 2 in (56 mm) tube. Because of this agreement, and because the larger tube proved more difficult to operate, ICI continues to favor the 2 in tube for inerting tests.

FMRC Explosion Burette (29,30). In two series of experiments, Factory Mutual Research Corporation investigated differences caused by venting the burette at the top vs venting at the bottom. The apparatus was a 57.5 mm i.d. x 135 cm long vertical tube. Gas mixtures were prepared from calibrated rotameters. The tube was thoroughly cleaned and purged with several volumes of gas mixture between determinations. The ignition source was an A-C spark of 20 to 30 J/sec at 40 Hz (in fact, the same type of spark coil that was used in the Du Pont and ICI burette tests). A criterion of 25% tube length flame propagation was applied in these tests. Six fuels were investigated (methane, propane, n-butane, i-butane, ethylene, and hydrogen) with both Halons 1301 and 1211. In the bottom-vented series, a fully open tube was first tried, but poor agreement with the ICI results were obtained. After further inquiry to ICI, the bottom opening was restricted to 21 mm dia. These results seemed to be more satisfactory. Data from this study are shown in Table XXIV.

TABLE XXIV

FACTORY MUTUAL RESEARCH CORP. 1973-1974

Effect of tube venting
57.5 mm i.d. x 135 cm long glass burette

Ignition source: A-C spark ~ 20-30 J/sec
25°C at 1 atm
25% flame propagation

Fuel	Agent	Flammability Peak Conc. of Agent, % Partially-Vented Botton	Fully-Vented Bottom	Vented Top
Methane	1301	4.0		6.8
Propane	"	5.25		8.0
n-Butane	"	5.0		7.2
i-Butane	"	4.75		7.2
Ethylene	"	8.3		10.5
Hydrogen	"	22.6		28.0
Methane	1211	3.6	2.3	4.3
Propane	"	3.6	1.6	7.0
n-Butane	"	6.25	-	6.3
i-Butane	"	4.8	1.6	6.3
Ethylene	"	7.0	-	12.0
Hydrogen	"	27.0		32.5

For all six fuels, the peak concentrations of both agents obtained with bottom venting were lower than those obtained with top venting. In the one case of n-butane with Halon 1211, approximately equivalent results were obtained with bottom and top venting. In general, the peak concentrations for Halon 1301 were somewhat higher than for Halon 1211, a reversal from the Purdue and ICI results. Neither set of Halon 1211 data agrees well with ICI's results. For both agents and both venting conditions, the values for ethylene and hydrogen are higher than for the hydrocarbons, confirming the relationship observed from previous tests. Results for the alkanes also show a surprising degree of internal inconsistency, compared to the Du Pont Mason jar and ICI data. Comparing the vented-top results to the Du Pont burette data, a curious agreement is noted. The FMRC data on methane and i-butane with Halon 1211 agree well with the Du Pont values with Halon 1301 for these same fuels.

Much of the inconsistency of the FMRC bottom-vented data is possibly due to venting procedure. The painstaking procedures developed by ICI to avoid disturbing the gas mixture in the lower portion of the tube while opening the vent do not appear to have been duplicated by FMRC.

Fenwal Large-Scale Tests (31). A study to observe the extent of flame propagation at various suppressant concentrations was conducted with Halon 1301 in a 5.31 ft i.d. x 17.22 ft high steel silo with propane and ethylene fuels. Both fuels were used at stoichiometric concentrations. With each fuel, the minimum concentration to prevent full-length flame propagation was first determined. Then, several higher agent concentrations were run, and the results observed photographically. Both maximum flame propagation distance upward from the ignition source and flame duration were determined from the film. These data are shown in Table XXV.

TABLE XXV
FENWAL, INC. 1973-1974

5.31 ft dia x 17.22 ft high silo (381 ft^3)

Ambient temperature at 1 atm
D-C spark ignition ∿100 J

Fuel and Concentration %	Halon 1301 Concentration %	Maximum Flame Propagation in	Flame Duration Sec
Propane 4.5%	0	>46	9.7
"	4	>46	7.1
"	5	>46	8.9
"	5.5	>46	9.1
"	6	>46	4.8
"	6.5	10	0.41
"	7	5	0.22
"	9	3	0.09
Ethylene 6.5%	8	>46	>7
"	8.5	>34	10.1
"	9	15	0.56
"	9.5	18	0.56
"	10	17	0.56

The minimum concentration of Halon 1301 to prevent full-length propagation was 6.5% for propane and 9% for ethylene. The propane value agrees well with the Du Pont burette data for isobutane, and the ethylene value agrees well with the Du Pont Mason jar data using the 27 J/sec ignition source. The ignition energy/unit volume in the Fenwal tests lies between the Du Pont Mason jar test and the Du Pont burette test with the 27 J/sec A-C spark. The Fenwal values are both lower than obtained by FMRC in the vented-top tests.

For a given fuel, the Fenwal tests demonstrated that increasing agent concentration beyond a minimum required to suppress full-length propagation reduced the amount of flame propagation obtained. However, a 40% increase beyond the minimum in the case of propane did not reduce the propagation to zero. With ethylene, increasing the Halon 1301 concentration from 9% to 9.5% to 10% did not significantly reduce the propagation distance. Observations of flame duration appear to vary directly with propagation distance.

Class A Fuel Evaluations. Halogenated suppressants, when used in total flooding systems, have been found to have a limited degree of effectiveness on certain types and configurations of Class A combustibles. Because the agents are applied to the fire as dilute gaseous mixtures in air, the cooling influence provided by agents such as water is missing. Also, Class A fuels often exhibit two types of combustion. One is surface flaming, producing flammable gases which are evolved from the solid fuel. This type of fire behaves much like a flammable liquid or gaseous fuel. The second type of combustion is the direct oxidation of carbon within the solid mass with little or no flames present. This type of fire is often called a "deep-seated" fire. In general, this type of combustion is more difficult to extinguish than surface flaming. Many fuels which exhibit this glowing combustion require high concentrations of halogenated suppressants and long soaking times to achieve complete extinguishment. In considering Class A fuels, not only must the fuel and its configuration be taken into account, but the agent concentration and soaking time must also be considered. The burning time preceeding application of agent may also determine whether the fire burns as a surface fire or becomes deep-seated.

Discussed below are eight series of tests in which Class A fuels have been examined with Halon 1301.

1. FMRC wood pallet tests - 1965
2. FMRC glovebox tests - 1967
3. Underwriters' Laboratories - 1971
4. Fenwal - 1971
5. Ansul - 1971
6. Du Pont - 1971
7. Safety First Products Corporation - 1972
8. Williamson 21.5 cu ft cabinet - 1972

Results from these test series are summarized in Tables XXVI and XXVII, and discussed in more detail below.

FMRC Wood Pallet Tests(32). Factory Mutual Research Corporation in 1965 conducted two tests in a 6,550 ft^3 test room using a stack of ten 4 ft x 4 ft hardwood pallets. Ignition was achieved with two gasoline-soaked torches inserted into the bottom of the stack. A 140°F heat detector was used to operate the Halon 1301 discharge system. The soaking time permitted in both tests was 60 min. In both tests the pallets were completely extinguished with 3% Halon 1301, but the cotton wadding of the torches reignited upon reopening the enclosure.

FMRC Glovebox Tests(33). A series of Halon 1301 extinguishment tests were conducted for Du Pont in 1967 in which wood, paper, and charcoal fires were examined in a 180 ft^3 glovebox.

TABLE XXVI
CLASS A FIRE TEST SUMMARY

Total Flooding with Halon 1301

Fuel	Investigator	Halon 1301 Conc.	Soaking Time Min	Result
10 Wood Pallets	Miller (FMRC) 1965	3.0	10	Extinguished
Wood Crib	UL 1971	3.88-6.09	10	Extinguished
Excelsior	UL 1971	6.00	10	Extinguished
Shredded Paper	UL 1971	<18%	10	Erratic
Shredded Paper	UL 1971	>18%	10	Extinguished
Roofing Brands	UL 1971	5.9	10	Extinguished
Corrugated Cardbd.	Williamson 1972	30%*	10	Extinguished
Low Density Ceiling Tile	Williamson 1972	30%*	10	Extinguished
Fiberglas Poly-ester Resin	Williamson 1972	3.5*	10	Extinguished
Plywood	Williamson 1972	20*	10	Extinguished
Masonite	Williamson 1972	25*	10	Extinguished
Poly-ether Foam	Williamson 1972	3.5*	10	Extinguished
Cellulose Acetate	Williamson 1972	3.3*	10	Extinguished
PVC Trim	Williamson 1972	3.3*	10	Extinguished
Charcoal	FMRC 1967	13*	10	Extinguished
Wood Blks.	FMRC 1967	12*	10	Extinguished
PVC Tubing	Du Pont 1971	2.6	10	Extinguished
5 lb Stacked Paper	Ansul 1971	5.1	10	Extinguished
5 lb Stacked Cards	Ansul 1971	Did not support combustion		

5 lb Paper and Cards (random)	Ansul 1971	>11.8	30	Extinguished
Poly-styrene and Poly-ethylene	Fenwal 1971	2.0*	10	Extinguished

*Threshold extinguishing concentration

TABLE XXVII
SAFETY FIRST PRODUCTS CORPORATION

Class A Fire Tests

Test No.	Fuel	Enclosure ft^3	Preburn Time min:sec	Halon 1301 Conc. %	Soaking Time Min	Results
1	Wood crib 24 pcs. 2x2x18, 4 rows	336	9:00	3.8	5	Extgsh'd, cold to touch
2	Wood crib, UL Std. 1A	336	10:00	3.8	5	Extgsh'd, cold to touch
3	3 lbs excelsior on floor	336	0:50	3.8	2:30	Extgsh'd, no reflash
4	3 lbs shredded paper on floor	336	0:15	4	0:30	Extgsh'd
5	5 lbs shredded paper in 2 ft dia wire basket	336	0:15	3.8	5:45	Extgsh'd
6	5 lbs shredded paper in 2 ft dia wire basket	336	0:35	3.8	5:00	Reignited when door opened
7	3 sheets fiberglass sheet 1/16 in x 16 in x 16 in	336	1:30	3.8	0:20	Extgsh'd
8	7 lbs PVC insulated wire in loose mass	336	1:40	3.5	0:30	Extgsh'd
9	2 lbs PVC wire in 27 ft^3 sub-enclosure	2000	0:20	3.8	5:00	Extgsh'd
10	1/8 in thick polyurethane sheets piled on floor 7 ft x 3 ft x 1.5 ft high	336	0:38	3.8	0:30	Extgsh'd
11	4 cardboard boxes of punch cards 40 lbs total	336	20:00	6.5	10:00	Slight wisps of smoke when room opened

12	Glazed fox fur with lining	336	0:30	3.8	5:00	Fur out, lining still smoldering slightly
13	Same as No. 12	336	0:20	7.6	5:00	Extgsh'd
14	21 lbs. corrugated cardboard	336	2:40	7.6	30:00	Reignited
15	4 pcs. (12 lbs) wood flake board	336	4:00	20	10:00	Reignited
16	15 pcs. (63 oz) wood fiberboard	336	2:00	20	10:00	Reignited
17	3 lbs cotton lint	336	0:10	3.8	1:05	Extgsh'd
18	1 pail each crumpled paper and carbon paper	336	0:15	3.8	5:00	Extgsh'd

The test enclosure is the same as reported previously in the FMRC flammable liquid fires. Of most interest were tests with charcoal and wood blocks.

In the charcoal fire tests, 21 oz of oven dried hardwood charcoal in 1 in to 1-1/2 in cubes were arranged in a wire basket measuring 6-1/2 in dia by 4-1/2 in deep. The charcoal was ignited by exposure to a gas-fired radiant heater for 4 min. The basket was placed on a scale in the glovebox and allowed to burn an additional 11 min. A predetermined quantity of Halon 1301 was discharged into the glovebox and allowed to remain until the basket temperatures and weight loss readings indicated the fire had been extinguished. The glovebox was then opened to the atmosphere and ventilation restored to determine whether reignition would occur. Approximately 15 determinations were made, with Halon concentrations ranging from 1.1% to 25%. Results of these tests are shown in Figure 15. Halon 1301 concentrations as low as 1-1/2 to 2% were effective in extinguishing these fires completely, provided minimum soaking times of 30-40 minutes were maintained. To keep the soaking time below 10 min, a Halon 1301 concentration of about 13% was required.

The wood block fires consisted of two oven-dried 2 x 4 fir blocks 1 ft long. The two blocks were mounted vertically with the 4-in faces parallel and separated by approximately 3/8 in. The blocks were ignited with a match and allowed to preburn 15 min before injecting agent. Halon 1301 concentrations were varied from 10.2 to 13.2%. Soaking times between 3 and 60 min were required. The curve represented by these four tests is also shown in Figure 15. Approximately 12% Halon 1301 is required for this fuel to keep the soaking time below 10 min.

Underwriters' Laboratories(34). Underwriters' Laboratories, Inc., compiled a set of fire test data accumulated from test programs for equipment listings. Their compilation included four types of Class A fires: standard 1-A wood cribs, shredded paper, excelsior, and roofing brands. Results from these tests are shown in Table XXVI. All fires except the shredded paper fires were extinguished with Halon 1301 concentrations of 6% or less maintained for 10-min soaking periods. With the shredded paper fires, Halon 1301 concentrations below 18% gave erratic results. With concentrations above 18%, extinguishment was achieved within the 10-min soaking time. In the shredded paper fires, either 5 or 10 lb of shredded paper was contained in an open wire basket elevated approximately 1 ft above floor level and ignited at the bottom center.

Fenwal, Inc., Polystyrene and Polyethylene Tests(35). As a part of an industry effort to examine fires which might typically occur in computer rooms, Fenwal, Inc., determined the minimum extinguishing concentration of Halon 1301 for polystyrene computer tape reels and tape cases and for polyethylene sealing

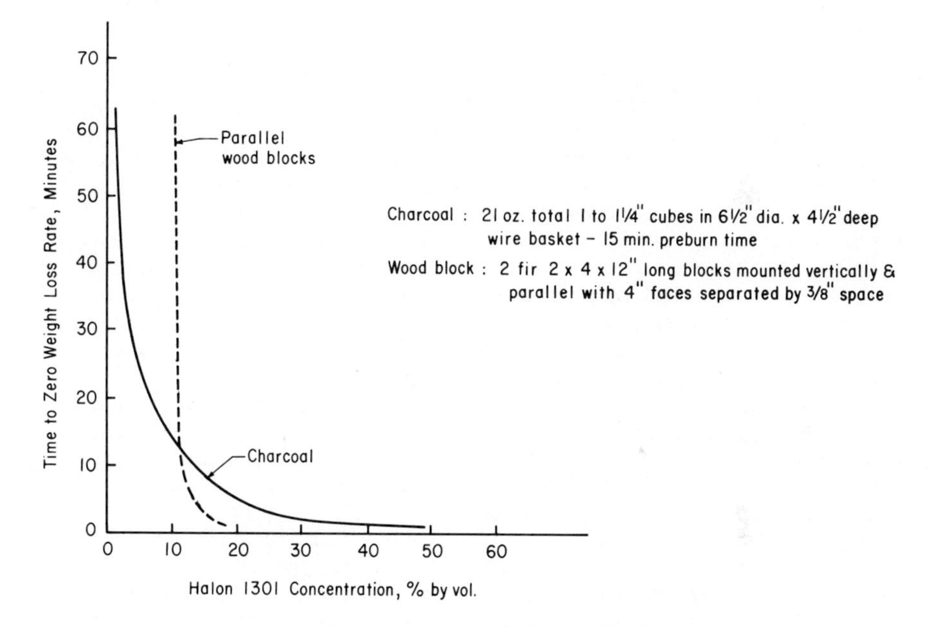

Figure 15. Soaking time vs. Halon 1301 concentration for charcoal and wood block test fires. FMRC (1967). 180-ft³ enclosure.

bands. The test enclosure measured 12 x 16 x 8 ft high and contained a volume of 1,536 ft^3. The test fire consisted of six empty polystyrene reels, four empty polystyrene tape cases, and six empty polystyrene reels sealed with polyethylene bands placed in each of the two lower shelves of a six shelf storage rack. The rack of components was ignited from 40 ml of n-heptane in a steel pan measuring about 8 x 16 x 1 in deep placed under the bottom row of components. Ten tests were performed in which Halon 1301 concentrations ranging from 2 to 6% by volume were applied using Fenwal commercial Halon 1301 components. In all tests in which the Halon 1301 concentration was 3% or greater, the fires were completely extinguished within a 10-min soaking time.

Ansul Cellulosic Tests (35). As a part of the same industry computer fire test program mentioned above, Ansul Company examined the extinguishment characteristics of cellulosic materials commonly used in computer rooms. Fuels used were a 5 lb stack of computer print-out paper, and 5 lbs (total) of paper punch tape, computer print-out paper, and data tabulating cards arranged randomly in an open wire basket. In tests containing the stacked print-out paper, one test was run in which the fire was ignited by placing an electrical heater wound around a single data tabulating card in the center of the stack. The ignitor was energized and the stack allowed to burn until a 0.5 lb weight loss had been observed. At that point, a Halon 1301 concentration of 5.1% was established in the 14.75 x 14.75 x 8 ft high (1,729 ft^3) enclosure. Following a 10-min soaking time, the fire was observed to be completely extinguished. In four tests performed on random arrangements of paper tape, computer print-out paper, and data tabulating cards, concentrations of 5.1% were found to extinguish flaming combustion but permitted the smoldering fire to continue throughout a 10-min soaking time. In two tests with concentrations of 11.8 and 21.0% Halon 1301, the fire was completely extinguished at the end of a 30-min soaking period.

Du Pont PVC Tests (35). As a final portion of the industry computer fire test program, Du Pont determined the threshold extinguishing concentration for Halon 1301 on 5 lbs of polyvinyl chloride tubing contained in an open wire basket. The tests were run in an enclosure measuring 12.5 x 15.5 x 8.75 ft high (1,695 ft^3). The fire was ignited from 25 ml of methanol in a 5 x 8 x 1 in deep aluminum pan placed directly on the floor beneath the fuel basket. Approximate burning time of the methanol was 1 min. Five tests were performed in which Halon 1301 concentrations were varied from 2.6% to 5% by volume. It was found that a 2.6% Halon 1301 concentration was sufficient to extinguish the polyvinyl chloride fire completely with a 1-min soaking period.

Safety First Class A Tests (36). Cholin in 1972 reported results from 24 tests conducted with Halon 1301 in 336 ft^3 and 2,000 ft^3 test buildings. Eighteen of these tests were conducted to determine the effects of Halon 1301 on various types of Class A fuels. The Halon 1301 was discharged into the test enclosures using commercial Safety First systems components. The results of this test series are shown in Table XXVII. Of the fuels examined, the following materials were completely extinguished with concentrations of 3.5 to 4% Halon 1301:

1. Small wood crib
2. UL size 1-A wood crib
3. Three pounds excelsior on floor
4. Three pounds shredded paper on floor
5. Three sheets fiberglass
6. PVC cable insulation
7. Polyurethane sheets
8. Cotton lint
9. Crumpled paper and carbon paper in solid wall pails

The following materials were not extinguished with low Halon 1301 concentrations for short soaking times:

1. Shredded paper in wire basket
2. Corrugated cardboard
3. Wood flake board
4. Wood fiberboard

The cardboard boxes of punches cards and the glazed fox fur with lining showed marginal results. Although neither fuel configuration was completely extinguished within the test conditions, the burning rate had been reduced to an extremely low level.

Williamson 21.5 ft^3 Cabinet Tests (37). Using the Cardox test cabinet described above, Williamson in 1972 determined the Halon 1301 extinguishing requirements for eight fuels under flow conditions. The fuel arrangement consisted of five 1 ft square panels mounted parallel and vertically with a 3/8-in space between each panel. The array was ignited from a small pan containing 40 ml of alcohol placed under the array which provided a burning time of about 1 min. Halon 1301 was added to the ventilating air inlet stream to establish the desired concentration within the cabinet. After a 10-min exposure period to the agent, the door to the cabinet was opened and the sample observed for continuing combustion. Results from the tests are given in Table XXVI. Corrugated cardboard, low-density ceiling tile, plywood, and masonite required concentrations between 20 and 30% for extinguishment. Four plastic materials investigated (fiberglass polyester resin, polyether foam, cellulose acetate, and polyvinyl chloride trim) required only 3.3 to 3.5% for extinguishment. These materials would not be considered to produce deep-seated fires.

Conclusion

From the foregoing analyses of individual test series, several "facts" may be extracted which hopefully will provide clues for a mechanism:

1. Halogenated agents containing bromine or iodine in the molecule are more effective extinguishants than those containing only chlorine or fluorine. Chlorine imparts a greater degree of effectiveness than fluorine. Compounds containing only fluorine do not show significantly greater effectiveness than inert gases. Fluorine atoms tend to impart stability to the compound.
2. One atom of an active halogen (Cl, Br, or I) in a compound gives a marked increase in extinguishing effectiveness over a similar compound which contains none of that halogen. Adding a second atom of the same halogen produces only marginal additional increase in effectiveness.
3. Required concentrations of a given halogenated agent to extinguish established fires of liquid or gaseous fuels depend primarily upon the fuel. Oxygenated or unsaturated hydrocarbon fuels (such as methanol, dimethyl ether, ethylene oxide, ethylene, and acetylene) require higher agent concentrations for extinguishment than do aliphatic hydrocarbon fuels. Hydrogen requires very high concentration of halogenated agents for extinguishment.
4. Halogenated agents added to the fuel side of a diffusion flame require higher concentrations for extinguishment than when added to the air side of the flame. In terms of the rate of agent injection for the total system, addition to the fuel side is probably more efficient than addition to the air side.
5. Halon 1301 concentrations to extinguish carbon monoxide diffusion flames are significantly lower than for most other fuels. With Halon 1001, carbon monoxide requires a higher concentration for extinguishment than do other fuels.
6. The extent to which minimum concentrations of halogenated agents to extinguish diffusion flames depend upon fuel temperature varies with type of fuel. Some fuels, such as alkanes, show little effect with temperature, while others, notably oxygenated fuels, show large effects.

7. Dynamic tests for determining flame extinguishment requirements of halogenated agents for a given fuel give somewhat higher threshold concentrations than do static tests. This is probably due to oxygen depletion or recirculation of combustion and decomposition products into the flame, which accompany static determination. For this reason, a dynamic procedure is preferred because of its conservative result.
8. Concentrations of halogenated agents to inert a given fuel/air mixture are significantly greater than to extinguish diffusion flames of that fuel.
9. Inerting requirements for halogenated agents vary with fuel type and temperature and conditions (ignition energy, container geometry, and amount of flame propagation permitted) of the system. These variables appear to operate as follows:
   a. The effect of fuel type is similar to that for extinguishment of diffusion flames. Paraffins require lower concentrations than olefins or oxygenated fuels. Hydrogen requires about the highest concentrations of any fuel.
   b. Increasing temperature produces increased flammability peak concentrations of agent. There is a suggestion that the extent of the dependence varies with fuel type.
   c. Increasing ignition energy increases the inerting requirement for a given agent/fuel system. The extent of this effect depends upon fuel type. Oxygenated fuels, such as dimethyl ether, exhibit greater effects than paraffinic hydrocarbons.
   d. Increasing the amount of flame propagation permitted in the system permits lower agent concentrations for inerting. This effect is greater when a high (>100 J) ignition energy is used.
   e. At high ignition energies, smaller scale equipment produces higher inerting requirements. At lower ignition energies, little dependence on scale is observed. (This assumes that the scale is not so small to

produce wall-quenching effects.)

10. Successful extinguishment of Class A (solid combustible) fuels with halogenated agents in total flooding systems depends upon the fuel and its configuration, preburn time, agent concentration, and soaking time. Thermoplastics, paper products in stacks or in solid wall waste containers, and wood arrangements which do not contain closely-spaced vertical surfaces, are completely extinguished with Halon 1301 concentrations of 5 to 6% and soaking times of 10 minutes or less. Wood or wood products containing closely-spaced vertical surfaces, corrugated cardboard, shredded paper in open wire baskets, and some types of thermosetting plastics, tend to develop deep-seated fires which require much higher agent concentrations and long soaking times for complete extinguishment. Application of low concentrations of Halon 1301 will "control" deep-seated fires during the soaking period.

The questions which these observations pose can serve as a check list to test proposed mechanisms. These relationships must be explained if a mechanism is to have practical significance to the fire protection engineer. While a valid mechanism will aid research for improved extinguishants, it must also act as a guide for application of existing agents to the variety of fuels, hazards, and conditions found in industry. The level of understanding necessary to formulate a workable mechanism cannot be programmed as an experiment, but must result from insight and inspiration. With this conference, we are taking the first step towards this understanding.

Literature Cited

1. Wharry, David and Ronald Hirst, Fire Technology: Chemistry and Combustion, Institution of Fire Engineers, Leicester, England (1974).
2. The Halogenated Extinguishing Agents, NFPA Q-48-8, National Fire Protection Association, Boston, Massahcusetts (1954).
3. Strasiak, Raymond R., The Development History of Bromochloromethane (CB), WADC Technical Report 53-279, Wright Air Development Center, Ohio, January 1954.
4. Standard on Halogenated Fire Extinguishing Agent Systems - Halon 1301. NFPA No. 12A - 1973. National Fire Protection Association, Boston, Massachusetts (1973).
5. Standard on Halogenated Fire Extinguishing Agent Systems - Halon 1211. NFPA No. 12B - 1973. National Fire Protection Association, Boston, Massachusetts (1973).

6. The Ansul Company, Research Department Project No. 147, Technical Report No. A-11, January 30, 1956.
7. Guise, A. B., "Extinguishants and Extinguishers", presented at 61st Annual Meeting, National Fire Protection Association, Los Angeles, May 1957.
8. Gorski, R. A., "Experimental 4 and 6B Extinguishers Containing 'Freon' FE 1301 Fire Extinguishing Agent", Technical Report KSS-2720A, E. I. du Pont de Nemours & Co., "Freon" Products Laboratory, November 9, 1960.
9. Gorski, R. A., "Experimental 6B Fire Extinguisher Containing 'Freon' FE 1301 Fire Extinguishing Agent", Technical Report KSS-3466A, E. I. du Pont de Nemours & Co., "Freon" Products Laboratory, November 17, 1961.
10. Stewart, R. L., "Design of F.P.L. 2:B Fire Extinguishing Head", Technical Report KSS-6269, E. I. du Pont de Nemours & Co., "Freon" Products Laboratory, May 27, 1968.
11. Mader, G. N., "Design Parameters for an FE-1301 Portable Fire Extinguisher with a 1-B Rating", Technical Report KSS-7289, E. I. du Pont de Nemours & Co., "Freon" Products Laboratory, October 25, 1971.
12. Owens, R. J., "Modification of a 50 lb. Carbon Dioxide Fire Extinguisher for Use with FE 1301 Fire Extinguishant", Technical Report KSS-7343, E. I. du Pont de Nemours & Co., "Freon" Products Laboratory, February 2, 1972.
13. Ford, Charles L., Unpublished Data, E. I. du Pont de Nemours & Co., Germay Park, Wilmington, Delaware, June 27-28, 1973.
14. Creitz, E. C., "Inhibition of Diffusion Flames by Methyl Bromide and Trifluoromethyl Bromide Applied the Fuel and Oxygen Sides of the Reaction Zone", J. Res. of Nat'l Bur. of Stds., 65, No. 4, 1961.
15. Fletcher, N., "Halon 1211 Systems: Determination of Extinction Concentrations in the Laboratory", PN-67-39-2, ICI Mond Division, Winnington Laboratory, February 4, 1970.
16. Booth, K., B. J. Melia and R. Hirst, "Critical Concentration Measurements for Flame Extinguishment of Diffusion Flames Using a Laboratory 'Cup Burner' Apparatus", ICI Mond Division, Winnington Laboratory, August 31, 1973.
17. Bajpai, S. N., "An Investigation of the Extinction of Diffusion Flames by Halons", Ser. No. 22391.2, Factory Mutual Research Corporation, Norwood, Massachusetts, November 1973.
18. Bajpai, S. N., "Extinction of Vapor Fed Diffusion Flames by Halons 1301 and 1211 - Part I", Ser. No. 22430, Factory Mutual Research Corporation, Norwood, Massachusetts, November 1974.
19. Ford, Charles L., "Intermediate-Scale Flame Extinguishment Tests with Halon 1301 and Halon 1211", E. I. du Pont de Nemours & Co., Wilmington, Delaware, May 8, 1974.
20. Fitzgerald, P. M. and M. J. Miller, "Evaluation of the Fire Extinguishing Characteristics of 'Freon' FE 1301 on

Flammable Liquid Fires", Ser. No. 16234.1, Factory Mutual Research Corporation, Norwood, Massachusetts, February 21, 1967.
21. Miller, M. J., "Determination of Design Criteria and Performance Testing of Prototype Fire Extinguishing Systems Using 'Freon' FE 1301", Ser. No. 16234.1, Factory Mutual Research Corporation, Norwood, Massachusetts, October 1968.
22. Fletcher, N., "Halon 1211 Systems: Threshold Concentration for Extinguishment of Various Fuels", PN-67-39-2, ICI Mond Division, Winnington Laboratory, February 3, 1970.
23. Wickham, Robert T., "Final Report on the Halon 1301 Threshold Fire Extinguishing Program", Wickham Associates, Marinette, Wisconsin, September 10, 1972.
24. "Final Report on Fire Extinguishing Agents for the Period September 1, 1947, to June 30, 1950", Contract No. W44-099-eng-507, Purdue Research Foundation, Lafayette, Indiana, July 1950.
25. Floria, Joseph A., "Du Pont FE 1301 Fire Extinguishant Fuel Inerting Concentrations by Mason Jar Technique", Technical Report KSS-7149, E. I. du Pont de Nemours & Co., "Freon" Products Laboratory, March 17, 1971.
26. Owens, R. J., "Determination of Inerting Concentrations of FE 1301 Fire Extinguishant in Intermediate Scale Equipment", Technical Report KSS-7310, E. I. du Pont de Nemours & Co., "Freon" Products Laboratory, November 10, 1971.
27. Floria, Joseph A., "Du Pont FE 1301 Fire Extinguishant Fuel Inerting Concentrations - Summary Report", E. I. du Pont de Nemours & Co., "Freon" Products Laboratory, March 20, 1972.
28. Lewis, D. J., "Provisional Draft: Comparison of Fuel-Air-BCF Flammability Limits in 2" and 4" Diameter Vertical Tube Apparatus - Part II", ICI Mond Division, Winnington Laboratory, September 13, 1973.
29. Bajpai, S. N. and J. P. Wagner, "Inerting Characteristics of Halogenated Hydrocarbons (Halons)", I & EC Product Research & Development, 14, No. 1, March 1975.
30. Bajpai, S. N., "Inerting Characteristics of Halons 1301 and 1211", Ser. No. 22391.3, Factory Mutual Research Corporation, Norwood, Massachusetts, August 1974.
31. Ford, Charles L., Unpublished Data, E. I. du Pont de Nemours & Co., Germay Park, Wilmington, Delaware, June 1973 - June 1974.
32. Miller, M. J. and R. L. Pote, "Fire Tests of Two Remote Area Fire Suppression System Concepts", Ser. No. 15974.1, Factory Mutual Research Corporation, Norwood, Massachusetts, November 29, 1965.
33. Miller, M. J., "Extinguishment of Charcoal, Wood, and Paper Fires by Total Flooding with 'Freon' FE 1301/Air and Carbon Dioxide/Air Mixtures", Ser. No. 16234.1, Factory Mutual Research Corporation, Norwood, Massachusetts, July 28, 1967.
34. Private Communication, W. A. Gawin (Underwriters' Labora-

tories, Inc.) to C. L. Ford (Du Pont), August 17, 1971.

35. Ford, Charles L., "Halon 1301 Computer Fire Test Program Interim Report", E. I. du Pont de Nemours & Co., Germay Park, Wilmington, Delaware, January 10, 1972.
36. Cholin, Roger R., "How Deep is Deep? (Use of Halon 1301 on Deep-Seated Fires)", Fire Journal 66, No. 2, January 1972.
37. Williamson, H. V., "Halon 1301 - Minimum Concentrations for Extinguishing Deep-Seated Fires", Fire Technology, 8, No. 4, November 1972.

DISCUSSION

W. D. WEATHERFORD, JR: It is of interest relative to mechanism clues to point out that the noneffectiveness of normally-liquid halon 2402 on cotton waste fires (observed by the Army Corps of Engineers) agrees with our observations with normally-liquid halon 1011 in diesel fuel. It has been shown by the Army Ballistic Research Laboratory that 5 liquid volume percent halon 1011 prevents burning of the bulk liquid diesel fuel, but data obtained at Southwest Research Institute demonstrates that this concentration of inhibitor does not retard the development of mist fire or wick-supported combustion. In fact, halon concentrations in excess of 25 liquid volume percent do not prevent wick-supported combustion. An apparently obvious explanation, based on enhanced surface area for evaporation (as in a kerosene latern) and consequent selection vaporization of the inhibitor does not completely explain these observations. On the other hand, dilution of the normally nonflammable mixture with air in a diffusion flame (after total vaporization) could cause the mixture to pass through the flammable zone (as could be illustrated on a diagram of inhibitor vapor concentration vs fuel concentration). Coupling of both phenomena could possibly explain observations.

J. DEHN: (In reply to comment of D. Weatherford that wicking fires with binary liquid mixtures of fuel and halons in air persist at halon concentrations sufficient for extinction in the absence of a wick.)

A wick evaporates liquids in a binary mixture essentially in proportion to their concentrations in the liquid phase, independent of their volatilities. In the case of surface evaporation without a wick, the more volatile component will be present in the vapor phase at a higher concentration than it is present in the liquid phase, thus, by choosing a liquid halon which is more volatile than a liquid fuel a smaller liquid volume percentage is required than might be supposed from a glance at vapor phase peak inert percentages as determined in explosion burettes, provided no wick is present. (cf. a paper to be published in _Combustion and Flame_ by J. Dehn on the subject of Flammable Limits over Liquid Surfaces.)

E. T. McHALE: How do the large-scale "Fenwal silo" results for peak percentage compare with smaller-scale data from the burette method?

C. L. FORD: The Halon 1301 value for propane in the Fenwal silo tests was 6.5%. This compares to about 5.25% (vented bottom) and 8.0% (vented top) from the Factory Mutual laboratory study. For ethylene, the Fenwal value of 9% compares to Factory Mutual values of 8.3% and 10.5%, respectively. The laboratory results with the vented-top apparatus give somewhat higher inerting concentrations that were observed at large scale.

R. G. GANN: In the early Ansul and Army tests which you reported, halon 2402 appears some 30% less effective than halon 1301 at pan fire suppression. This is contrary to flame speed reduction measurements, such as those of Miller, Evans and Skinner, (Comb. and Flame,(1963), 7, 137) which showed that halon 2402 is far more effective. As with the poor showing of halon 1211 in these tests, can the results be attributed to the experiment design and agent delivery technique?

C. L. FORD: Yes, I think so. Optimization of discharge nozzle design is very important in the design of a portable extinguisher. In the early tests, the agents were evaluated in portable extinguishers which were designed for another type of agent, namely CO_2. I think that if the nozzles had been optimized, Halons 1211 and 2402 would have showed superior performance then, as they are doing now. Tests by Montedison in Italy show Halon 2402 to be very effective in portable extinguishers.

R. W. BILGER: How does the fire go out? Does it blow off or gradually decrease the intensity of fire? Please comment on both real fires and the dynamic laboratory tests.

C. L. FORD: The effects are somewhat different in a total flooding room application than in a laboratory cup burner. In the room situation, for liquid fuels, sub-extinguishing concentrations have little effect on the apparent burning rate of the fire. The flame height is about the same as with no inhibitor present. Then, as slightly more inhibitor is added, the fire suddenly goes out. A person seeing this for the first time is usually astonished.

In the laboratory cup burner, as the extinguising concentration is approached, the flame lifts off the cup and may burn quite stably several centimeters above the fuel surface. Increasing the agent concentration by only a few tenths of a percent will drive the flame further up the tube, until is eventually goes out.

C. HUGGETT: The data shown indicate that Halon 1301 is much more effective than methyl bromide in extinguishing a carbon monoxide flame. It is well known that the combustion of carbon monoxide is accelerated by traces of hydrogen, presumably through participation of hydrogen-containing radicals in a chain reaction process. The commonly accepted mechanism for the action of the halon extinguishing agent involves reaction with these same radicals. The concentration of hydrogen will be very low in the carbon monoxide flame, arising from water vapor and other trace impurities. Only a small amount of extinguishing agent is needed to interact with the small concentration of hydrogen-containing radicals present. The methyl bromide will introduce large quantities of hydrogen into the flame, increasing the concentration of hydrogen-containing radicals and greatly increasing the inhibitor requirement.

This suggests that the study of carbon monoxide flames containing small, controlled amounts of hydrogen might provide a very useful tool for the study of the mechanisms of halon extinguishment processes.

A. S. GORDON: There is a surprisingly large effect of halon 1301 on the extinguishment of charcoal. Has the effect of N_2 been compared under the same experimental conditions?

C. L. FORD: Nitrogen was not examined, but carbon dioxide was. To give comparable effects, the concentrations of CO_2 were about 6 to 7 times higher than Halon 1301. For a 10-minute soaking time, about 65% by volume of CO_2 would be required.

R. G. GANN: Let me provide some details on recent work performed at NRL. Fielding, Woods and Johnson (J. Fire and Flam. (1975), 6, 37) flowed premixed 10% halon 1301 in air through a heated charcoal bed. The smoldering combustion was not suppressed and the

CO/CO_2 ratio in the effluent gas was much higher than that observed when normal air was flowed through the bed. They interpreted their result as indicating little effect of CF_3Br on the surface generation of CO and significant effect on the gas phase oxidation of CO to CO_2.

C. L. FORD: The Factory Mutual work was performed in a quiescent atmosphere, with no forced air flow through the bed. Some work which Cardox has performed in a flow cabinet suggests that higher concentrations are required when a flow is present. From tests with other fuels, concentrations of Halon 1301 perhaps as high as 30% might be required for extinguishment under flow conditions.

2

The Relevance of Fundamental Studies of Flame Inhibition to the Development of Standards for the Halogenated Extinguishing Agent Systems

MYRON J. MILLER

Factory Mutual Research Corp., Norwood, Mass. 02062

The principal responsibility of National Fire Protection Association (NFPA) committees is to provide unambiguous guidelines for the design and evaluation of safe, effective systems. Some of the factors that must be considered in the design of Halon Systems are shown as a Fault Tree in Figure 1.

The behavior of the agent in preventing or suppressing fire and explosion and the effect of the agent on personnel who might be exposed have received a large amount of attention over the past twenty years.

About two years ago inconsistencies were noted in extinguishing and inerting data which had been accumulated since the NFPA Halogenated Fire Extinguishing Agent Systems Committee was established in 1966. Of particular concern was the lack of a systematic relationship between fire extinguishment and inerting data and increasing discrepancy between the results obtained by different investigators and by different experimental methods.

It would be satisfying if the data available could be tested against a theoretical ideal. However, fire (being the varied event that it is) has so far frustrated efforts to develop an all-encompassing theoretical model. There have been promising advances in the past ten years toward an understanding of fire, and it is my purpose here to present some of the data developed by members of the NFPA Halon Committee in a way that will highlight our conclusions to date and perhaps stimulate further investigation responsive to the questions that remain.

Perhaps the best way to convey the information is to organize it into segments based upon our principal concerns regarding the Precision, Accuracy, Predictability and Applicability of the data. Since it is my contention that flame extinguishment, inerting and explosion suppression concentration requirements should all be related in some logical way, I will avoid breaking the discussion into these categories. We have also made an attempt to publish all experimental work conducted by the Factory Mutual Research Corporation. These and other sources are referenced, and I will avoid discussing the details of these investigations. (1-39)

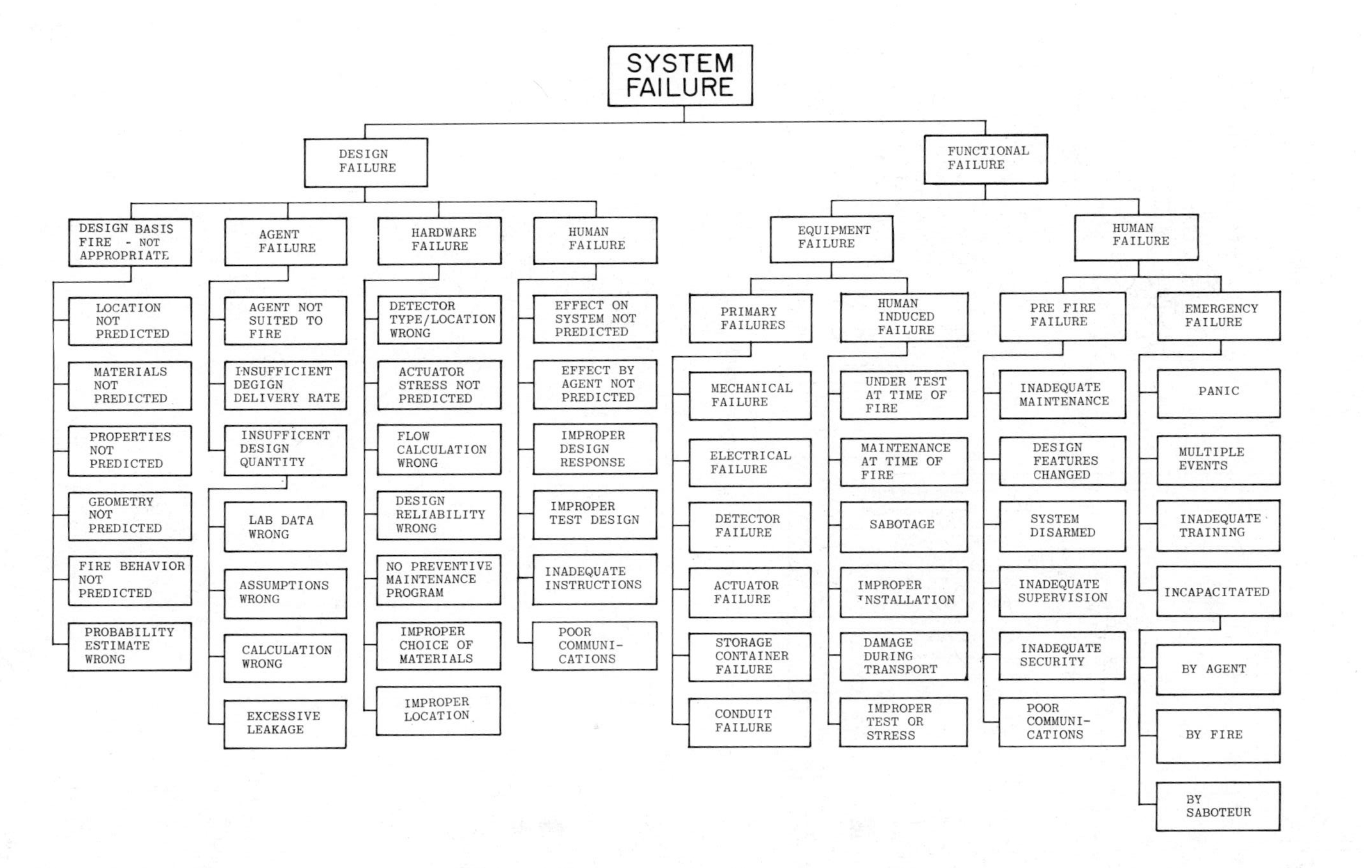

Figure 1. System effectiveness considerations

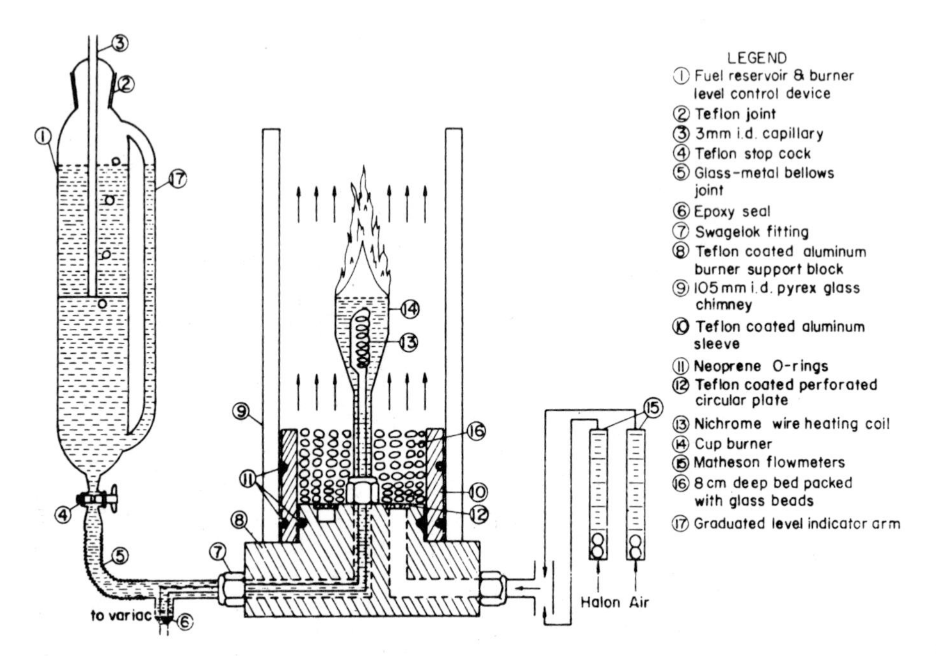

Figure 2. Experimental setup showing "cup burner"

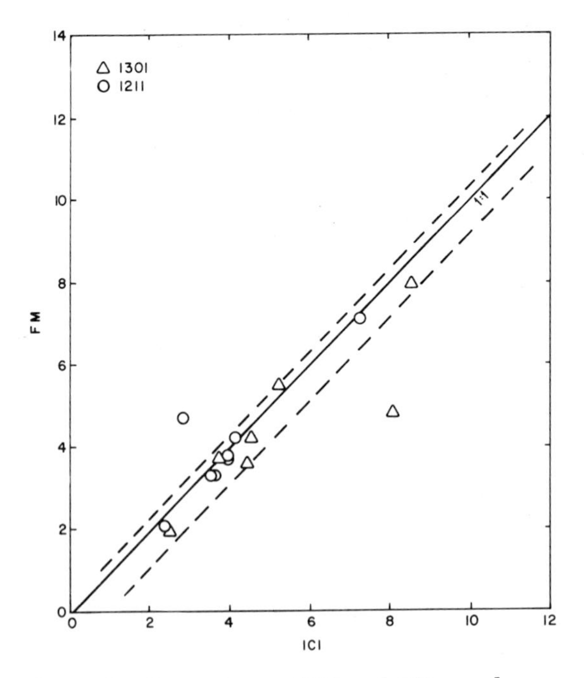

Figure 3. Comparison of FM and ICI cup burner data

Mr. C.L. Ford of DuPont has discussed a number of these in the preceding paper.

Precision

In fire research we are often confronted with the need to produce results which are approximately correct rather than precisely wrong from the standpoint of the ultimate application. It is more appropriate from an applicability standpoint to conduct full-scale simulations and apply a safety factor to the highest value obtained. Precise, reproducible experimental data must be available, however, to test theoretical models and form a basis for establishing confidence in the predictability of the results.

Since individual investigators are quite likely to reproduce their own experimental data, we have initiated an interlaboratory comparison of essentially equivalent experiments. Figure 2 shows the general arrangement of a laboratory "cup burner" experiment from which data were obtained. Figure 3 shows the relationship between results obtained using Halons 1211 and 1301 in which the fuel was heated to maintain bulk fuel temperature at the boiling point. A 1:1 equivalence line is shown for reference. Constructing an envelope and comparing possible answers at a common value of 5% indicates a potential error of up to 20% without considering whether the test accurately represents a realistic application condition. The points outside the envelope represent a research question, why? It would be quite useful to users of data if individual investigators would provide realistic assessments of the precision of their own experiments when only one laboratory is involved.

Accuracy

To some extent predictability is a measure of the accuracy of the experiment. While predictions based on fundamentals will be covered in the next section some comparisons of agents and experimental systems will be covered here.

A comparison of Halons 1211 and 1301 in various flame extinguishment and inerting (explosion prevention) experiments is shown in Figure 4. The vapor burner for which data are presented is shown schematically in Figure 5. The inerting data were obtained for both agents using an open top, closed bottom, 2-in. diameter tube with ignition at the bottom. (14)

Comparison of flame extinguishment and inerting data in general shows a systematic relationship as shown in Figures 8 (ICI) and 9 (Du Pont). The Du Pont "Mason Jar" Technique is described in reference (23).

To date there is little published data on the use of Halons for explosion suppression, and the data available show little systematic relation to inerting or extinguishment data (Figure 10).

A comparison of inerting data obtained in open top versus

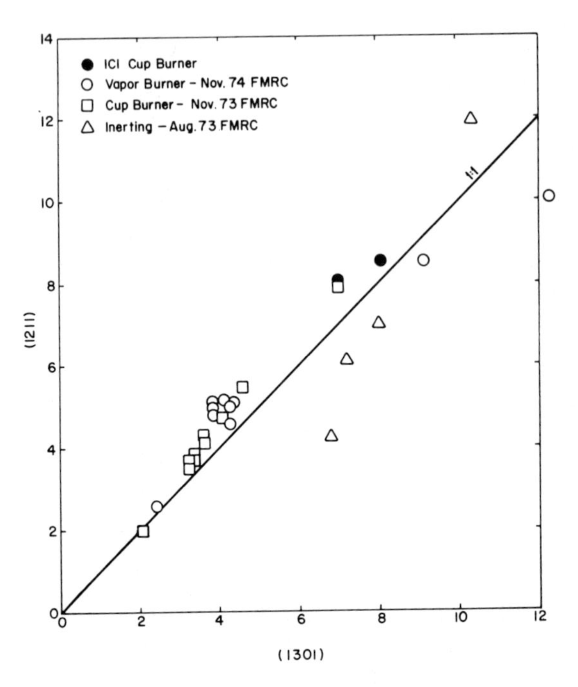

Figure 4. Relationship between Halons 1211 and 1301

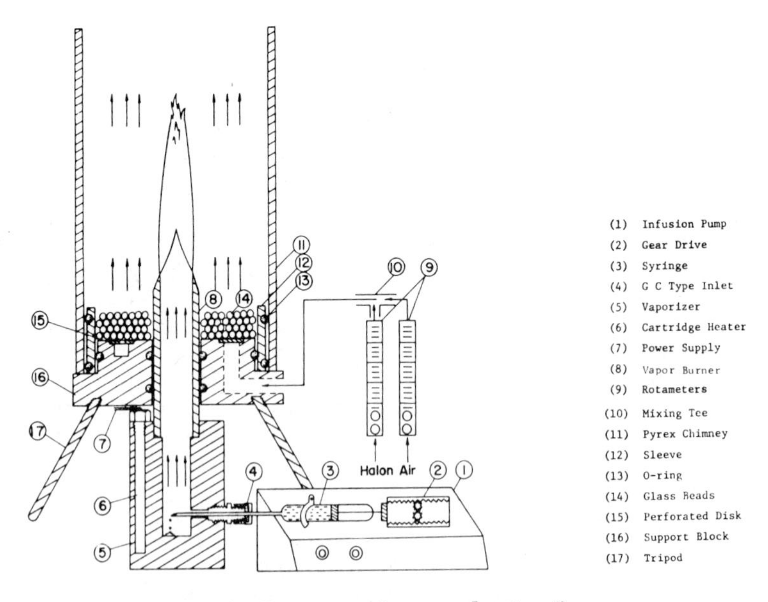

Figure 5. Experimental setup showing "vapor burner"

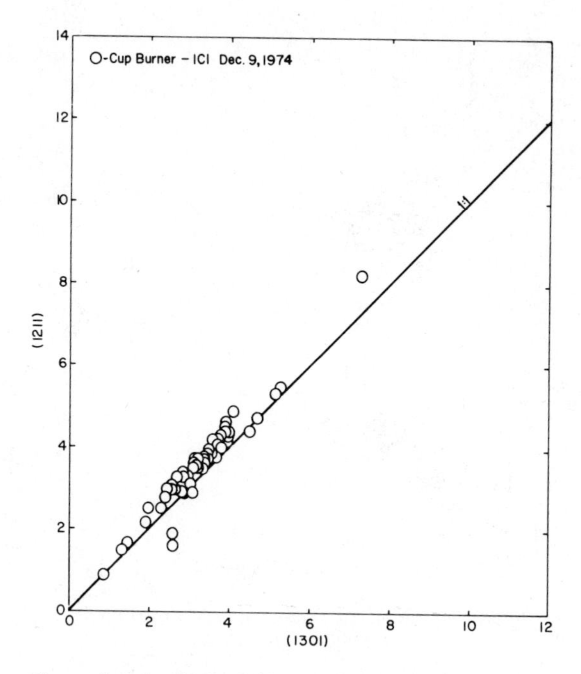

Figure 6. Analysis of ICI cup burner data for Halon 1211 and Halon 1301 with fuel at room temperature

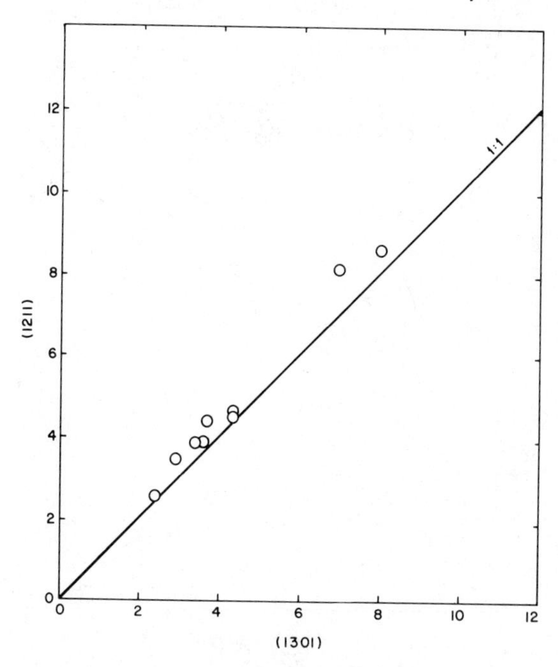

Figure 7. Analysis of ICI cup burner data with fuel at boiling point

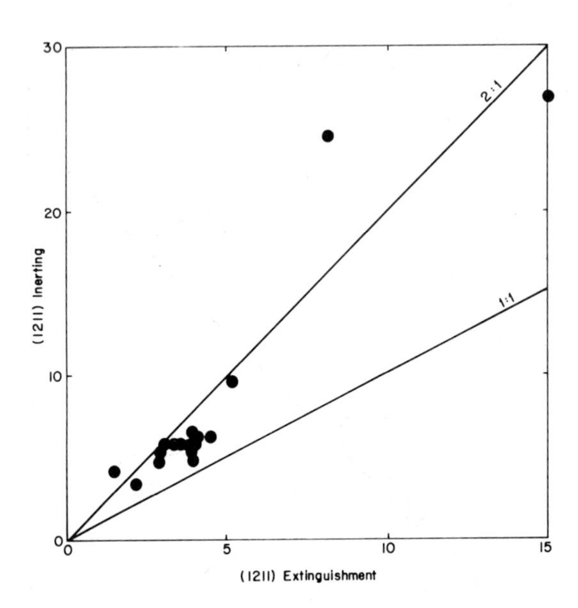

Figure 8. Relation between extinguishment and inerting (ICI data)

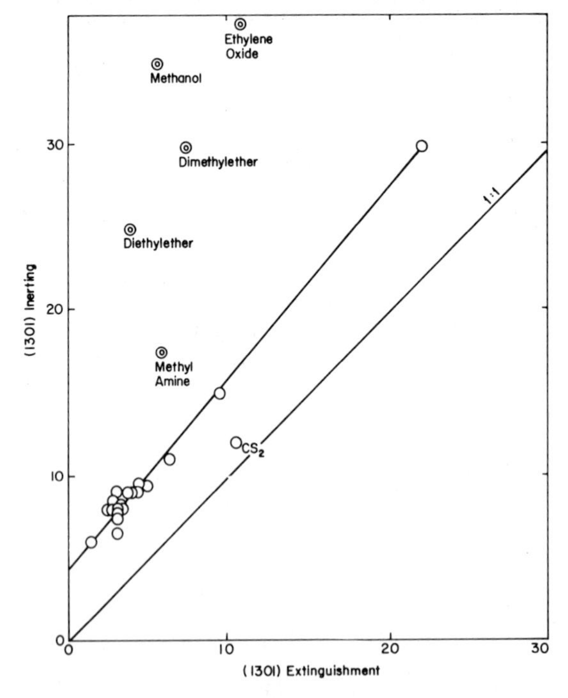

Figure 9. Analysis of Dupont "Mason jar" data for flame extinguishment and inerting with Halon 1301

open bottom tubes shows a reasonably tight correlation (Figure 11). Added experimental work is presently being conducted. Data comparing a porous glass burner (Figure 12) show a better correlation (Figure 13) with cup burner data for liquids than for gases, which give higher results due to greater preheating of the fuel.

Correlation of these laboratory experiments with the results of larger scale tests will be covered in the section on Applicability.

Predictability

Maximum confidence in design data would require that all values be entirely predictable and that these predictions be verified by an experiment that adequately represents all aspects of a predictable event. A complete model based on the material properties and the chemistry and physics of the situation is still somewhat beyond us; however, Tewarson (22) has made significant strides in developing a theoretical model of the radiation augmented oxygen index type experiment shown in Figure 14. Data for the extinguishment of plastics using Halon 1301 (with oxygen mole fraction held constant) are shown in Figure 15 and compared with oxygen index values for the same materials. This and other papers show that energy balance has a significant effect on results. Additional evidence of significant thermal effects on some materials is shown in Figure 16 which shows the effect of temperature on the extinguishing concentrations for ethylene glycol obtained by ICI using their heated cup burner. A similar effect on inerting concentrations was obtained by Lewis of ICI for n-Hexane and Halon 1211, as shown in Figure 17.

Since we have mentioned the oxygen index type of experiment it is noted that a systematic relationship seems to exist between Halon extinguishing and inerting requirements and the oxygen index values for the same materials as shown in Figure 18. Creitz (Figure 19) (30) has also suggested that the amount of inhibitor required for extinguishment is related to the amount of oxygen above that required for combustion (the oxygen index). This also suggests that there is a relationship between the Halons and the inert gases. If the inerting values for hydrogen, ethylene, propane-butane, and methane are plotted as a function of the heat capacity, a general trend is shown (Figure 20). Larsen (29) develops a thermal explanation of inhibition even further by calculating the total heat capacity of the system. A general trend of agent requirements is shown with fuel-air system parameters such as the lean limit equivalence ratio (Figure 21) and adiabatic flame temperature (Figure 22) reported by Hertzberg (24). Bajpai (14) in his paper also proposes a correlation based on flame temperature (Figure 23). Consideration of other materials using values for flame and autoignition temperatures reported by Woinsky (Figure 24) (25) throws some doubt on this approach. His intrinsic safety "Group Classification Temperature" also seems to

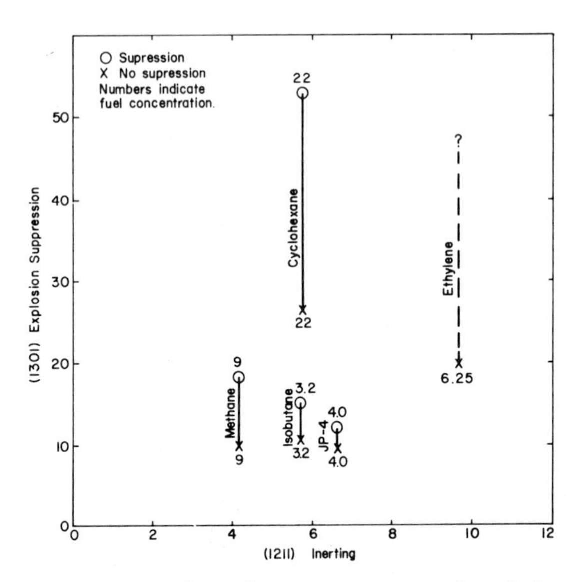

Figure 10. Relation between inerting and explosion suppression

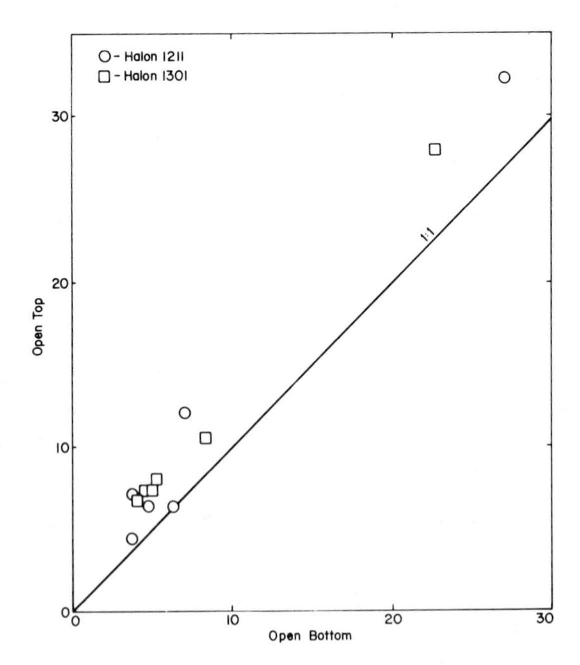

Figure 11. Comparison between Factory Mutual Research Corp. inerting experiments using two different techniques

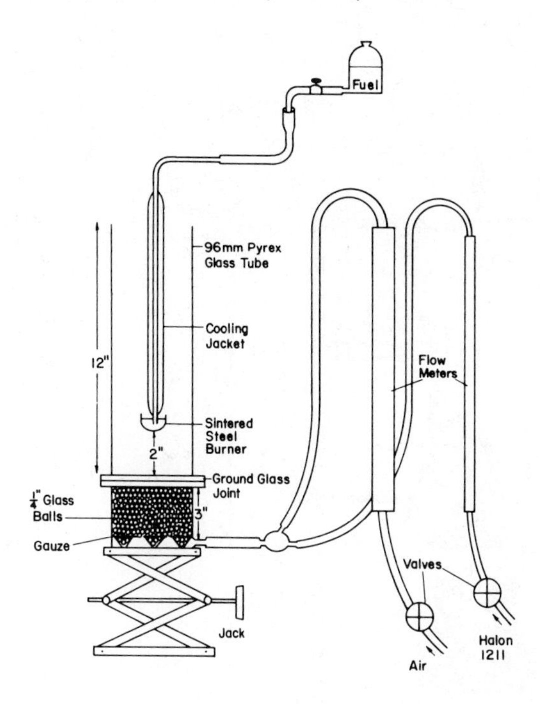

Figure 12.—ICI porous burner

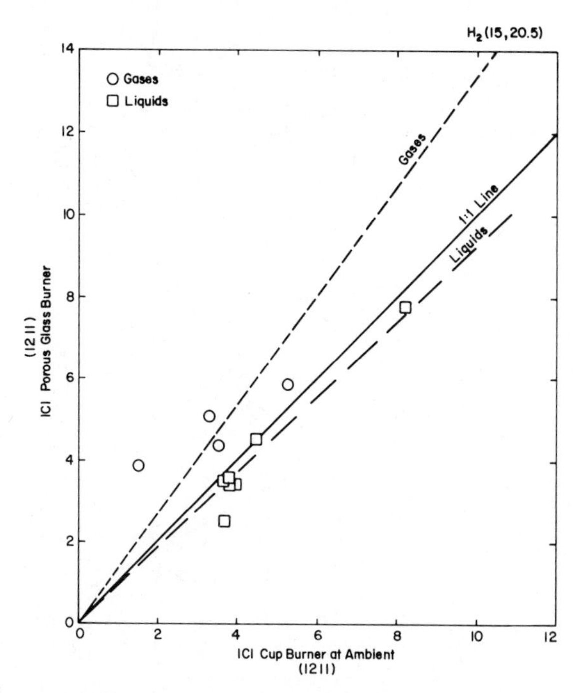

Figure 13. Comparison of ICI cup and porous burners

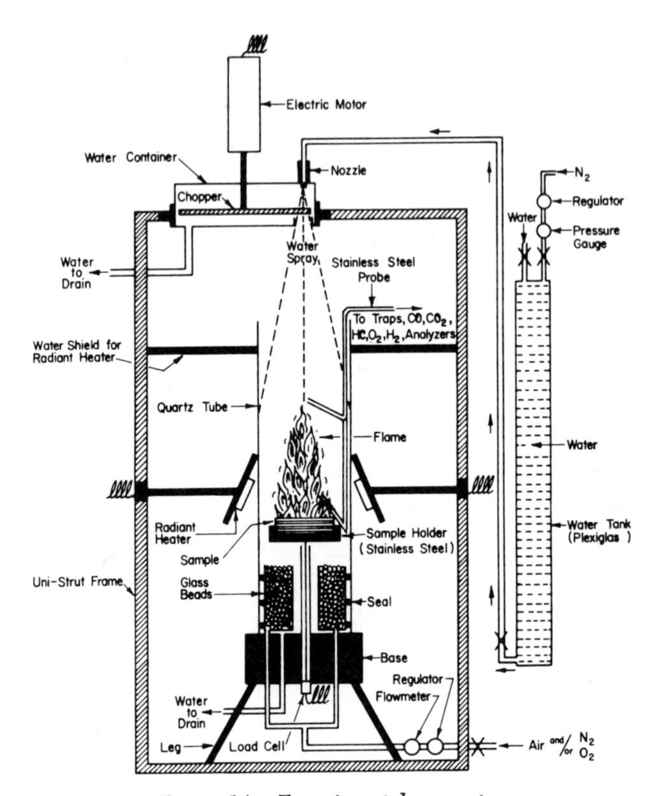

Figure 14. Experimental apparatus

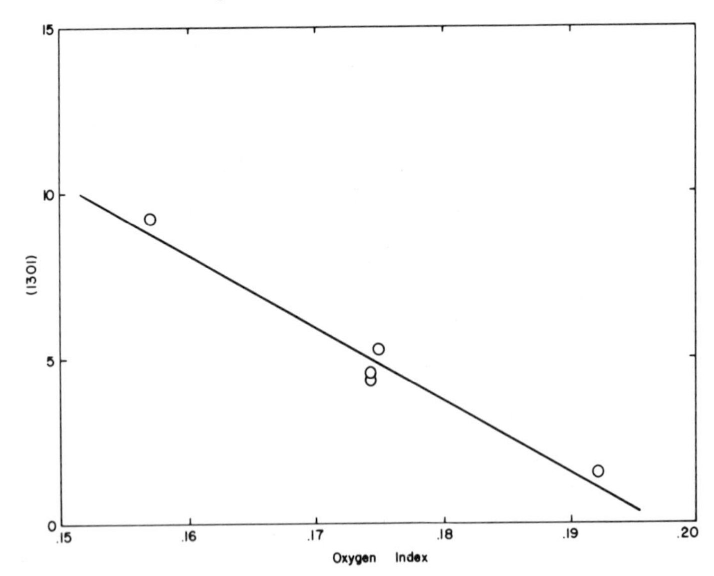

Figure 15. Comparison of oxygen index and Halon concentration (at constant 0.21 oxygen) (Data from Tewarson, 22)

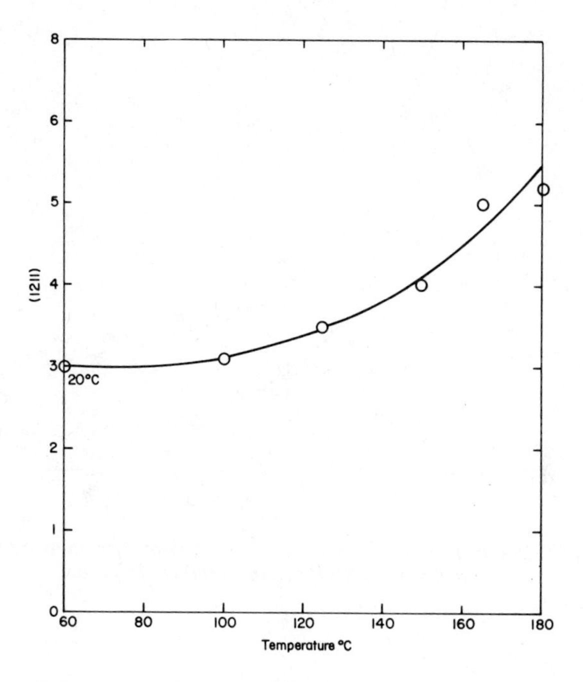

Figure 16. Effect of temperature on ICI cup burner data for ethylene glycol

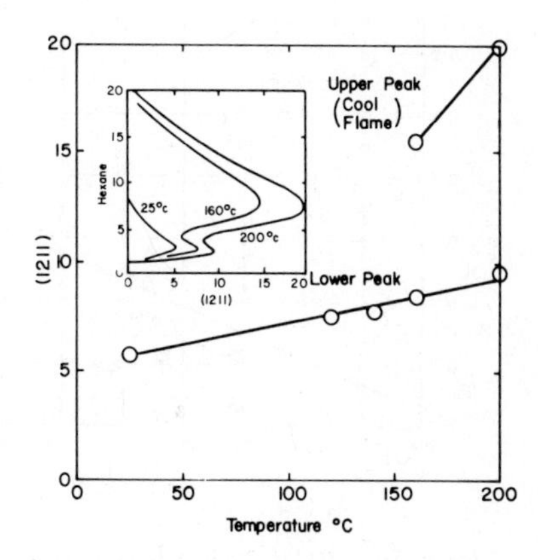

Figure 17. Inerting concentration for n-*hexane* vs. *temperature*

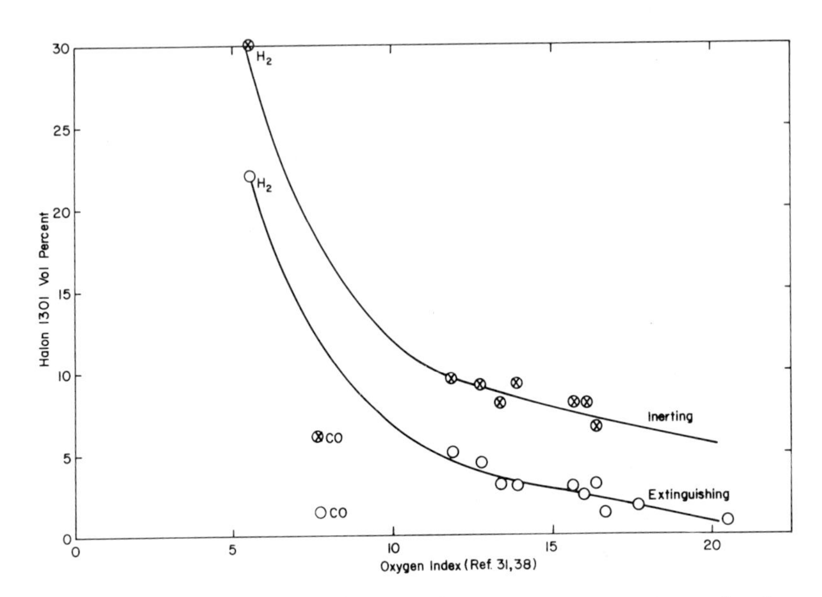

Figure 18. Correlation of Halon 1301 concentrations for inerting and extinguishment with the oxygen index of the fuel

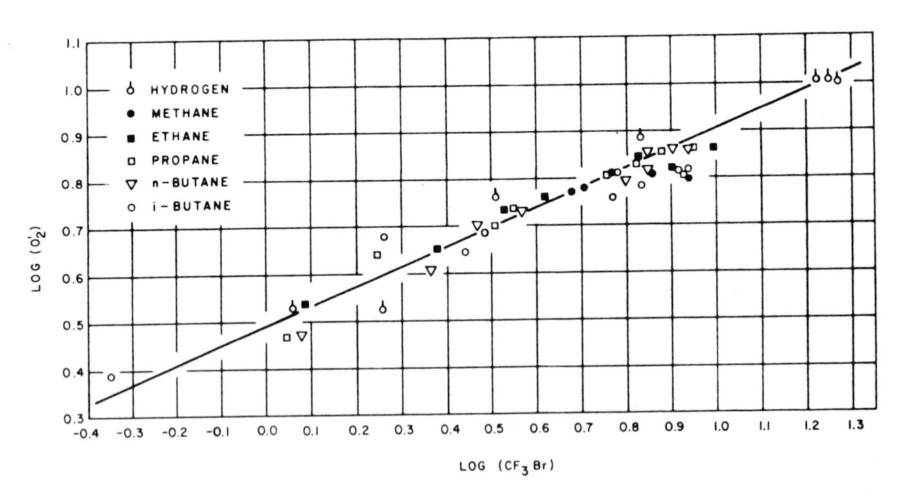

Figure 19. Relationship between oxygen concentration above the minimum required for combustion and the amount of CF_3Br required for extinction of a 2.4-cm high diffusion flame $[O_2'] = [O_2]$ (total)—$[O_2]$ (required for combustion) (data from Creitz, 30)

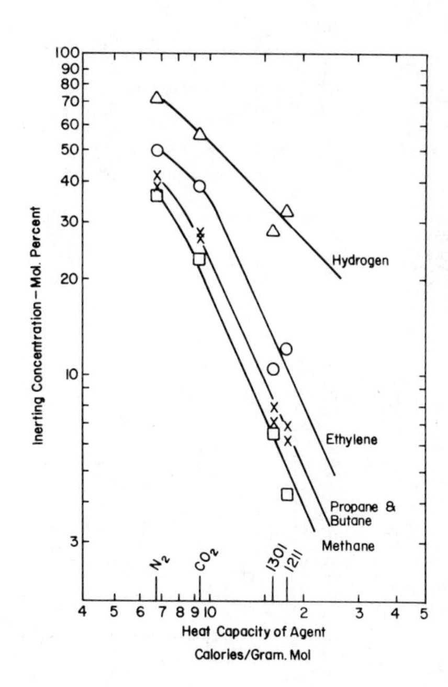

Figure 20. Inerting concentration vs. *heat capacity of agent*

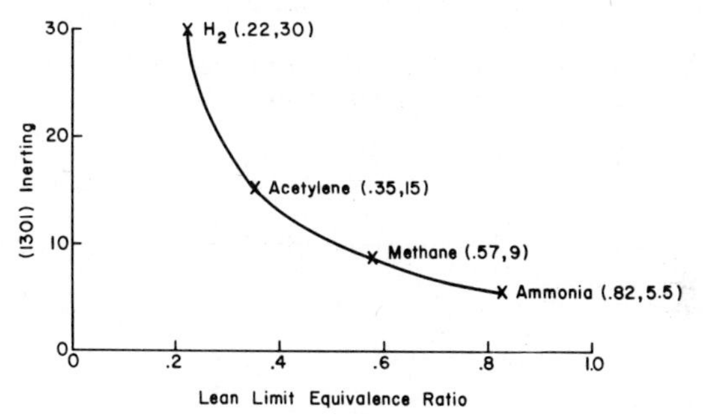

Figure 21. Relation of agent requirements to combustion properties of fuel

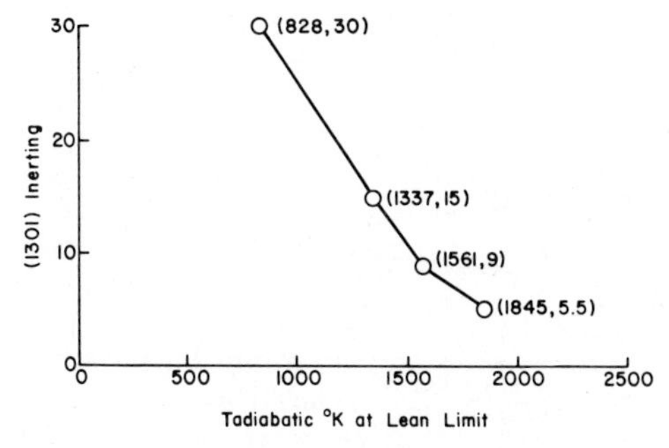

Figure 22. Relation of agent requirements to flame properties (24)

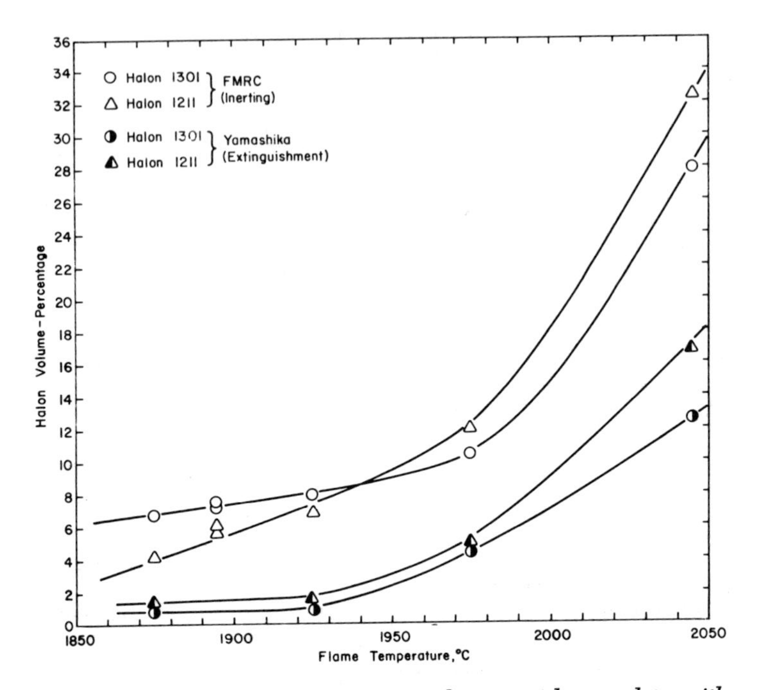

Figure 23. Correlation of inerting and extinguishment data with flame temperature (14)

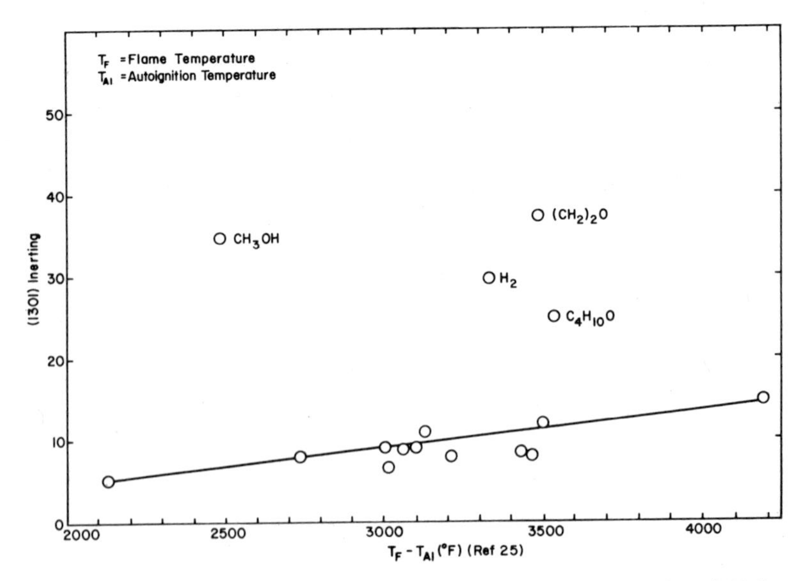

Figure 24. Relationship between combustion properties of fuel and Halon 1301 inerting concentrations

be of little use in correlating inerting data (Figure 25) and the reciprocal nature of flammability and extinguishability is questionable. If the analysis is limited to simple paraffinic hydrocarbons, data can be correlated using fuel boiling point (Figure 26), limit flame temperature (Figure 27), etc. However, when more complex fuels are added, simple correlations, such as with the number of carbon atoms, produce trends only within natural groups as shown in Figure 28. Use of "B" number or other parameter based on heat of combustion, boiling point or other thermal properties does little to improve the situation since they ignore chemical kinetic effects. Such combustion system parameters as flame velocity which contain both thermal and kinetic effects provide another possible basis for comparison (Figure 29) and directly relatable effects such as quenching distance provide a basis for correlating data on a wide variety of fuels as shown in Figure 30. Other combustion system parameters which include both thermal and kinetic effects, e.g., minimum ignition energy and ignition delay should provide useful correlations.

Forman Williams (26) has proposed the use of the Damköhler Number for correlating fire extinguishment data. This has yet to be demonstrated, and this or any other approach needs to be tempered by consideration of its practical as well as theoretical utility and the ease with which numerical values can be obtained. I have belaboured this point principally because there is a need for improved understanding to provide added confidence that unexpected effects will not nullify our design assumptions and create catastrophic consequences.

Applicability

The ultimate test of all laboratory data is its pertinence to the fire event which actually occurs. In this respect, even full-scale fire extinguishment tests represent only one set of conditions, although it is hoped that these conditions are representative of most, if not all, fire events that might occur. For instance, a test program oriented toward computer rooms might include fires in various configurations of plastics and cellulosic materials but not consider a spray fire caused by a leaking hydraulic system for tape drive units. Often these are considered "improbable" events without considering the actual numerical probability.

Most of the laboratory extinguishment studies of flammable liquids have been oriented toward applications in which a two-dimensional fuel surface is assumed. In the absence of unusual heating effects there is generally a rather good correlation between laboratory, intermediate and larger scale experiments as shown in Figures 32-35.

By varying experimental apparatus and technique some materials, such as nitromethane, ethylene glycol and isopropyl nitrate have been shown to be thermally sensitive in that modest in-

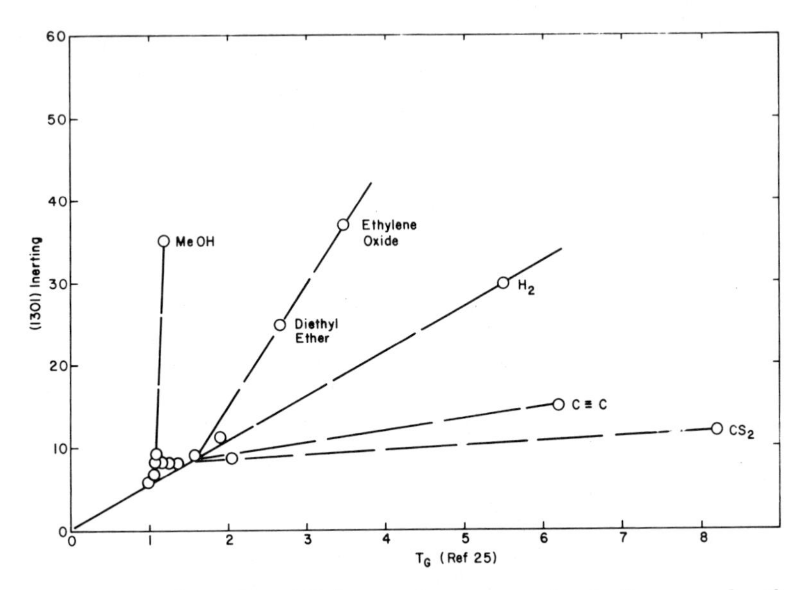

Figure 25. Relationship of Halon inerting concentration to the group classification temperature of the fuel

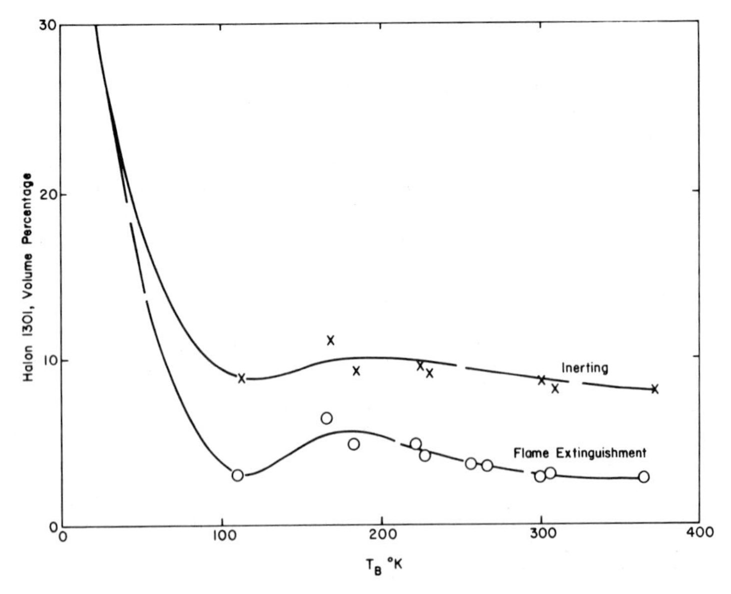

Figure 26. Relationship of agent requirements to boiling point of fuel

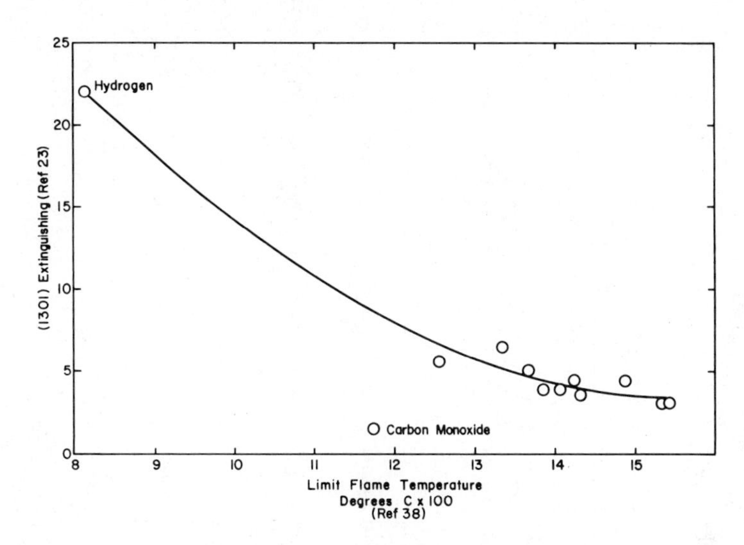

Figure 27. Comparison of flame extinguishment concentration with limit flame temperature

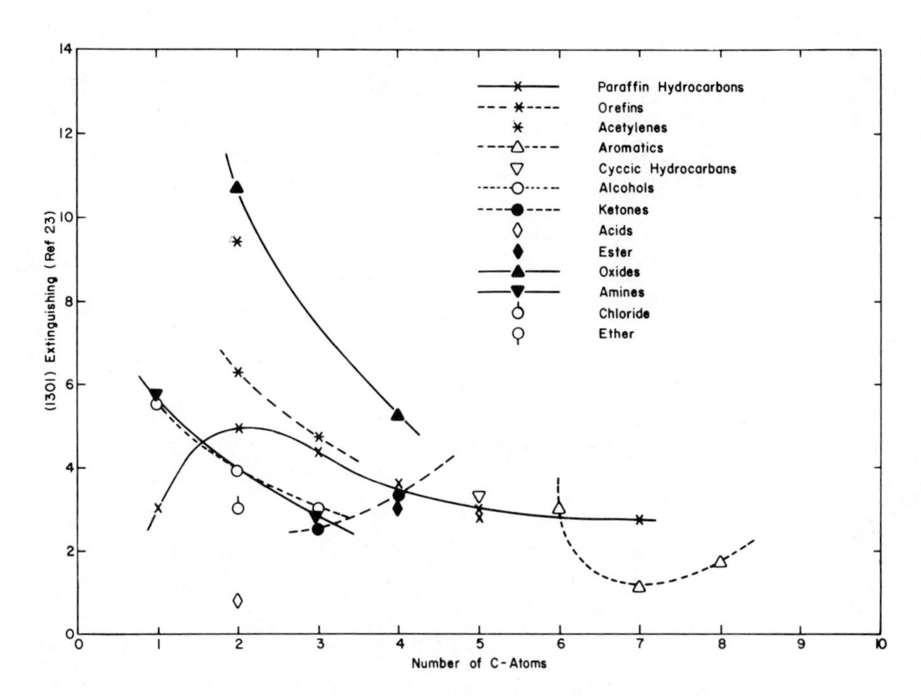

Figure 28. Relation of Halon concentration to number of carbon atoms in fuel (36)

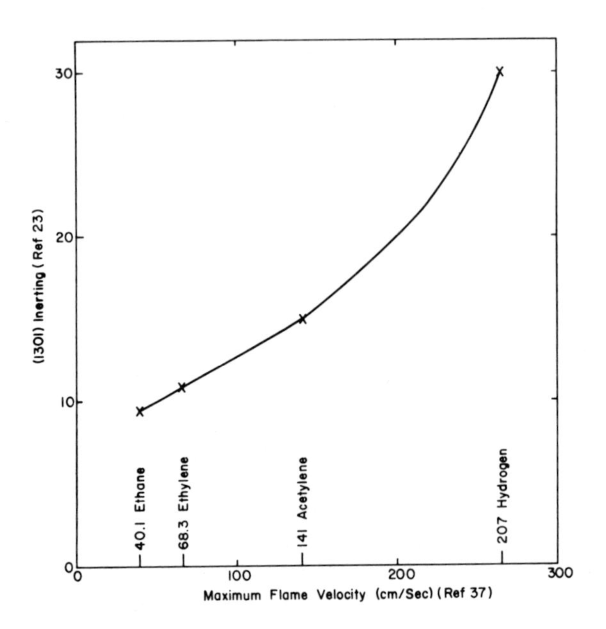

Figure 29. Halon concentration vs. *flame velocity*

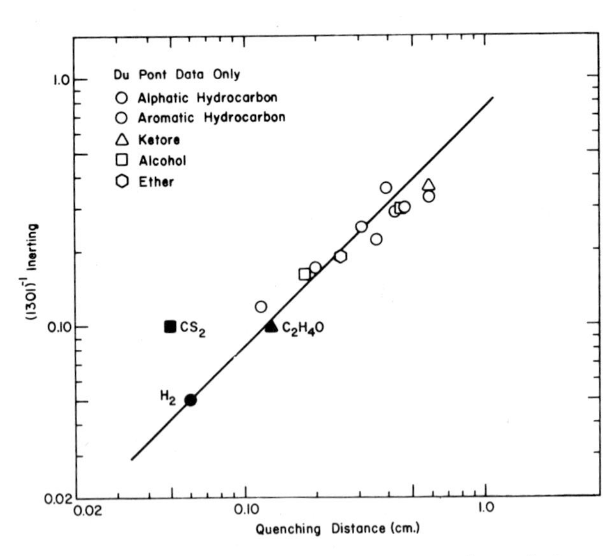

Figure 30. Halon concentration vs. quenching distance (22)

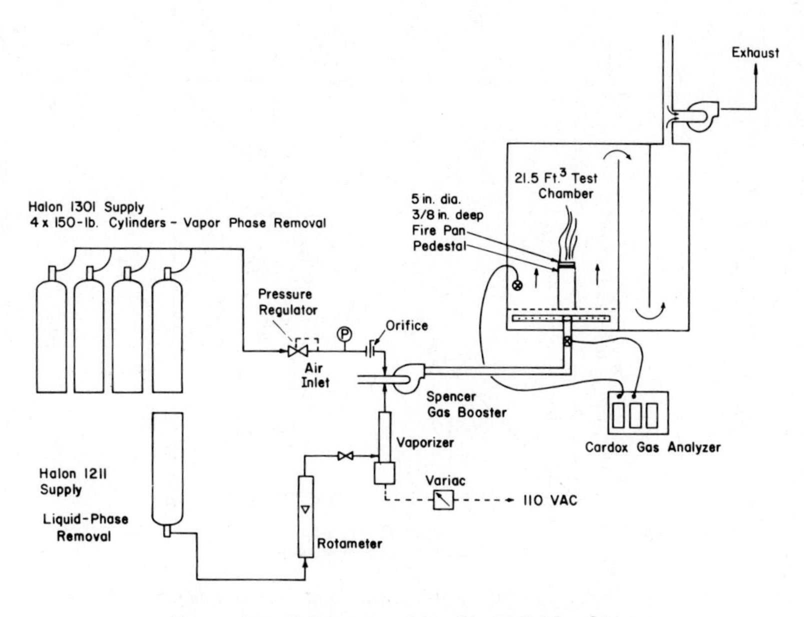

Figure 31. Schematic of Cardox 21.5-ft³ cabinet

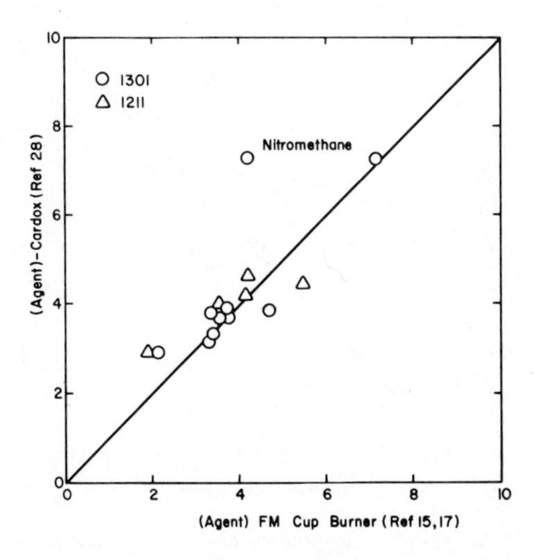

Figure 32. Results of Factory Mutual cup burner and Cardox intermediate scale tests

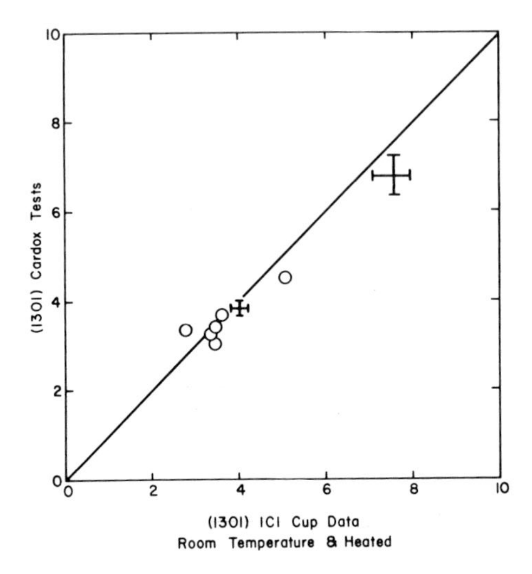

Figure 33. Laboratory and intermediate scale extinguishment tests

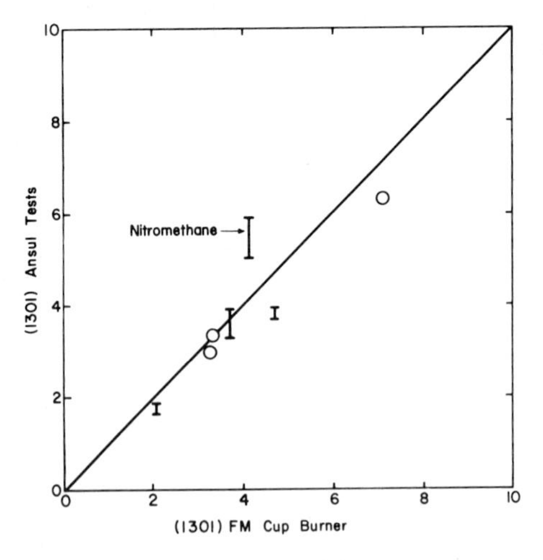

Figure 34. Laboratory and larger scale tests

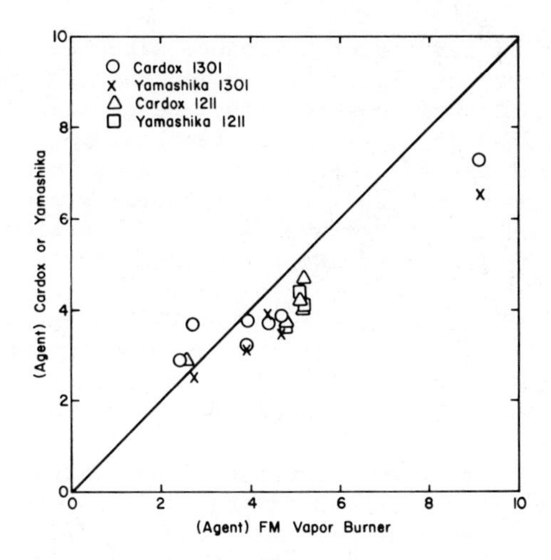

Figure 35. Factory Mutual vapor burner data vs. large-scale tests of Cardox and Yamashika

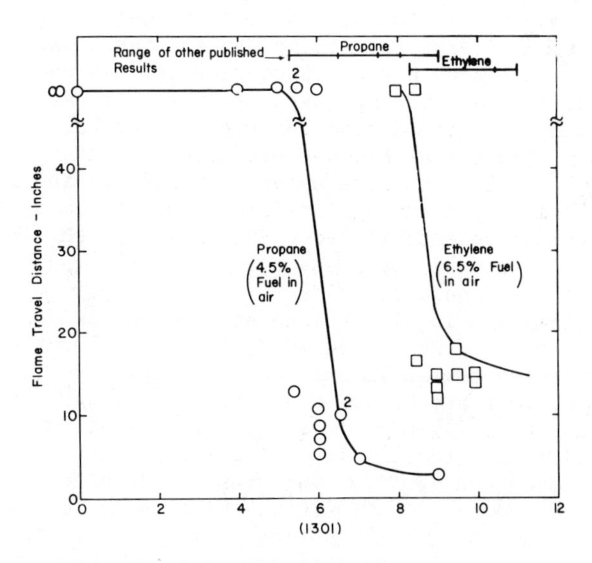

Figure 36. Large-scale inerting tests of Fenwal vs. *other published results*

creases in initial temperatures produce sizeable changes in the results. For instance, Figure 16 showed that the Halon 1211 extinguishing concentrations for ethylene glycol nearly doubled (from 3 to 6%) when going from room temperature to 180°C. Inerting concentrations for n-Hexane were shown to nearly double between room temperature and 200°C at a near stoichiometric mixture and, in fact, at about 150°C a second reaction zone in fuel rich mixtures was indicated which required nearly four times the concentration of agent for inerting (Figure 17).

In concentrating on two-dimensional fuel surfaces we have made a definite choice to limit the investigation to obtain a solid base of information. In using heated fuels and later in prevaporizing the fuels we have attempted to minimize thermal scaling effects.

There are, of course, flammable liquid fires in which the fuel surface is other than two dimensional; however, the quantities of fuel involved (see NFPA 12A/B par. A-2300) in this type of fire would most likely require the higher inerting concentration. There is still, however, some uncertainty as to whether inerting concentrations are necessarily pertinent to these three-dimensional fire situations since only a limited number of experiments have been conducted. For instance, in perusing the data which has been tabulated by Norm Alvares (27) the results for Marine Diesel Oil show a range of 1.1% to 4.4% Halon 1301 required for spray fires. Other investigations show a range of 3.3 to 5.6%, whereas the single inerting concentration reported is 4.4%. Additional consideration is obviously needed unless large safety factors are applied - and even these become questionable with only limited understanding. Both mixing and thermal effects influence the results and these effects must not only be defined but the probability spectrum of fire situations and the resulting stresses must be predicted with some degree of confidence. An understanding of the flammability and burning characteristics of materials is a critical element in the design and analysis of Halon systems.

Various fire scenarios must ultimately be described, the characteristics of these fire events defined and the requirements for extinguishment, inerting (prefire or postfire) and/or explosion suppression established. Elements of both fire and explosion can conceivably be present in the same event and the sequence must be considered.

In a flammable liquid storage area an initial spill could conceivably heat a drum of liquid causing it to vent hot vapor. We may extinguish the spill but how about the vapor jet? Can this jet, in turn, heat another drum and create a cascading effect? If the fire is extinguished will sufficient fuel be vaporized to create local if not totally homogeneous explosive mixtures? Are the temperatures and ignition energies predictable, defined, and their effects known? If a fuel rich mixture is formed can an explosive mixture be created during venting due to the influx of fresh air? Is an explosion suppression system needed as back-up if this situation is possible?

Certainly, NFPA Standards will never cover every situation. Probably no design guide ever will, but designers and users should be aware of potential effects and recognize the need for caution and perhaps added experimental data where questions exist. Since three-dimensional flammable liquid fires do exist, additional tests are needed using a fire array such as is being proposed by the International Standards Organization (ISO) Committee concerned with fire extinguishing agents (ISO TC 21/5). Other exploratory tests are needed to determine if common events such as leaking pipe may preheat the fuel and require higher agent concentration. Little work has been done with dusts, aerosols, sprays and stabilized flames. Electrical and electronic systems are using a wide variety of synthetic materials and experimental data on fire behavior and extinguishability is still scant. The use of Halons in storage and office areas is conceivable and a wide variety of materials may be involved. Other investigators are generating data of various types. If parallels can be drawn between oxygen index, quenching diameter or other measurements, perhaps it will not be necessary to completely duplicate all measurements using Halons. This does, however, require a considerably broader perspective than has been prevalent to date.

Conclusions

Through cooperative research we now have a fairly large body of data with which to assess precision, accuracy, predictability and applicability. Guidelines for many common materials and situations can now be developed with greater confidence.

More importantly, however, these investigations are showing that thermal and other effects are significant and indicate caution in all design exercises. Future design problems should include a more detailed consideration of the nature of the combustion and the sequence of events leading to its extinction.

Experimental data should be applied with caution and where uncertainty exists, this data should be confirmed with realistically scaled experiments pertinent to the situations that might be expected.

A better understanding of combustion and the agents used in its inhibition will not only increase our confidence but provide the theoretical models and data upon which more complete engineering analyses can be based.

Added fundamental and empirical research is not so much needed to discover new agents as to permit us to employ intelligently those that do exist. While very specific studies of very specific systems are needed, we must also ultimately establish the fundamental relationships between diverse systems. Only then will we be able to design with maximum confidence.

Acknowledgments

The author appreciates the support of DuPont, ICI, Ansul, Cardox, Fenwal and the Factory Mutual System in making available fire extinguishment, inerting and explosion suppression data for analysis. In addition, DuPont has provided financial support for much of the experimental work conducted at the Factory Mutual Research Corporation by S.N. Bajpai and others.

The conclusions to date are largely the author's and final resolution remains a function of the data review subcommittee consisting of Messrs. C.L. Ford, DuPont; J. Novotny, ICI; J. Riley, Ansul; and G. Grabowski, Fenwal with the author as chairman. Revised guidelines are expected to be incorporated into the appropriate NFPA Standards in 1976.

Abstract

This paper addresses the Precision, Accuracy, Predictability and Applicability of experimental data for fire extinguishment, inerting and explosion suppression using Halons 1301 and 1211.

Graphical comparisons of data showing the effect of experimental systems, technique, agent, temperature and scale on the results provide a guide to safety factors required in system design.

Several proposed correlation schemes based on simple fuel properties and flame temperatures are shown to break down when a variety of fuels is considered.

Correlations are also shown which indicate that agent requirements are predictable from a knowledge of oxygen index or quenching diameter.

The purpose of this paper is to review graphically current data and ideas and stimulate further investigation toward a more complete understanding of fire and explosion suppression phenomena of practical importance to the design engineer.

Literature Cited

(1) Smith, J.B., Cousins, E.W., Slicer, J.S., Miller, M.J., Pote, R.L., Jr., "Packaged, Self-Contained Fire Suppression System for Use in Remote Areas Where A Normal Water Supply Is Not Available for Structural Fire Fighting," NTIS-AD No. 466 694, May 1965.

(2) Smith, J.B., Miller, M.J., Pote, R.L., Jr., "Fire Tests of Two Remote Area Fire Suppression System Concepts," NTIS-AD No. 475 343, November 1965.

(3) Fitzgerald, P.M., Miller, M.J., "Evaluation of the Fire Extinguishing Characteristics of "Freon" 1301 on Flammable Liquid Fires," Factory Mutual Research Corporation Report RC67-TC-8, February 1967.

(4) Miller, M.J., "Extinguishment of Charcoal, Wood & Paper Fires by Total Flooding with "Freon"1301-Air and CO_2-Air Mixtures," Factory Mutual Research Report RC67-TC-9, July 1967.
(5) Troup, E.W.J., "Nomographs for Design of "Freon" 1301 Total Flooding Systems," Factory Mutual Research Corporation Report RC67-TC-10, November 1967.
(6) Miller, M.J., "The Effect of Gaseous Inhibition in Homogeneous Combustion Systems," Factory Mutual Research Corporation Report RC67-TC-11, December 1967.
(7) Fitzgerald, P.M., Miller, M.J., "Evaluation of the Fire Extinguishment Characteristics of "Freon" 1301 on Flammable Liquid Fires," Factory Mutual Research Corporation Report RC67-TC-12, February 1967.
(8) Livingston, W.L., Miller, M.J., Troup, E.W.J., "A Study of the Industrial Market for "Freon" 1301 Fire Extinguisher Systems," Factory Mutual Research Corporation Report RC67-TC-15, November 1967.
(9) Yao, C., ""Freon" 1301 Fire Protection System Total Flooding and Local Application," Factory Mutual Research Corporation Report RC67-P-16, November 1967.
(10) Yao,C., Smith, H.F., "Convective Mass Exchange Between a "Freon" 1301-Air Mixture in an Enclosure and the Surrounding Through Openings in Vertical Walls," Factory Mutual Research Corporation Report RC67-T-2, January 1968.
(11) Crandlemere, W., Miller, M.J., "Analysis and Toxicity of Gases Produced by Extinguishing Fires with Bromotrifluormethane," Factory Mutual Research Corporation Report RC68-TC-6, February 1968.
(12) Miller, M.J., "The Extinguishment of Flammable Liquid by $CBrF_3$," Factory Mutual Research Corporation Report RC68-TC-7, March 1968.
(13) Buckley, J.L., "Evaluation of a Novel, Slurry-Type Fire Extinguishing Agent (Final Technical Report) (WPAFB), Factory Mutual Research Corporation Report RC71-T-31, August 1971.
(14) Wagner, J.P., Bajpai, S.N., "Study of Inerting Characteristics of Halons 1301 and 1211," Factory Mutual Research Corporation Report RC73-T-26, August 1973.
(15) Bajpai, S.N., "An Investigation of the Extinction of Diffusion Flames by Halons," Factory Mutual Research Corporation Report RC73-T-37, November 1973.
(16) Bajpai, S.N., "Measurement of the Extinction of Direct Fuel-Vapor-Fed Diffusion Flames by Halons," Factory Mutual Research Corporation Report RC73-P-39, November 1973.
(17) Bajpai, S.N., "An Investigation of The Extinction of Diffusion Flames By Halons," Factory Mutual Research Corporation Report RC74-TP-4, February 1974.
(18) Bajpai, S.N., "Inerting Characteristics of Halons (Du Pont)," Factory Mutual Research Corporation Report RC74-T-32, July 1974.

(19) Bajpai, S.N., "Inerting Characteristics of Halogenated Hydrocarbons (Halons)," Factory Mutual Research Corporation Report RC74-TP-43, August 1974.
(20) Bajpai, S.N., "Extinction of Turbulent Jet Diffusion Flames By Halons 1301 and 1211," Factory Mutual Research Corporation Report RC74-T-52, November 1974.
(21) Bajpai, S.N., "Extinction Concentration Measurements of Halons 1301 and 1211 For Vapor-Fed Diffusion in Flames," Factory Mutual Research Corporation Report RC74-P-53, November 1974.
(22) Tewarson, A., "Flammability of Polymers and Organic Liquids, Part 1 - Burning Intensity," Factory Mutual Research Corporation Report RC75-T-6, February 1975.
(23) Floria, J.A., "Du Pont FE1301 Fire Extinguishant, Fuel Inerting Concentration-Summary Report," E.I. du Pont de Nemours & Co., Inc., Wilmington, Delaware, March 1972.
(24) Burgess, D. and Hertzberg, M., "The Flammability Limits of Lean Fuel-Air Mixtures - Thermochemical and Kinetic Criteria for Explosion Hazards," ISA AID 74414 (61-70), 1974.
(25) Woinsky, S.G., "Predicting Flammable-Material Classifications," Chem. Eng. pp 81-86, November 27 , 1972.
(26) Williams, F.A., "A Unified View of Fire Suppression," J. Fire and Flammability Vol 5 (January 1974), pp 54-63.
(27) Alvares, N. J., This Symposium.
(28) Ford, C.L., "Intermediate Scale Extinguishment Tests With Halon 1301 and Halon 1211," Report of Work Done at The R&D Division of Cardox, May 8, 1974.
(29) Larsen, Eric, R., "Halogenated Fire Extinguishants: A Physical Mechanism of Fire Suppression," Paper No. INDE 054 at The 166th National ACS Meeting.
(30) Creitz, E.C., "Extinction of Fires by Halogenated Compounds - A Suggested Mechanism," in Fire Protection by Halon's UFPA Publication No. SPP-26, (1975) pp 68-77.
(31) Nelson, G.L. and Webb, J.L., "Oxygen Index of Liquids - Technique and Application," J. Fire and Flammability, Vol 4, July 1973, pp 210-226.
(32) Booth, K., "Critical Concentration Measurements for Flame Extinguishment of Diffusion Flames at Ambient and Elevated Bulk Fuel Temperatures Using a Laboratory 'Cup Burner' Apparatus," ICI, RES/KB/NSS, December 9, 1974.
(33) Anon, "Halon 1211 Systems: Determination of Extinction Concentrations in The Laboratory," ICI Technical Report 6 RES/NF/RH, February 2, 1970.
(34) Lewis, D.J., "Provisional Draft: The Effect of Initial Temperature of n-Hexane - BCF - Air Mixtures on BCF Peak Concentration Values," ICI - Mond Division, March 13, 1974.
(35) Lewis, D. J., "Provisional Draft: Summary of Results for BCF Inerting Obtained in a 2-in. Vertical Tube Apparatus Up to September 1973," ICI Mond Division, September 13, 1973.

(36) Kanury, M., Internal Factory Mutual Research Corporation Memo to M. J. Miller dated October 11, 1972.
(37) Metzler, Allen J., "Minimum Ignition Energies of Six Pure Hydrocarbon Fuels of the C_2 and C_6 Series, NACA RM E52F27, August 11, 1952.
(38) Vanpee, Marcel, "The Ignition of Combustible Mixtures by Laminar Jets of Hot Gases," USBM RI6293 (1963).
(39) Kubin, R. F. and Gordon, A. S., "Inhibition of Liquid Methanol Fuel Flames," Presented at The Seminar on Experimental Methods in Fire Research, May 9-10, 1974 at Stanford Research Institute.

DISCUSSION

C. L. FORD: The comment was made that in the small-scale inerting tests perhaps the influence of varying ignition energies was simply an influence of temperature increases.

The data which I presented seems to reinforce this observation, and further to suggest that some fuels are more sensitive to this effect than others. In the DuPont explosive burette data (Table XXI in my paper). The hydrocarbon fuels are affected only moderately by varying the ignition energy from 27J/sec to 176J. Dimethyl ether, however, is drastically affected. This would indicate that perhaps each fuel has an "activation energy" which would produce similar results to dimethyl ether.

N. J. BROWN: The match (as an ignition source) is described in energy units while the electrical ignition is described in units of energy/time. Comparison is only possible between qualities in identical units.

A. D. LEVINE: In comparing inerting characteristics with explosion inhibiting characteristics for given agents, one must consider geometry, pressure, etc. as crucial factors. Has any effort in this direction been attempted?

M. J. MILLER: Differences in pressure will certainly affect the results. Both temperature and pressure effects need to be defined in more detail. Questions of geometry become less significant if the definition for "inert" permits little flame travel from the point of ignition.

J. C. BIORDI: You mentioned apparent anomalies in results obtained from experiments carried out in top open or bottom open configurations. Aren't the results understandable from buoyancy considerations, as, e.g., the differences in flammability limits for upward and downward propagation?

M. J. MILLER: The effects are explainable. In both cases flame propagation is upward and the differences arise due to the way in which burned gases are vented.

A. S. GORDON: A mix which is sparked has flame propagation as a result should appear identical to the unburned mix as does the normal flame propagation. Yet the inerting mix requires as much as twice the concentration of quenchers as does the extinguishment situation. This is very disconcerting to me.

K. L. WRAY: It seems to me that some of your attempts to correlate the agent concentration required with various parameters are misleading and tend to obscure the phenomenology which you are trying to understand. The correlations work for hydrocarbons but fail when the agent molecule includes oxygen. This is not surprising, it merely shows the importance of the details of the chemistry. Most chemical and physical properties of hydrocarbons scale reasonably well with chain length, hence, your correlations do not shed much light on what is going on.

M. J. MILLER: I have attempted in the paper to show both the breakdown of some proposed correlation schemes when a wide variety of fuels is considered, and the potential success of others. Both are presented as a basis for further discussion and to highlight those fuels which vary most widely from the norm. In most cases these are fuels containing oxygen and an extended investigation of these fuels may lead to better explanations of the mechanisms involved and permit predictions based on basic fuel properties and the oxygen required for combustion.

3

CF_3Br Suppression of Turbulent, Class-B Fuel Fires

NORMAN J. ALVARES[+]

Stanford Research Institute, Menlo Park, Cal. 94025

Based on tests conducted by the Naval Ship Engineering Center at the Philadelphia damage control training center, CF_3Br[*] was selected as the agent for extinguishing combination bilge and oil spray fires in engine rooms. Consequently, 1301 total flooding systems are now being designed to protect spaces aboard Navy ships that contain quantities of flammable liquids and gases.

Investigation of literature pertaining to concentration of agent required to extinguish a variety of fuels was disappointing, since many of the flammable liquids and gases common to Navy ships were not included in the tables of 1301 suppression data. Furthermore, the testing methods, criteria, and critical test parameters were not easily identifiable or the procedures were not entirely satisfactory for large compartment fires, e.g.;

1. Most quantitative data were obtained using laminar fires; no attempt was made to extrapolate these data to large turbulent fires.

2. In general, only a particular set of environmental conditions was used to establish a single value for extinguishment concentrations.

3. In large-scale tests critical parameters such as oxygen depletion, ventilation, agent distribution, and fuel burning rate were generally ignored.

[+]The work presented in this paper was supported by the Naval Surface Weapons Center, White Oak, Maryland.

[*]From here on in this paper we will use 1301 instead of CF_3Br.

Because of these problems the following objectives were established for this task:

1. Examine existing techniques for delineating 1301 extinguishment concentrations and summarize available data.

2. Identify or design a satisfactory test method to establish the agent's critical concentration for extinguishment (CCE) for various flammable liquids and gases of interest to the Navy.

3. Demonstrate the adequacy of the selected test method.

4. Measure the CCE for Navy flammable fluids and gases over a specified range of ventilation and fuel temperature conditions.

Status

The available-pertinent data are summarized in Tables I and II. Table I is a summary of both extinguishing and inerting concentrations for large and small scale tests. These data are derived from reports and papers currently available in the open literature and from commercial and governmental sources. The first row under the tabular headings contains a code that identifies the test method used for obtaining these values. This code, a brief description of the test methods, and the agencies using these methods, are contained in Table II.

In Table I it is apparent that the agreement in either inerting or extinguishing results is fairly good. However, there appears to be considerable scatter in the CCE's from large-scale tests, which is not apparent in the small scale. This non-uniformity can be attributed to several factors. First, there is the inherent difficulty in monitoring the parameters and controlling the environment of large-scale tests, and second, the method by which the agent is applied to the fire and the equilibrium concentration of the agent at the time of extinguishment are difficult to define. Generally, the agent is distributed by an array of nozzles; the application velocity of the agent, the entrainment characteristics of the fire, the burning rate, the air temperature, the convection cells, air leaks, and internal obstacles all interact to modify agent distribution. Thus, at the time of extinguishment, the agent concentration can vary from zero to much greater than the design concentration, depending on where (and how) the measurement is secured.

Table I

COMPARISON OF 1301 CONCENTRATION FOR EXTINGUISHMENT AND INERTING OF VARIOUS FUELS BY CURRENT ESTABLISHED TESTING PROCEDURES
(Volume Percent)

Fuel Parameters					Extinguishment values										Inerting Values			Exting. Summary		Inert Summary		Ratio
	SRI	NFPA* 12-A	Fire Research Institute Japan	DuPont (Hot fuel)	Ansul (for DuPont)	ICI America Inc.	Factory Mutual (for DuPont)	DuPont	U.C. San Diego	China Lake	USCG Mobile	USN DCTC Philly	NRL	NFPA* 12-A	DuPont Dupont	Factory Mutual (for DuPont)	Ansul (for DuPont)	Average Exting. Conc.	Maximum Exting. Conc.	Average Inert Conc.	Maximum Inert Conc.	Inerting (avg) Extinguish (avg)
Table 2, Technique Code	I	J	E	F	J	G	G	I	H‡	G	K	K(-)	I	A	B and C	A						
Acetone			3.0	2.8	3.0-3.63	3.5	3.7							5.3	8.0			3.28	3.7	6.65	8.0	2.0
Acetylene			7.0	8.5											15.0			7.75	8.5	15.0	15.0	1.9
Benzene	2.7		3.1	3.5		2.9	3.3		1.9					4.3	6.5			2.9	3.5	5.0	6.5	1.6
Carbon Disulphide		12.0	7.8	10.0		2.6								12.0	12.0			8.1	12.0	12.0	12.0	1.5
Carbon Monoxide		1.0	2.0	1.3											6.0			1.43	2.0	6.0	6.0	4.2
Diethyl Ether			3.9	3.5	3.34-5.04		4.3							6.3	25.0			4.01	5.04	15.65	25.0	3.9
Ethanol		4.0	3.5	3.5		3.9	4.0							4.0	9.0			3.78	4.0	6.5	9.0	1.7
Ethylene		7.2	4.4	5.8										11.0	11.0	10.5	5.4-6.05	5.8	7.2	8.78	11.0	1.5
Hydrogen		20.0	12.7	20.0						18.0				20.0	30.0	28.0		17.7	20.0	26.0	30.0	1.5
Methane		2.0	0.9	2.8		1.5				1.7				6.3	9.0	6.8		1.8	2.8	7.36	9.0	4.1
Methanol		4.0	6.5		6.33-7.06	7.3	8.1		7.8						35.0		7.3-7.6	6.73	8.1	16.63	35.0	2.5
Pentane			3.6	2.8		3.3									8.5			3.23	3.6	8.5	8.5	2.6
Propane		3.2	1.9	4.0		3.1								6.5	9.0	8.0		3.05	4.0	7.83	9.0	2.6
N-Butane		2.9		3.3		3.0				2.0					9.0	7.2		2.8	3.3	8.1	9.0	2.9
i-Butane		3.3		3.3		3.0									9.0	7.2		3.2	3.3	8.06	9.0	2.5
Ethane				4.5											9.5			4.5	4.5	9.5	9.5	2.1
N-Heptane		3.7		3.0		3.4	3.7		3.45					8.0	8.0			3.45	3.7	8.0	8.0	2.3
Toluene	2.0		2.0	1.8		1.9	2.7											2.1	2.7	-	-	-
Linseed Oil			1.9															1.9	1.9	-	-	-
Kerosene		2.8	2.5	3.3														2.87	3.3	-	-	-
Marine Diesel Oil (Bunker C)				3.8	5.65		3.3	1.1 Spray 4.4			3.4-5.6				4.0			3.89	5.65	4.0	4.0	1.0
Lube Oil	2.0																					
Naval Distillate	2.35			3.5	3.16-4.18		3.1					1.0-7.0	2.0					3.42	7.0	-	--	-
Hydraulic Jack Oil	2.0			2.3														2.3	2.3	-	-	-
Transformer Oil				3.0														3.0	3.0	-	-	-
JP-4				3.5										6.6				3.5	3.5	6.6	6.6	1.9
JP-5	2.5			3.0											8.0			3.0	3.0	8.0	8.0	2.7
Gasoline	2.5		3.5	3.3		3.5									8.0			3.43	3.5	8.0	8.0	2.3
Deisel	2.7																					

*10% Safety factor for design concentration included (these data also listed in DuPont Advertisements).
+Data range for spray- bilge- and spray plus pilge fires.
‡Maximum value at lowest air flow rate.

Table II

DESCRIPTION OF TESTS USED FOR DETERMINING INERTING AND EXTINGUISHING CONCENTRATIONS FOR VARIOUS LIQUID AND GASEOUS FUELS

Key	Scale/Purpose	Designer/User	Description
A	Small, inerting	Bureau of Mines Factory Mutual	Premixed gas mixture Explosion Burette ≃ 1.5 meters long Spark ignition source.
B	Small, inerting	DuPont Factory Mutual (for DuPont)	Premixed gas mixture, one quart Mason jar. Spark or electrically initiated ignition source.
C	Intermediate, inerting	DuPont	Premixed or stirred gas mixture, 55 gallon drum. Spark or electrically initiated match ignition source.
D	Large-scale 10^3 ft^2 or larger Inerting	Factory Mutual DoD Ansul, Fenwall	Premixed and/or circulated gas mixture established in test chamber, attempted ignition by various sources (e.g., spark, flare, and diffusion torch).
E	Intermediate, inerting	Fire Research Institute of Japan	Apparently, premixed mixture into which a diffusion flame is introduced. Test conditions not completely explained in English abstract.
F	Small to intermediate scale extinguishing	DuPont Factory Mutual (for DuPont)	Premixed gas mixture, one quart Mason jar and 55 gallon drum. An aliquot of burning fuel is plunged into a pre-mixed gas mixture--depth of plunge at extinguishment used as test criteria.
G	Small,extinguishing	ICI Corporation Factory Mutual Naval Weapons Lab (China Lake, Calif.)	Dynamic-continuous flow system. Similar to oxygen index flammability test.
H	Small, extinguishing	University of California San Diego	Stagnation point flow system. Liquid Fuels only--research system.
I	Intermediate, 6 ft^3	Naval Research Laboratories Washington, D.C.	Constant pressure test chamber fuel area = 10 in^2, ignition by spark, agent discharged after preburn period.
J	Large-scale, 10^3 ft^3 or larger	Factory Mutual, Ansul, Fenwall, Fyr-Fyter, Underwriters Lab.	Various size pan fires (class B fuels) Ignition by various means, agent discharged after preburn period.
K	Full-scale, extinguishing	U.S.N.D.C.T.C., Philadelphia, Pa. U.S.G.G.-F.S.T.F. Mobile, Alabama	Bilge Fires--various surface areas ignition by torch. Agent discharged after preburn period.

In small-scale tests, particularly the dynamic techniques, the environmental air velocity, fuel, and geometric parameters can be easily controlled. However, there are fundamental differences between the physical and chemical characteristics of small laminar flames and large turbulent diffusion fires. How these differences interact to modify the CCE has not been established. There is ample evidence to show that method G (the dynamic small-scale technique) gives repeatable results. It also offers a degree of versatility not demonstrated by the other methods. If a similarly well controlled method, using a naturally turbulent and scalable fire, would yield results similar to method G, then the effect of scale on CCE would be resolved. Thus, it would be possible to use this small-scale technique to determine 1301 CCE's for the wide variety of flammable liquids and gases resident aboard Navy vessels.

An intermediate scale test reactor has been constructed, and the major emphasis is directed to determining CCE valves for the fuels listed in Table III. Small-scale tests using the same fuels have been initiated at the Naval Weapons Center, China Lake, and the results from the large- and small-scale dynamic tests will be compared as soon as they are available. The rest of this paper will be devoted to describing the design criteria for the large-scale turbulent fire test reactor, and reporting the results obtained thus far.

Design Criteria

To establish the validity of the test method, and to obtain results that have meaning with respect to full-scale fires, the first requirement is to provide a naturally turbulent and scalable fire. Once this is established, it is then necessary to provide independent regulation of the controlling variables, such as fuel burning rate, ventilation, agent concentration and agent distribution so that the effects of 1301 can be determined as a function of the fuel and environment parameters.

Apparatus Design

To assure scalability, the fuel area must be large enough to produce a turbulent diffusion flame, i.e., no less than a foot in diameter. In addition, it is desirable that the burning rate of the fuel per unit surface area is independent of size. For this condition to be met, the total heat feedback to the surface must

Table III

FLAMMABLE LIQUIDS AND GASES FOR TESTING

Naval Fuels	Mil Spec
Aviation gasoline	G-5572
JP-4	T-5624
JP-5	T-5624*
Diesel	F-16884*
Naval Distillate (Foreign and Domestic)	F-24397*
NSFO (East and West Coast)	F-859E
Bunker C (Grade 6)	WF-815
Hydraulic Fluids	
H-580	H-22072 (or) H-19457
H-575	F-17111*
H-515	H-5606
Lubricating Oils	
1010	L-6081
0-156	L-23699
2075TH	L-17672
2110TH	L-17672
2135TH	L-17672
2190 TEP	L-17331
2100AW	H-24459
9250	L-9000*
4065	L-15019
5190	W-L-1071
5230	W-L-1071
Flammable Gases	
Propane	
Butane	
Chlorine	
Hydrogen	
Acetylene	
Ammonia	

*Fuels tested to date.

be constant, Consequently, for smoke-producing hydrocarbon fuels, the fuel bed should be no less than 2.0 ft in diameter. (1)(2)

The apparatus shown schematically in Figure 1 was designed with this size for the basic fuel reservoir. The outer plenum, 6 ft in diameter and 6 ft in height, encloses the inner combustion chamber, which is 4 ft in diameter and 8 ft high. Adequate access is provided to the interior of both the chambers. Air is pumped to the outer chamber at a rate that can be varied from 500-2000 cfm. The volume between the outer and inner chamber acts as a plenum so that the air can be uniformly distributed radially inward through the holes in the inner chamber. The distribution membrane made up of several layers of steel screen provides a pressure drop of about 0.1 inch of water for an air velocity of about 8.0 ft/min/ft^2 at 500 cfm. This velocity is comparable to the air entrainment velocity of natural convective pool fires.(3)

The advantage of this technique over simply introducing air at the bottom is that the entrainment flow of the flame is not perturbed by an imposed directional flow field. This could be a very important factor when the air is loaded with 1301, e.g., for air entering at the bottom there is a potential for the flame to lift off the surface of the fuel but continue to burn flame gases at some distance upstream from the fuel. (This behavior is not uncommon in small-scale flow experience.) The radial impingement of 1301-laden air all along the entrainment zone would not have this effect. Further, since the 1301 will be attacking the flame uniformly over its surface area, less agent should be required per unit of air volume, thus providing a truer lower concentration limit for the agent. This value should be close to the values found during large-scale testing, since the mechanism of agent application would be more nearly approximated.

The air flow is measured by the venturi meter installed upstream from the pump. The agent is introduced downstream from the venturi. Complete mixing is ensured by the transit of the 1301 + air mixture through the air pump. Controlled flow of 1301 is achieved by metering the liquid agent through a calibrated rotometer. Agent application is controlled by a fast acting valve located within 10 inches of the agent exit nozzle in the air duct, allowing a single-phase liquid at high pressure throughout the system. The agent reservoir is maintained at a constant temperature (110°F) in a regulated water bath to ensure accuracy of the flow measurement and calculations. A schematic of the agent supply systems is shown in Figure 2.

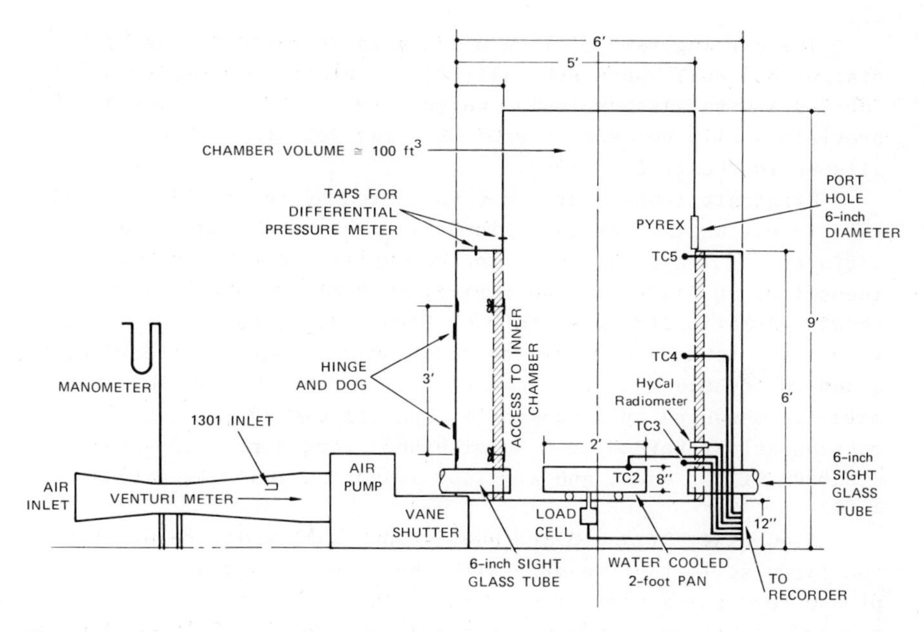

Figure 1. Large scale dynamic testing apparatus for determining critical concentration of 1301 for extinguishment of flammable fluids and gases

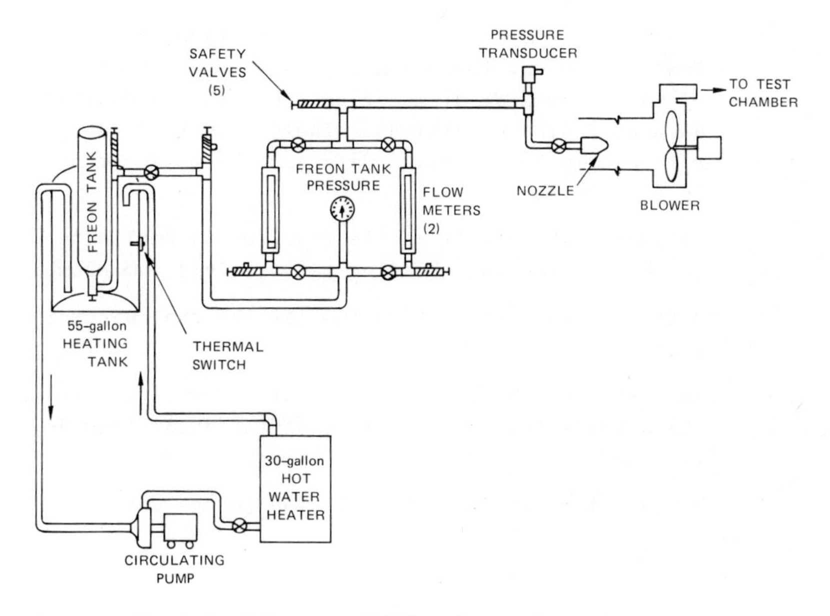

Figure 2. Schematic of 1301 pre-heat and metering system

The burning rate of liquid fuels is measured by the hydrostatic load cell shown schematically in plate a of Figure 3.(4) For tests with gaseous fuels, which have not been conducted yet, provisions will be made to provide a gas jet symmetrically aligned in the test chamber.

Strategic temperatures are monitored by thermocouples, and time to extinguishment is defined by both a thermocouple and a radiometer. The location of thermocouples and the radiometer is identified in Figure 1, and typical thermal sensor circuitry is shown schematically in plate b of Figure 3. External parameters (i.e., air temperature and relative humidity) are monitored by standard techniques. The volumetric air flow through the venturi meter is measured by a manometer, and all electrical signals are continuously recorded by a multichannel recorder. The tested accuracy of the agent and air flow measuring instrumentation is $\pm$ 5%.

Figures 4, 5, and 6 are photographs that show, respectively, the large-scale test reactor with the external doors to the outer plenum removed, a view of a calibration fire test, for fully ventilated conditions, and the apparatus completely sealed and ready for testing.

Test Procedure

The procedure described below was employed for each fuel tested to date.

1. Measure the "free standing" burning rate of the fuel at ambient temperature. This was accomplished by using the 2-ft-diameter pan in the open during quiescent wind conditions. These values of burning rate supply the baseline data for the measurements in the chamber.

2. Establish the air flow rate required to duplicate the "free standing" burning rates within the chamber.

3. Determine the CCE for the fuel at the chosen burning rates.

4. Measure the change in burning rate of fuel and flame characteristics as a function of the agent concentration.

5. Measure the effect of ventilation on the CCE.

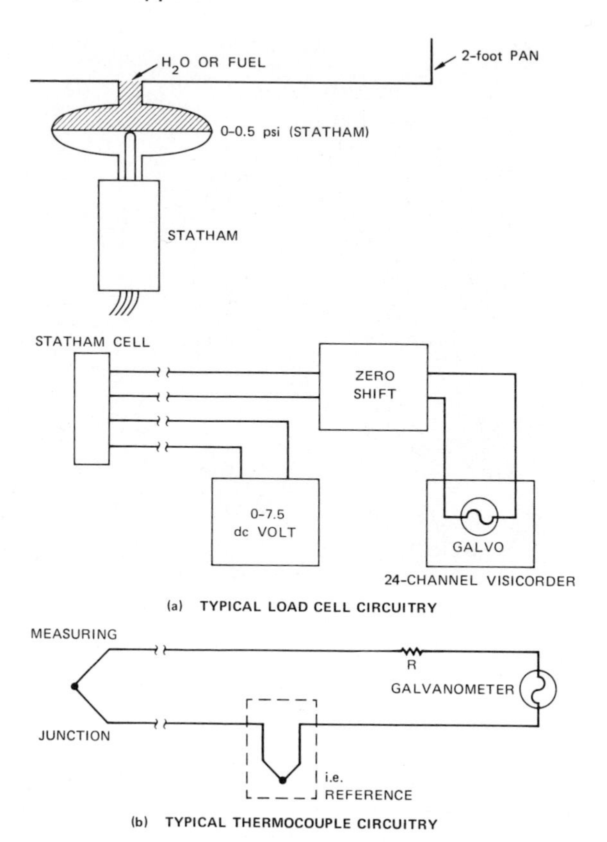

Figure 3. Details of hydraulic load cell and temperature measuring circuitry

Figure 4. Testing apparatus with outer plenum walls removed

Figure 5. Diesel oil fire in 2-ft pan

Figure 6. Large scale testing apparatus in test ready condition

Results

Tables 1 through 5 of the appendix collect the pertinent data resulting from tests conducted in the large-scale turbulent fire reactor. Figures 7 through 9 are curves of burning rate versus 1301 concentration for concentrations that are less than or equal to CCE. Figure 7 contains data for "calibration fuels." These fuels were selected for testing because they were well documented in 1301 suppression literature, as can be seen from Table I. The first column in Table I contains the CCE as indicated by the 1301 concentration where the burning rate falls to zero. For toluene and benzene the CCE compares favorably with results listed in the table. (Note: for these fuels only, results from small or intermediate tests have been published.) For methanol it appears that the burning rate varied directly with 1301 concentration over the concentration range covered in these tests. Because of the high consumption of 1301, tests with methanol to establish the CCE were not continued.

Figure 8 shows the burning rate versus 1301 concentration for four propulsion fuels. As expected, the fully ventilated burning rate for gasoline is significantly higher than the rates for the heavier fuels. However, the CCE for all these fuels is less than 3%, which is considerably less than the small-scale results collected in Table I. Note that large scale results obtained with the naval distillate generally indicate lower values than those obtained by small-scale techniques. (The solid curve through the heavy fuel data was drawn as a visual average for all the heavy fuels, while the dashed curve simply represents an outer bound for these data.)

Figure 9 shows burning rate versus 1301 concentration for a typical naval engine oil and hydraulic fluid. For the hydraulic fluid we were able to generate a data curve, but for the engine oil, the burning rate with even minimal 1301 concentrations was so low that it was difficult to determine whether or not the fire was consuming fuel, i.e., the radiometer indicated fire, but the burning rate was imperceptible. Consequently, I have drawn a straight, dashed line to the CCE concentration.

Figure 10 indicates the effect of air flow rate on CCE. That the CCE decreases or even increases with increasing ventilation is not surprising in light of small-scale results from both Williams (5) and Bajpai (6), which show significant variations in CCE with both increased ventilation and fuel temperature

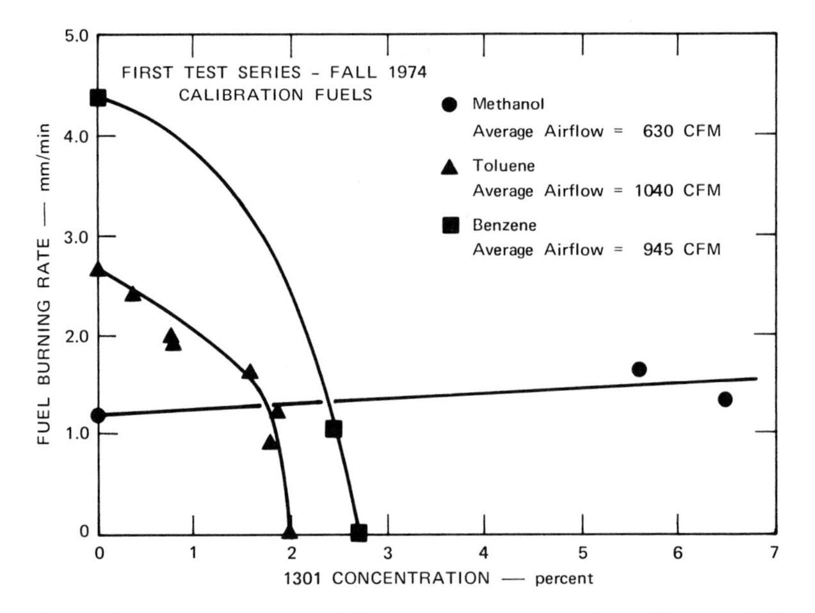

Figure 7. Burning rate vs. *1301 concentration in air for calibration fuels (methanol, toluene, benzene)*

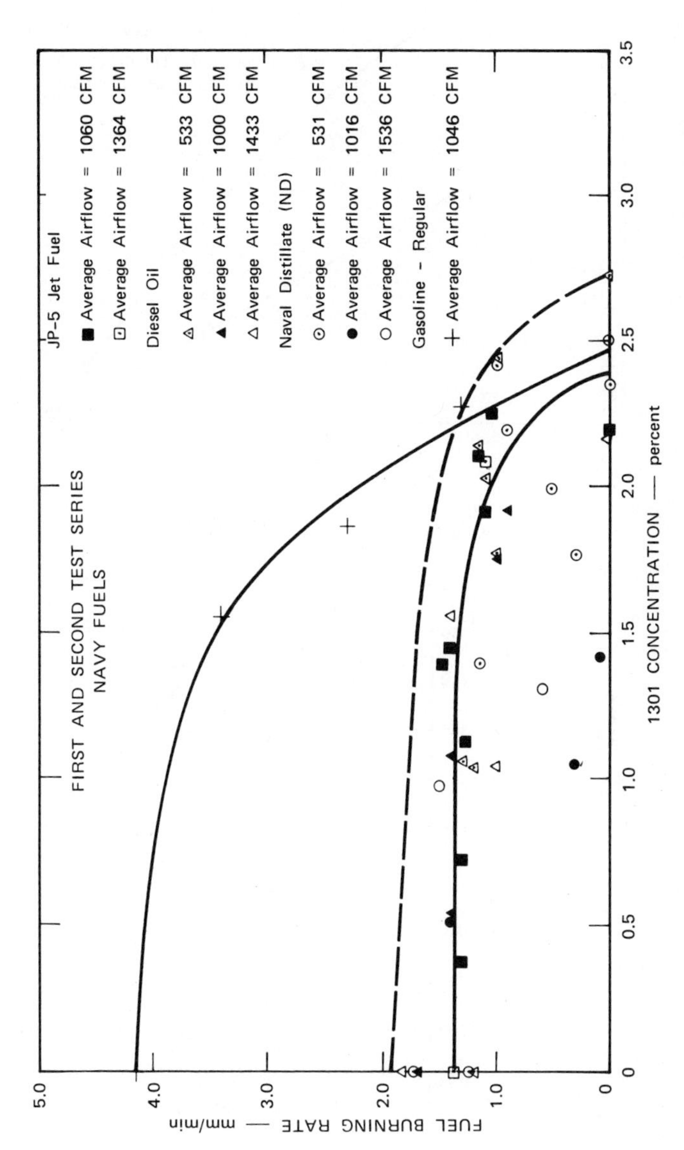

Figure 8. Burning rate vs. *1301 concentration in air for navy propulsion fuels (JP-5, diesel oil, naval distillate, gasoline)*

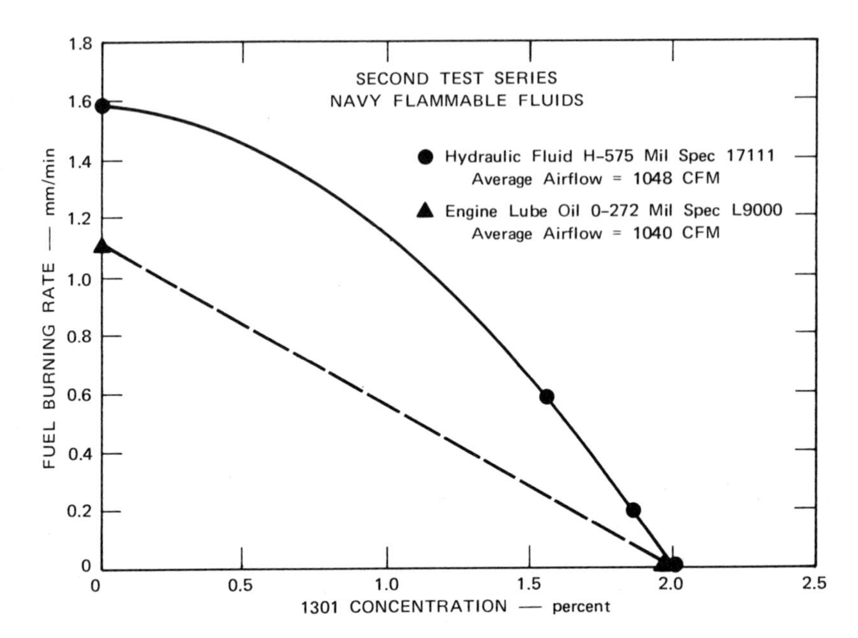

Figure 9. Burning rate vs. *1301 concentration in air for navy flammable fluids (hydraulic fluid-H-575, Enbine lube oil-O-272)*

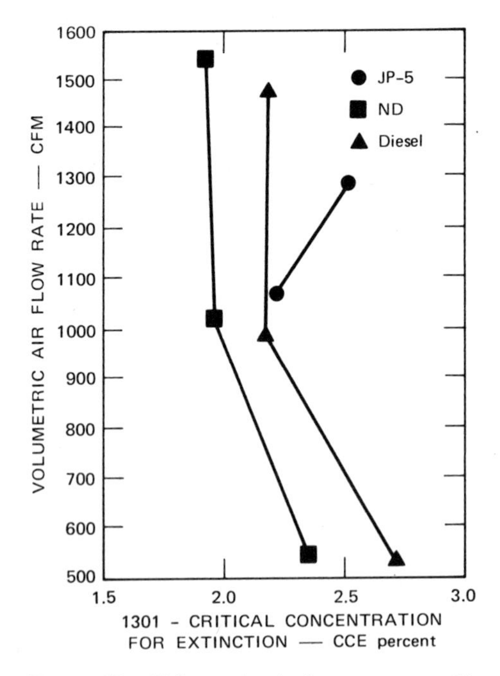

Figure 10. Volumetric air flow rate vs. *critical concentration for extinction for navy propulsion fuels*

The fact that the CCE for JP-5 increases with ventilation while the ND and Diesel CCE's decrease is disconcerting since they are so similar in other respects.

Discussion

The first column of Table I contains the CCE for flammable fluids tested in the intermediate scale apparatus described in this paper. It is to be noted that all fires conducted in this apparatus were fully turbulent and that ventilation rates and 1301 concentrations were critically monitored. In general, it appears that these CCE values are less than the lowest CCE's found during the collection of data for the table. Exceptions to this observation are small-scale toluene and large-scale naval distillate results.

From these data, it appears that the CCE's for large turbulent fires are less than CCE's for laminar flames. We can speculate that this observation can be explained by consideration of the enhanced mixing that is induced during turbulent conditions as opposed to the streamline flow patterns that are established during laminar conditions. It is possible that the turbulent mixing of agent in the combustion zone provides a more efficient environment for the suppressive action of the agent.

All of the CCE data reported in the table are single value results. However, there is ample evidence to show that CCE's are variable with respect to ventilation rate, fuel burning rate, mode of flaming (i.e., laminar or turbulent), and possibly the temperature and moisture state of the ambient air. This information should be available to fire protection systems designers, since such data are critical to the establishment of safety factors for design concentrations of 1301 systems. Until such information is readily available, one will have to rely on existing data and his own ingenuity for conjuring of safety factors. This practice is both wasteful in terms of money and potentially dangerous in terms of the toxic potential of the agent and its pyrolysis products.

Several fuels, including methanol, showed a slight increase in burning rate with low concentrations of 1301. While this effect was not dramatic, it was indeed a valid observation. Similarly, the radiance of the fires, where burning rates increased, was measurably higher both by visual and radiometric observations. What this observation means, in terms of suppression mechanism, is beyond the scope of the project, and is included only to demonstrate the sensitivity of the apparatus to recordable phenomena.

The repeatability of the results from the intermediate-scale apparatus and the reasonable agreement from other test methods are encouraging with respect to establishing the validity of the method. However, this procedure is expensive both in time and in consumption of 1301. Therefore, we look with anticipation to comparative results obtained on the same fuels by the dynamics small-scale apparatus of the Naval Weapons Center, China Lake.

Literature Cited

1. Hottel, H. C., Fire Research Abstracts and Reviews, 4 No. 2, p. 41 (1959).
2. Wood, B. D., and Blackshear, Jr., P. L., "An Experimental Study of the Heat and Mass Transfer to the Surface of a Burning Array of Fuel Elements," Combustion Laboratory Tech. Report No. 7, University of Minnesota, Nov. 1969.
3. Ricou, F. P. and Spaulding, D. B., "Measurements of Entrainment by Axisymmetrical Turbulent Jets," Journal of Fluid Mechanics, Vol. 9 (1961).
4. Alger, R. S., and Nichols, J. R., "A Mobile Field Laboratory for Fires of Opportunity," NOLTR73-87, Oct. 3, 1973.
5. Kent, J. H., and Williams, F. A., "Effect of CF_3Br on Stagnation-Point Combustion of a Heptane Pool," Preprint, Western States Combustion Institute, Fall Meeting, Oct. 29-30, 1973.
6. Bajpai, Satya N., "An Investigation of the Extinction of Diffusion Flames by Halons," FMRC Serial No. 22391.2, Factory Mutual Research Report, Nov. 1973.

Table A1

FIRST TEST SERIES--FALL 1974

Calibration Fuels

Date/Test	Air CFM	1301 CFM	1301 Conc. (%)	Fire Duration (sec)	Fuel $\dot{M}$ Preburn (mm/min)	Fuel $\dot{M}$ 1301 (mm/min)	Ambient Temperature F°	Relative Humidity	Observation
Alcohol									
16 September									
1	580	38	6.5	-	1.1	1.35	77.5	56	Flames visible
2.	685	38	5.6	-	1.3	·1.65	78	55	Same
$\dot{M}$ (Avg) = 1.2									
Toluene									
17 September									
4	1100	17.15	1.56	-	2.75	1.62	87	39	
5	1025	32.9	3.21	18.5	2.70	-	90.5	35	
6	1010	25.2	2.49	25	2.73	-	92	34	
18 September									
7	990	7.64	0.77	-	2.55	1.9	81	54	
2	1045	23.2	2.22	20	2.7	-	86	44	
3	1090	19.31	1.77	-	2.78	0.9	86.5	45	Almost out
4	1038	19.31	1.86	-	2.84	1.2	88.5	42	Almost out, gusty wind
5	1065	21.21	1.99	18	2.38	-	88	43	Gusty wind
6	1036	21.21	2.13	19	2.52	-	88	43	Gusty wind
19 September									
1	1010	7.48	0.74	-	2.47	1.98	70	70	Cool fuel
2	1025	3.71	0.36	-	2.82	2.4	72.5	67	
3	1010	19.76	1.96	25	2.85	-	77	59	
CCE = 1.96% $\dot{M}$ (Avg) = 2.67									
Benzene									
20 September									
11	930	30.9	3.3	19.5	4.14	-	85.5	44	
12	950	23.1	2.43	-	4.52	1.06	83.5	43	
13	950	26.9	2.83	20	4.52	-	84	43	
14	950	25.6	2.69	37	4.54	-	83.5	43	
CCE = 2.69% $\dot{M}$ (Avg) = 4.4									

Table A2

FIRST TEST SERIES--FALL 1974

JP-5 Jet Fuel

Date/Test	Air CFM	1301 CFM	1301 Conc. (%)	Fire Duration (sec)	Fuel $\dot{M}$ Preburn (mm/min)	Fuel $\dot{M}$ 1301 (mm/min)	Ambient Temperature F°	Relative Humidity	Observation
19 September				Medium Air Flow Rate ~ 1000 CFM					
4	1065	38.4	3.6	14	1.30	-	82.5	52	Fuel oscillates
5	1065	26.88	2.52	19	1.23	-	83	53	Fuel oscillates
6	1100	15.36	1.4	-	1.23	1.47	82.5	52	Fuel oscillates
7	1100	21.11	1.92	-	1.28	1.1	86	48	Fuel oscillates
8	1080	24.71	2.29	23	1.3	-	81	52	Fuel oscillates
9	1047	7.7	0.73	-	1.28	1.3	83	56	Fuel oscillates
10	1010	3.8	0.38	-	1.28	1.32	83	55	Fuel oscillates
20 September									
1	1065	22.42	2.11	-	1.16	1.14	67	75	Cool fuel
2	1065	24.42	2.27	22	1.33	-	70	72	Fuel oscillates
3	1065	23.80	2.20	27	1.26	-	72	69	Fuel oscillates
4	1045	15.19	1.45	-	1.3	1.41	76	60	Fuel oscillates
9	1025	11.60	1.13	--	1.3	1.28	86	47	Fuel oscillates
10	1025	23.20	2.26	-	1.3	1.04	86	47	Fuel oscillates
				High Air Flow Rate ~ 1300 CFM					
5	1585	38.3	2.42	-	1.42	1.0	81	51	Almost out
6	1292	34.61	2.68	14.5	1.36	-	84	52	
7	1290	27.10	2.10	-	1.34	1.1	87	43	
8	1290	32.50	2.51	48	1.36	-	88.5	43	

$CCE_{low} = 2.2\%$ $\dot{M}_{low} = 1.27\%$

$CCE_{high} = 2.5\%$ $\dot{M}_{high} = 1.38\%$

Table A3

SECOND TEST SERIES--SPRING 1975

Navy Fuels--Diesel Oil

Date/Test	Air CFM	1301 CFM	1301 Conc. (%)	Fire Duration (sec)	Fuel $\dot{M}$ Preburn (mm/min)	Fuel $\dot{M}$ 1301 (mm/min)	Ambient Temperature F°	Relative Humidity	Observation
				Low Air Flow Rate ~ 500 CFM					
20 March									
1	540	19.7	3.65	22	1.1	-	54.5	51	
2	540	16.2	3.01	34	1.2	-	55	52	
3	560	9.95	1.78	-	1.2	1.0	54	53	
4	520	10.4	2.03	-	1.15	1.1	57	55	
5	520	5.4	1.04	-	1.15	1.2	57	55	
6	510	10.9	2.14	-	1.2	1.15	59	47	
7	510	5.4	1.06	-	1.2	1.3	59	47	
8	580	12.6	2.18	-	1.15	-	56	51	Runt out of B01
21 March									
1	520	12.6	2.42	-	1.2	1.0	50	56	Gusty wind
2	540	16.1	2.98	-	1.3	-	51	70	Gustier
25 March									
1	510	14.42	2.84	-	1.25	1.0	57	77	1301 duration too short
2	530	14.42	2.72	68	1.15	-	57	71	
3	545	13.31	2.44	-	1.15	1.0	57	50	
				CCE = 2.72%	$\dot{M}$(Avg) = 1.19%				
				Medium Air Flow Rate ~ 1000 CFM					
25 March									
1	1000	31.0	3.1	24	1.55	-	58	37	
2	1000	29.0	2.9	23	1.65	-	58	37	
3	970	25.45	2.6	24	1.6	-	58	37	
4	940	21.8	2.32	32	1.6	-	57	40	
5	1020	18.05	1.77	-	1.75	1.0	57	40	
6	1025	10.8	1.08	-	1.7	1.4	56	42	
7	1025	5.5	0.54	-	1.7	1.4	56	42	
26 March									
1	990	21.45	2.17	32	1.7	-	53	41	
2	1025	19.65	1.92	-	1.85	0.9	52.5	38	Almost out
				CCE = 2.17%	$\dot{M}$(Avg) = 1.63%				
				High Air Flow Rate ~ 1500 CFM					
26 March									
1	1480	32.21	2.18	64	2.0	-	55.5	36	Close
2	1450	15.0	1.04	-	1.65	1.0	56	40	
3	1400	21.8	1.56	-	1.9	1.4	57	45	
4	1400	32.5	2.32	30	1.8	-	56	39	
				CCE = 218%	$\dot{M}$(Avg) = 1.84%				

Table A4

SECOND TEST SERIES--SPRING 1975

Navey Fuels--Naval Distillate (ND)

Date/Test	Air CFM	1301 CFM	1301 Conc. (%)	Fire Duration (sec)	Fuel $\dot{M}$ Preburn (mm/min)	Fuel $\dot{M}$ 1301 (mm/min)	Ambient Temperature F°	Relative Humidity	Observation
27 March									
1	580	12.8	2.2	-	1.4	0.9	58	50	Re-flash (wick)
2	490	12.8	2.6	54	1.2	-	57	45	
3	545	12.8	2.35	55	1.2	-	59	45	
4	520	7.2	1.4	-	1.23	1.15	57.5	43	
5	520	10.7	2.0	-	1.23	0.5	57.5	43	
				CCE = 2.35%	$\dot{M}$(Avg) = 1.25%				
				Medium Air Flow Rate ~ 1000 CFM					
27 March									
1	1010	21.77	2.15	30	1.6	-	55	49	
2	1025	17.95	1.75	-	1.66	-	54.5	42	Almost out-wick
3	1020	19.9	1.95	38	1.9	-	54	48	
4	1010	5.27	0.52	-	1.75	1.4	54	48	Establish air flow, change 1301 concentration as indicated
5	1010	10.6	1.05	-	1.75	0.3	54	48	
6	1010	14.3	1.42	-	1.75	0.1	54	48	
				CCE = 1.95%	$\dot{M}$(Avg) = 1.73				
				High Air Flow Rate ~ 1500 CFM					
28 March									
1	1530	32.2	2.1	32	1.65	-	51	50	Cool fuel
2	1530	27.1	1.77	-	1.75	0.3	51	47	Almost
3	1540	15.05	0.98	-	2.00	1.5	54	42	
4	1540	19.65	1.31	-	2.00	0.6	54	42	
5	1540	29.6	1.92	34	1.65	-	56.5	39	
				CCE = 1.92%	$\dot{M}$(Avg) = 1.76				
				Navy Fuels--Regular Gasoline--Leaded					
				Medium Air Flow Rate ~ 1000 CFM					
28 March									
1	1050	30.4	2.93	20	4.55	-	56.5	42	
2	1060	26.3	2.48	21	4.0	-	57	40	
3	1060	19.85	1.87	-	3.6	2.3	57.5	48	Extinguish with 3% 1301 CO_2 does not work here
29 March									
4	1030	23.42	2.28	-	4.4	1.3	53	47	
5	1030	16.1	1.56	-	4.15	3.4	54	46	
				CCE = 2.48%	$\dot{M}$(Avg) = 4.14%				

Table A5

SECOND TEST SERIES--SPRING 1975

Navy Flammable Fluids
Hydraulic Fluid--H575--MIL SPEC 17111

Date/Test	Air CFM	1301 CFM	1301 Conc.(%)	Fire Duration (Sec)	Fuel $\dot{M}$ Preburn (mm/min)	Fuel $\dot{M}$ 1301 (mm/min)	Ambiant Temperature F°	Relative Humidity	Observations
Medium Air Flow Rate ~ 1000 CFM									
29 March									
1	1070	21.85	2.04	36	1.73	-	59	44.5	
2	1040	16.32	1.57	-	1.45	0.6	61.5	47	
3	1040	19.45	1.87	-	1.50	0.2	61	46.5	Almost out
4	1040	20.85	2.00	46	1.65	-	61.	46.5	
CCE = 2.0% $\dot{M}$(Avg) = 1.58%									
NAVAL FLAMMABLE FLUIDS--LUBRICATION OIL--ENGINE 0-272-MIL SPEC L9000									
29 March									
1	1040	22.0	2.1	25	1.25	-	62	47	
2	1050	16.4	1.56	-	1.00	-	63	46	Almost out
3	1030	20.17	1.97	54	1.05	-	65	44	
CCE = 1.97% $\dot{M}$(Avg) = 1.10%									

DISCUSSION

R. W. BILGER: It appears from your data that extinguishment occurs by decreasing the intensity of the fire as halon concentration increases rather than a sudden blow-off as occurs in the ICI cup test. Is this a correct description of the extinction phenomenon?

N. J. ALVAREZ: Yes. In tests where the 1301 concentration is close to the CCE, the burning rate of fuel decreases monotonically to zero. We observe no "blow-off" phenomona or radiation peak upon application of the agent.

A. S. GORDON: We have noted a large effect on burning rate of methanol pool fires of the distance of the surface below the lip. Have you made similar measurements in your apparatus? If you are in the diameter regime where the burning rate is independent of diameter, there should be little effect on the rate when the distance is varied.

N. J. ALVAREZ: All of our tests are conducted with the fuel level one inch or greater below the water cooled lip. We have noted no effect on burning rate as the fuel level recedes. If the fuel level is close to the lip, we observe migration of fuel over the lip and down the sides of the pan. This does perturb the burning rate measurement.

R. FRISTROM: Would you comment on the problems connected with separation phenomena in mixtures due to differing vapor pressures of the constituents - both time dependent and steady state effects?

N. J. ALVAREZ: With the heavier fuel oils and greases, (those with low API Gravities such as Bunker C fuel oil) there is a definite fractionating of the fuel resulting in a succeedingly more viscous residue during natural diffusion burning. In testing these fuels to determine CCE we intend to use a large volume of fuel, and internal stirring to reduce the effect of change in fuel character on the burning rate. Since our pre-burn times are relatively short, this procedure should be sufficient to insure reasonable accuracy in determining the CCE. We have not made burning rate measurements with these fuels as yet, but we have burned and noted the qualitative character of fuel behavior with time.

4

Mechanistic Studies of Halogenated Flame Retardants: The Antimony–Halogen System

JOHN W. HASTIE and C. L. McBEE

Inorganic Chemistry Section, Institute for Materials Research, National Bureau of Standards, Washington, D. C. 20234

Introduction

In addition to the prominent use of various halocarbons as flame extinguishing agents there exists another class of halogenated flame retardants which, at present, incorporates primarily the additional elements of Sb and P. Despite the inconclusive mechanistic understanding for halogen induced flame inhibition in general, there is reason to believe that the mechanistic action of halocarbons, *e.g.*, CF_3Br, and halogenated Sb- or P-containing species, *e.g.*, $SbCl_3$, $SbBr_3$ or $POCl_3$, has a common molecular basis (1). For both retardant categories, the primary function of the non-chloride or bromide moiety appears to be its ability to serve as a convenient carrier for the flame inhibiting chloride or bromide component to the flame front. However, some synergistic effects are observed. For instance, the reduction in burning velocity is far greater for species such as $SbCl_3$ than for the equivalent amount of Cl (2). Similarly, for CF_3Br the observed degree of flame inhibition exceeds that expected for the equivalent amount of Br. In order to understand such synergistic systems it is necessary to separate the individual mechanistic effects for each active component. The present study attempts to define the flame inhibiting mechanism for the synergistic interaction of antimony and halogens, with particular emphasis on systems involving [Sb_2O_3] and chlorinated hydrocarbons (square brackets denote solid state compounds).

There are, we believe, two main factors governing the flame retarding performance of the halogenated Group V systems in commercial use. First, in practical retardancy formulations the thermodynamic and chemical kinetic properties of the substrate, whereby volatile fuel and inhibitor species are produced, are of great significance. Thus, for example, the *in situ* interaction of HCl or HBr with a polymer incorporant of [Sb_2O_3] to yield volatile species such as SbOBr, $SbBr_3$, $SbCl_3$, and SbOCl provides the means for introducing flame inhibiting species to the flame front. The thermodynamic and kinetic parameters for such substrate interactions will therefore be considered in the discussion which follows. The second factor; namely, the mechanistic action of these volatile halogen-containing species within the flame, will also be investigated. The laboratory flames of interest include primarily the atmospheric premixed systems of H_2-O_2-N_2 and CH_4-O_2-N_2, burning under fuel rich conditions. Such flames are considered reasonable model systems of "real-fire" chemical kinetic phenomena, as has been argued elsewhere (1).

Apparatus and General Experimental Procedure

Very specialized experimental techniques are required to characterize the basic molecular processes where both moderate temperature pyrolysis and high temperature flaming combustion occur. We utilize a combination of mass spectrometric and optical spectroscopic techniques for the characterization of flame inhibition phenomena. A detailed description of our mass spectrometric apparatus and procedures for sampling reactive species from atmospheric pressure flames, has already been given elsewhere (3,4). Figure 1 provides a schematic of this system in its flame sampling mode of operation. A Knudsen reactor system is used as the primary means of characterizing substrate pyrolysis phenomena. For the present study, established optical spectroscopic procedures are used to monitor the species H, OH and SbO in atmospheric pressure flames.

The Knudsen Reactor System. A convenient technique for the study of pyrolysis phenomena utilizes a so-called Knudsen effusion reactor in combination with the molecular beam mass spectrometer. A Knudsen cell serves as a temperature- and pressure-controlled inert container for the reacting system. The Knudsen cell is heated radiatively by a resistance furnace as shown in Figure 2. Substrate samples are contained in an alumina cup within the Knudsen cell. Note (in Figure 2) the presence of a gas inlet line to the Knudsen cell. This allows for the external control of gaseous components such as HCl and H_2O. Providing the total gas pressure within the cell does not exceed about 10^{-3} atm and an orifice area of $< 10^{-2}$ cm^2 is used, an effusing molecular beam may be generated from such a cell. This beam having, by definition,a composition representative of that for the vapor contained by the cell can then be conveniently analyzed using the line-of-sight mass spectrometric detector. The molecular beam is directed along the center axis of the vacuum space denoted as region II in Figure 1. Upon entering the mass spectrometer chamber (see Figure 1) the molecular beam is chopped by a mechanical wheel, partially converted to positive ions by electron impact, mass selected and detected using frequency-dependent lock-in phase sensitive detection methods as described elsewhere (3).

From the theory of Knudsen effusion mass spectrometry *e.g.*, see Grimley (5), the magnitude of the detected ion signal is related to the partial pressure for the precursor species by

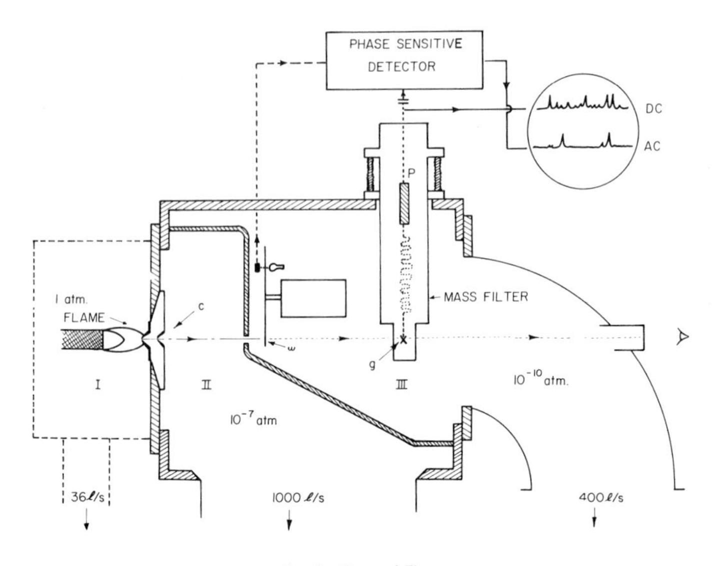

Combustion and Flame

Figure 1. Schematic of molecular beam mass spectrometric system for sampling 1-atm flames (indicated) and effusing vapor from a Knudsen cell (not shown but normally located in region II) (3)

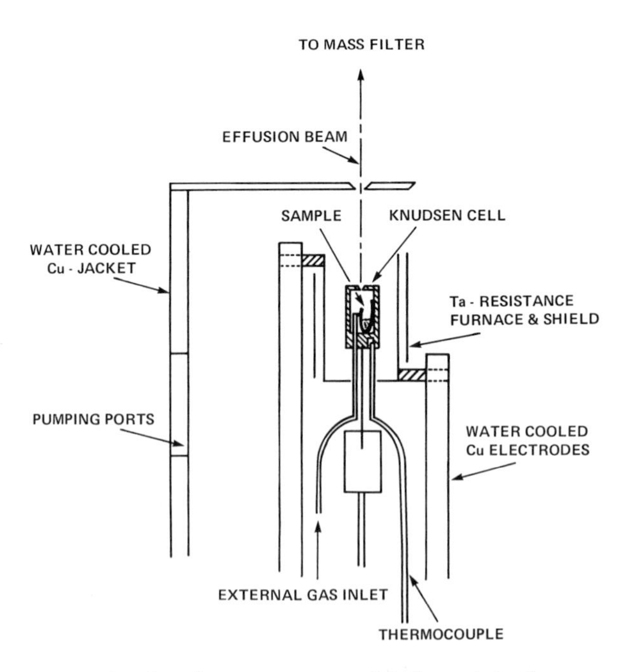

Figure 2. Knudsen reactor assembly located in the vacuum space denoted as region II in Figure 1

$$P = kI\,T,$$

where P is the partial pressure for the species within the Knudsen cell, I is the intensity of the mass analyzed detected positive ion signal, T is the temperature of the Knudsen cell, and k is a calibration constant indicative of the instrument sensitivity and the nature of the electron impact process. Typical values of k for our apparatus fall in the region of 10^{-11}-10^{-12} atm/μV K (for 10^7 Ω). In our studies k is determined by two separate methods. Either I is measured for a species of known P{*e.g.*, Sb_4O_6 (6)}, or use is made of the relation:

$$k = \frac{2.2557\times10^{-2} \times w}{AM^{1/2}\Sigma(IT^{1/2}\Delta t)} \text{ atm/}\mu\text{V K},$$

where w is the total weight loss in gm, A is the Knudsen orifice area in cm^2, Δt is the time interval in s, M is the gm molecular weight, and T is the temperature in K, *e.g.*, see Drowart and Goldfinger (7).

A further consideration in making the conversion of ion intensity to partial pressure data involves the assignment of ions to their proper neutral precursors. This is a relatively straightforeward process for systems where a single vapor species is present, *e.g.*, for Sb_4O_6. But for a system containing both SbOCl and $SbCl_3$, ions such as $SbCl^+$ and Sb^+ could be derived from each species. In order to resolve such ambiguous ion intensity data we adjust the reaction conditions of HCl pressure and temperature to favor only the presence of $SbCl_3$, for instance. A summary of our ion-precursor assignments is given in the Appendix. We assume that the total ionization cross sections are equivalent for each molecular species of interest. An uncertainty of up to a factor of two in species partial pressure is likely to result from this unavoidable simplification.

As the mass spectrometer serves to identify the various gaseous components of a reacting system, the determination of P for each species allows equilibrium constants and hence reaction free-energies to be calculated. Furthermore, as the temperature is a controlled variable, the observed temperature dependence of the equilibrium constant allows reaction enthalpy and entropy data to be determined. To summarize the basic thermodynamic concepts, the equilibrium constant K is a function of the reactant and product species partial pressures, and:

$$\frac{d\ell nK}{d(1/T)} = \frac{-\Delta H}{R}$$

$$\Delta F = -RT\ell nK = \Delta H - T\Delta S,$$

where ΔH, ΔS and ΔF are the reaction enthalpy, entropy and free energy respectively at the temperature T; R is the universal gas constant.

It should be noted that the residence time for a molecular species in our Knudsen cell reactor is typically in the region of 10^{-4}-10^{-5} s. During this time the gas species undergo approximately 10^{9} collisions with the sample surface. Only in exceptional cases are these conditions unfavorable to the attainment of thermodynamic equilibrium within the cell. A convenient test for the presence of equilibrium within the cell is provided by the obtainment of partial pressure and apparent equilibrium constant data that are independent of the Knudsen orifice diameter. We use two separate hole diameters of 0.1 cm and 0.025 cm for this test. As a further test, the equilibrium constant should be independent of pressure for constant temperature conditions.

The Optical Spectroscopic System. Optical spectroscopic techniques may be used to monitor flame temperatures and certain atomic and diatomic species (8). Our apparatus and procedures are similar to those used by other workers (9,10).

The method of monitoring H atom concentrations is an indirect although well established one (10). Basically we rely on the presence of the following balanced reaction in Li containing flames:

$$Li + H_2O = LiOH + H$$

for which the equilibrium constant is known, as is the H_2O concentration. The concentration ratio Li/LiOH may be experimentally determined by comparing the resonance emission intensities for Na and Li where the total Na and Li concentrations are made equal. Hence the H atom concentration can be derived. We also make use of an alternative method, involving the measurement of CuH emission (428.3 nm) intensities as a measure of relative H atom concentrations (9).

Flame temperatures are determined with the same apparatus using the Na D-line reversal technique.

Flame retardant species are introduced to the flame either by transpiration of vapor with the premixed combustion gases, or by entrainment of nebulized aqueous solutions. The nebulizer used is of the type described by Mavrodineanu and Boiteux (11).

Experimental Results for Substrate Reactions

Reaction of Ethylene Chlorinated Polymer (40% Cl) with $[Sb_2O_3]$. A powdered mixture (50/50 by weight) of an ethylene chlorinated polymer with $[Sb_2O_3]$ can be taken as a representative example of commercial retardant formulations involving the interaction of a halogen source with $[Sb_2O_3]$ or related antimony compounds. Similar mixtures have been characterized by Touval (12), among others, using thermogravimetric analysis (TGA) and differential thermal analysis. Also the effectiveness of such mixtures as flame retardants is empirically well known, *e.g.*, see the review of Pitts (13).

From the TGA results (12), it is known that about 76% of the initial $[Sb_2O_3]$ component is lost by a halogen induced vaporization process. Using the Knudsen reactor mass spectrometric technique we can establish the molecular details of this process.

Typical experimental results are shown in Figure 3, together with the TGA curve of Touval (12) for comparison. Note that the loss of $[Sb_2O_3]$ from the substrate mixture occurs primarily over the temperature interval of 250-450 °C and in the form of $SbCl_3$ and SbOCl molecular species. In this experiment the initial mixture had a composition equivalent to a 3.3 mole ratio of HCl/ $[Sb_2O_3]$, and from an analysis of the end-of-run residue (primarily $[Sb_2O_3]$) it was found that the corresponding mole ratio lost by vaporization was 5.0. By comparison, the similar TGA data indicated about a 4.3 mole ratio loss of HCl to $[Sb_2O_3]$.

From our observation of $SbCl_3$ and SbOCl as the principal antimony-containing vapor species, and from the thermodynamic properties of these species (see Appendix), the *overall* vapor-forming gas-solid reactions can be written as:

$$6HCl + [Sb_2O_3] = 2SbCl_3 + 3H_2O \quad (1)$$

$$2HCl + [Sb_2O_3] = 2SbOCl + H_2O \quad (1a)$$

Thus if these reactions proceeded completely to the right, reactions (1) and (1a) would result in the utilization of $HCl/[Sb_2O_3]$ mole ratios of 6 and 2, respectively. Under conditions of complete reaction, the observed ratio of 5 would require reaction (1) to be three times more effective than reaction (1a) in transporting antimony to the vapor phase. A check on these mass-balance considerations is provided by the areas under the SbOCl and $SbCl_3$ pressure-temperature curves of Figure 3. These curves indicate an overall ratio of $SbCl_3$/SbOCl $\sim$2.0 which compares reasonably with the mass balance prediction of 3.0, particularly as the reactions are not completely to the right.

In order to express the pressure-temperature data, contained in the curves of Figure 3, in terms that may readily be transferred to other experimental or even "real-fire" conditions it is necessary to define apparent equilibrium constants for the various possible reactions. Thus, for reaction (1) we have:

$$K_1 = \frac{(PSbCl_3)^2 (PH_2O)^3}{(PHCl)^6 \ a[Sb_2O_3]},$$

where P represents species partial pressure and $a[Sb_2O_3]$ is the thermodynamic activity of condensed $[Sb_2O_3]$ which, by definition, is unity for the pure $[Sb_2O_3]$ phase. For the present, we consider K_1 as an "apparent" equilibrium constant since we have yet to establish whether reaction (1) is at equilibrium. Also, $[Sb_2O_3]$ is assumed for the moment to be present at unit activity. Values of K_1 as a function of temperature are given in Figure 4. Similarly,

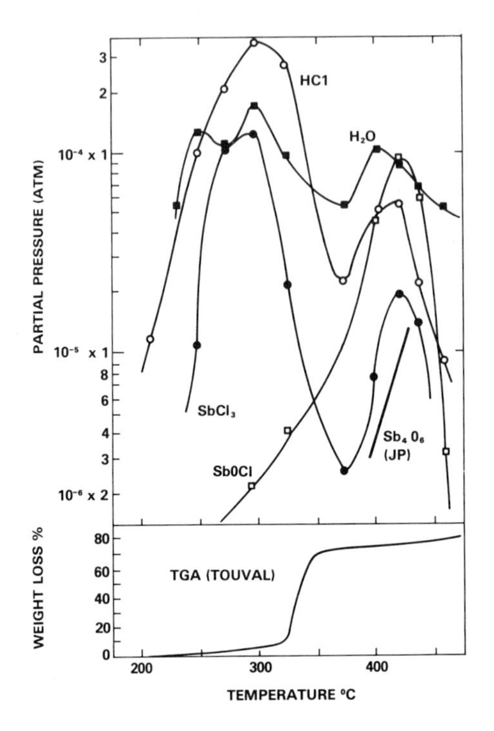

Figure 3. Partial pressure of volatile species resulting from the thermal interaction (heating rate 2°C/min) of an ethylene chlorinated polymer (40% Cl, Borden Chemical) with an equal weight of [Sb_2O_3]. Thermogravimetric analysis curve of Touval (12) is also shown for comparison. Curve for Sb_4O_6 is based on absolute pressure data given by Jungermann and Plieth (6).

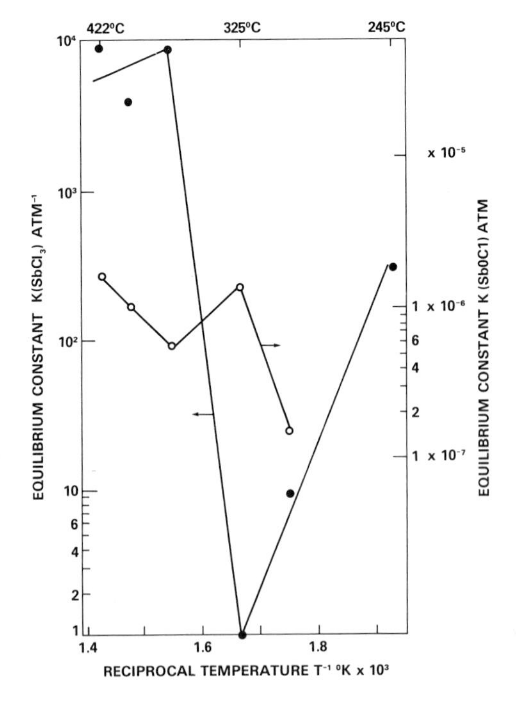

Figure 4. Apparent equilibrium constants for production of $SbCl_3$ and SbOCl as a function of reciprocal temperature, based on data in Figure 3 and assuming that reactions (1) and (3) hold, respectively.

values of K_3 for the homogeneous reaction:

$$SbCl_3 + H_2O = SbOCl + 2HCl \quad (3)$$

are also indicated in the Figure 4. Data for K_{1a} have not been plotted as this reaction is merely another form of reactions (1) and (3). Note that K_1 is particularly sensitive to the temperature and the absence of a single monotonic dependence of K_1 on the reciprocal temperature is indicative of the presence of additional reactions, other than reaction (1), involving the vapor species $SbCl_3$, HCl and H_2O.

Reaction of HCl with [Sb_2O_3]. In order to determine the nature of the additional reactions inferred above, it is necessary to control the total pressure of HCl externally rather than by the internal release mechanism provided by the pyrolysis of the chlorinated polymer. Figure 5 indicates the nature of the K_1 temperature dependence for total HCl pressures in the region of 10^{-5}-10^{-6} atm over an initially pure [Sb_2O_3] substrate. Note that K_1 reaches a maximum value in the region of 420°C which is a similar result to that obtained for the chlorinated polymer -[Sb_2O_3] system. Again, more than one process for the production of $SbCl_3$ is evident from the slope change of the log K_1 versus T^{-1} curve. It is noteworthy that the slope of this curve between temperatures of about 420 and 520°C yields a reaction enthalpy of -20 kcal/mol, in agreement with a value based on the literature heats of formation for the various reaction components (see Appendix).

Figure 5 also indicates the dependence of K_3 on temperature for the same system.

The molecularity, or stoichiometric dependences, of the reaction(s) leading to the formation of $SbCl_3$ and SbOCl can be inferred from the dependence of the products partial pressure on the reactant pressure for isothermal conditions. For instance, if reaction (1) is predominant in forming $SbCl_3$ and K_1 is pressure independent (*i.e.*, the system is at equilibrium) then the pressure of $SbCl_3$ should vary as the 6/5 power of the HCl partial pressure. This result assumes that H_2O is formed primarily by reaction (1) and is not extensively involved in secondary processes such as reaction (3). As is shown in Figure 6 the pressure of $SbCl_3$ shows the expected 6/5 power dependence with HCl pressure thus indicating reaction (1) to be the main *overall* reaction for these particular conditions.

The reactions (1) and (3) may not be in a state of general thermodynamic equilibrium. A condition of general thermodynamic equilibrium requires K to be pressure independent for constant T and that the variation of K with T yield a second law reaction enthalpy in agreement with the corresponding third law value. In practice, we find that K_1 decreases with increasing P(HCl) where T < 420°C. The value of K_3 increases with increasing P(HCl) where T <420°C. At higher temperatures, K_1 and K_3 are essentially constant

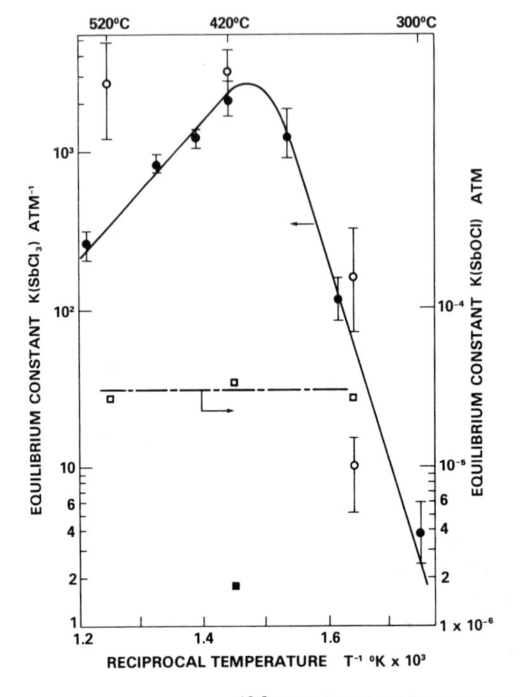

Figure 5. Variation of K_1 (= $KSbCl_3$) and K_3 (= $KSbOCl$) with reciprocal temperature, for interaction of HCl gas with a pure [Sb_2O_3] substrate sample. The closed circle, or square, data were obtained with a Knudsen cell orifice area of 7.75×10^{-3} cm^2 and the open circle, or square, data with an area of 5.6×10^{-4} cm^2. Vertical bars are variation in K_1 that results from a controlled change in the HCl pressure by about 2 orders of magnitude. Closed circle, or square, data were obtained using an EAI quadrupole mass filter and the open circle, or square, data with an Extranuclear Laboratories quadrupole mass filter (note: the specification of commercial instrumentation does not imply endorsement by NBS).

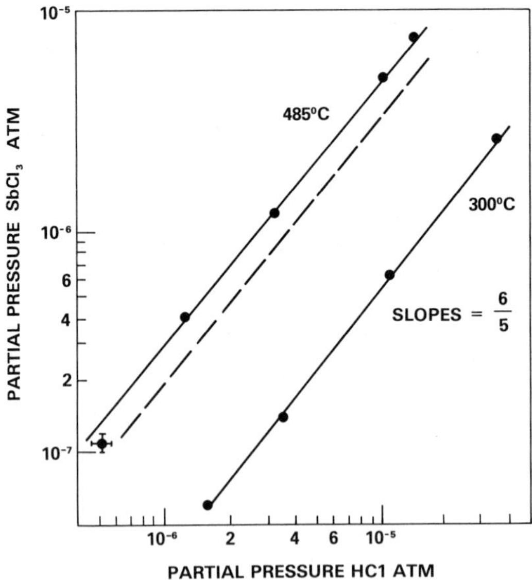

Figure 6. Variation of $SbCl_3$ partial pressure with HCl partial pressure for fixed temperature conditions in the HCl–[Sb_2O_3] system. Broken line indicates hypothetical effect of an order of magnitude decrease in K_1 (485°C) on the $SbCl_3$–HCl pressure dependence.

with HCl pressures ranging from 10^{-7}-10^{-5} atm. Also at T > 420°C the variation of K_1 with T yields a second law reaction enthalpy which is in good agreement with the literature value, as noted above. However, in practical systems most of the flame inhibiting action is expected to occur over the temperature regime of 300 < T< 420°C where it appears that the production of $SbCl_3$ is less efficient than would be expected from reaction (1).

A key assumption in process (1) is the presence of $[Sb_2O_3]$ at unit activity. Hence, by definition, the partial pressure of Sb_4O_6--the free molecular form of $[Sb_2O_3]$ -- in the HCl + $[Sb_2O_3]$ system should be equal to that for the standard state $[Sb_2O_3]$ system. By monitoring the partial pressure of the Sb_4O_6 species we find that for T > 420°C, $[Sb_2O_3]$ is at unit activity where P(HCl) = 10^{-7}-10^{-4} atm. However, at a temperature of 378°C, and lower, we find $P(Sb_4O_6)$ to be dependent on P(HCl) as shown in Figure 7. These data indicate that the thermodynamic activity of $[Sb_2O_3]$ decreases with increasing P(HCl). The apparent equilibrium constant K_1 can be corrected for this changing activity, *i.e.*,

$$K_1' = \frac{K_1}{a[Sb_2O_3]}$$

and the results are indicated in the Figure 7. With this correction the pressure dependence of K_1 is virtually eliminated. This strongly suggests the low apparent efficiency of reaction (1) in the foreward direction to be associated with a lowering of the $[Sb_2O_3]$ thermodynamic activity.

Vaporization of [SbOCl] and $[2SbOCl \cdot Sb_2O_3]$ Substrates. The observed reduction in the thermodynamic activity of $[Sb_2O_3]$ can be accounted for by the formation of new antimony oxide condensed phases containing a halogen component. The existence of two such phases; namely, [SbOCl] and $[2SbOCl \cdot Sb_2O_3]$ has been reported in the literature. A third phase of probable formula $[SbOCl \cdot Sb_2O_3]$ is also believed to exist, *e.g.*, see the review of Pitts (13). Some limited thermodynamic data is available for the first two mentioned oxychloride solids (see the Appendix). This data, although approximate in nature, suffices to establish *a priori* that the reaction of HCl and $[Sb_2O_3]$ to yield each of these oxychloride products should be very favorable for our temperature-pressure conditions. The key question, then, is how does the presence of such phases modify the release of $SbCl_3$ (and SbOCl) to the vapor phase over a mixed chloride-antimony oxide substrate?

Figure 8 summarizes the results of a mass spectrometric analysis of the vapor phase in the presence of the three oxychloride solids and at various temperatures. These data were obtained using [SbOCl] or $[2SbOCl \cdot Sb_2O_3]$ separately as starting materials. A controlled vaporization of these phases was allowed to occur by increasing the temperature using heating rates of 0.2°/min for [SbOCl] and 1.3°/min for $[2SbOCl \cdot Sb_2O_3]$. Eventually, the halide

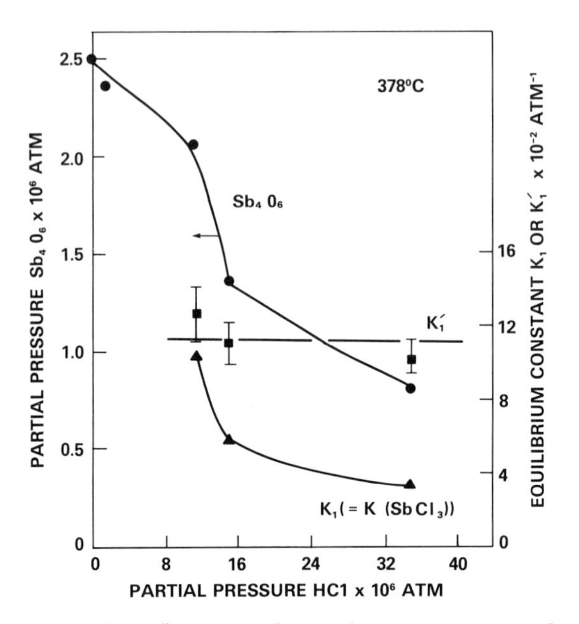

Figure 7. Effect of HCl partial pressure on partial pressure of Sb_4O_6 and apparent equilibrium constants K_1 and K_1' at a fixed temperature of 378°C.

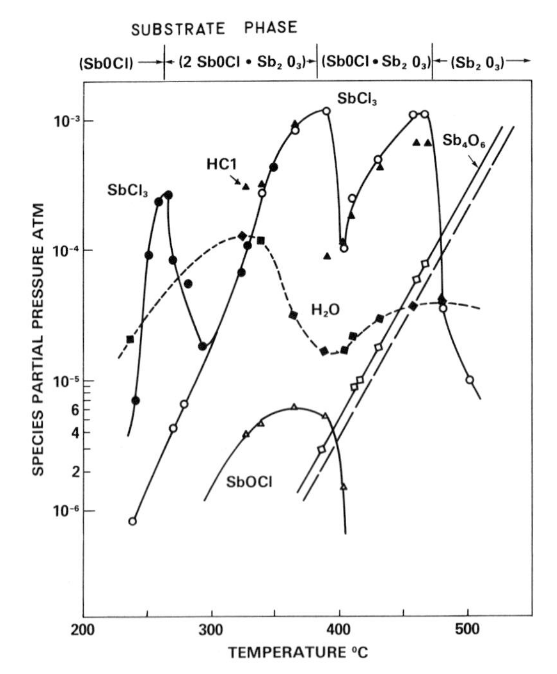

Figure 8. Composition of vapor phase resulting from controlled separate heating of SbOCl (closed circle data) and $2SbOCl \cdot Sb_2O_3$ (open circle data). Broken line, representing the calculated partial pressure curve for the less stable valentine form of [Sb_2O_3] (14), is given for comparison with senarmonite curve. Smaller diameter Knudsen orifice was used for these measurements.

component of these phases was depleted and only [Sb_2O_3] remained in the substrate. The temperature regimes whereby the respective phases exist are indicated in the Figure 8. It is informative to note also the temperatures of maximum and minimum $SbCl_3$ partial pressure that result from the presence of the various oxychloride phases. In particular, the pressure minimum in the region of 300° C corresponds to the temperature of minimum K_1 found for the chlorinated polymer -[Sb_2O_3] substrate system (see Figure 4) and for the HCl - [Sb_2O_3] system (see Figure 5). Thus it appears that solid [$SbOCl \cdot Sb_2O_3$] is formed in each of these systems. Note that the presence of free unit activity [Sb_2O_3] is indicated by the presence of Sb_4O_6 in the vapor at its saturation pressure for temperatures as low as 387°C, as shown in Figure 8.

The production of HCl and SbOCl in these systems results from the presence of bound H_2O in the initial substrate materials. Reaction (3) is found to be the main process leading to SbOCl formation in this system and the production of HCl results from the hydrolysis of the oxychloride substrates with their conversion to [Sb_2O_3]. We should recall the presence of a significant H_2O concentration in the chlorinated polymer - Sb_2O_3 system (see Figure 3). Hence the observed strong interaction of H_2O with these condensed oxychloride phases should also be of significance in the more practical retardancy formulations where H_2O is usually present in appreciable quantity.

Summary of Main Reactions in the Antimony Oxide-Halogen Substrate Systems. A summary of the main reactions indicated from the various Sb_2O_3 - halogen substrate pyrolysis experiments is given in Table I. Also indicated in the Table are the "apparent" thermodynamic properties of these reactions and the corresponding literature thermodynamic data. The differences between apparent and literature thermodynamic data are the result of two factors; namely, a less than equilibrium degree of reaction of HCl with [Sb_2O_3] for the lower temperature conditions and, in some cases, the large uncertainty in the basic literature thermodynamic data.

A measure of the significance of the various reactions listed in Table I can be given in terms of the reaction free energies and these are expressed as a function of temperature in Figure 9. The most favored reactions are those having the most negative free energy change. From this set of data it is possible to calculate the amount of halogen and antimony entering the vapor phase of a flame retardant system for any given conditions of substrate composition and temperature, as shown below.

Table I

Summary of main reactions in the antimony oxide-halogen substrate systems.

	Reaction	ΔF_T Obs.	ΔF_T Lit.[l]	ΔH_T Obs.	ΔH_T Lit.	ΔS_T Obs.	ΔS_T Lit. [T=298]
1.	$6HCl + [Sb_2O_3] \rightarrow 2SbCl_3 + 3H_2O$	-11 [T=693]	-20	-20[T=693]	-20		2
2.	$2HCl + [Sb_2O_3] \rightarrow 2[SbOCl] + H_2O$		-26[T=500]		-23		~ 0[n]
2a.	$2HCl + 2[Sb_2O_3] \rightarrow [2SbOCl \cdot Sb_2O_3] + H_2O$		-17[T=600]		-17		~ 0[p]
2b.	$2HCl + 3[Sb_2O_3] \rightarrow 2[SbOCl \cdot Sb_2O_3] + H_2O$		-14[T=700]				
3.	$H_2O + SbCl_3 \rightarrow SbOCl + 2HCl$	18.7[T=693]	40	0[T=693]	63	27[T=693]	33
4.	$5[SbOCl] \rightarrow SbCl_3 + [2SbOCl \cdot Sb_2O_3]$	8.7[T=538]	~25	76&30[T=526]	25		~ 0[q]
4a.	$4[2SbOCl \cdot Sb_2O_3] \rightarrow SbCl_3 + 5[SbOCl \cdot Sb_2O_3]$	8.2[T=663]		36 [T=590]	[m]	42[T=590]	
4b.	$3[SbOCl \cdot Sb_2O_3] \rightarrow SbCl_3 + 4[Sb_2O_3]$	8.2[T=743]		23 [T=693]	[m]	19[T=693]	
5.	$H_2O + [2SbOCl \cdot Sb_2O_3] \rightarrow 2HCl + 2[Sb_2O_3]$	4.5[T=640]	~16		16		~ 0[r]
6.	$[SbOCl] \rightarrow SbOCl$				64		~ 35[s]
7.	$[Sb_2O_3] + SbCl_3 \rightarrow 3SbOCl$	32 [T=693]	97		168[t]		103
8.	$2[Sb_2O_3] \rightarrow Sb_4O_6$	15.8[T=693][u]	15.8	44[T=693]	44	41[T=693]	41

Footnotes for Table I

l. All literature data are based on the thermodynamic parameters given in the Appendix.
m. From the observed ΔH_T for reactions 4(a) and (b) we can show that for these data to be self consistent, $\Delta H_f[SbOCl \cdot Sb_2O_3] = -356$ kcal/mol; this is within the uncertainty for the reported literature value (see the Appendix).
n. Based on an estimated $S_{298}[SbOCl] = 37 \pm 5$ cal/deg mol.
p. By analogy with above reaction 2.
q. From $S_{298}[SbOCl] = 71.6 \pm 0.8$ cal/deg mol (calc.) and $S_{298}[SbOCl] = 37$ cal/deg mol (est.).
r. Based on ΔS for reactions 2 and 2a and the literature S_{298} for $SbCl_3$, HCl, $[Sb_2O_3]$, and H_2O.
s. Based on ΔS for reaction 2a.
t. Using $\Delta H_f[SbOCl] = -28.7$ kcal/mol (see Appendix).
u. Our $P(Sb_4O_6)$ data, determined using the total vaporization method of calibration for the k-factor, were found to be in satisfactory agreement with the vapor pressure data given by Jungermann and Plieth (6) for senarmonite.

Mass Transport of Retardant to Flame Front

Controlling Factors. Under actual fire conditions the release and transport of a flame inhibiting species, or some suitable molecular precursor, from a decomposing polymer substrate to the flame will be determined primarily by:

- the partial vapor pressure for the inhibitor at the polymer surface, and
- the formation of a diffusive barrier or boundary layer at the surface-gas interface which limits the transport of retardant to the flame.

These flame inhibition process-determining-factors will in turn be dependent upon the temperature, the rate of bulk gas flow across the surface, the effective thickness of the flame front, and the thermodynamic activity of retardant in the polymer substrate.

Consider first the possible limitation of $SbCl_3$ and HCl transport from the substrate to the flame due to boundary layer formation between the vaporizing surface and the bulk gas, as shown schematically in Figure 10. The effect of such a barrier would be to reduce the flux of retardant reaching the bulk gas and hence the flame front. From Fick's law of mass diffusion (15) and the accepted concept of a boundary layer we have:

$$J(SbCl_3) = (DM/\delta RT)[P(\text{surface}) - P(\text{bulk})],$$

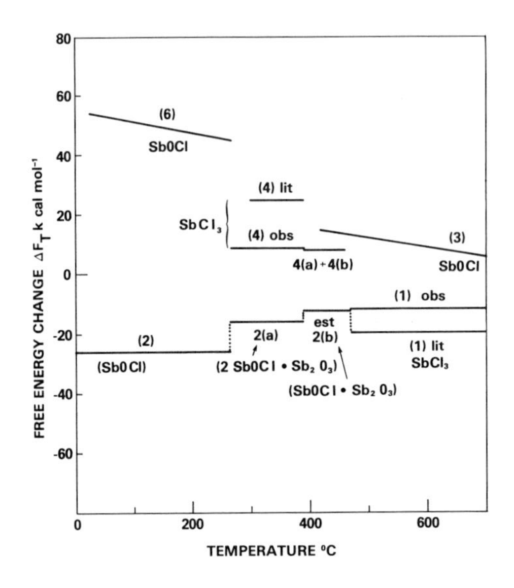

Figure 9. Reaction free energy diagram showing the main reaction products at various temperatures, based on data in Table I. Condensed phases indicated by parentheses.

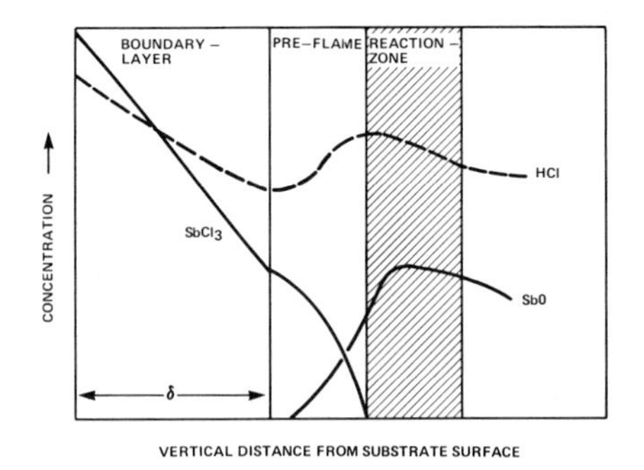

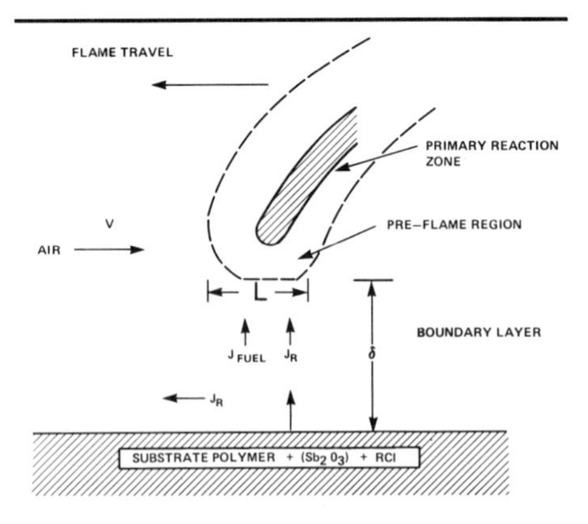

Figure 10. Idealized fire system including a boundary layer between substrate and flame

where $J(SbCl_3)$ is the flux of $SbCl_3$ (units g/cm^2-sec),

D is the diffusion coefficient of $SbCl_3$ in the atmospheric gas (mainly N_2), in units of $cm^2 sec^{-1}$,

δ is the effective thickness of the boundary layer (cm)

P(surface) = $P(SbCl_3)$ at the surface, and

P(bulk) = $P(SbCl_3)$ in the bulk gas (units atm).

In the absence of such a diffusion limited boundary the flux would be given by the Langmuir expression for free molecular vaporization:

$$J = P/[2.557 \times 10^{-2} (T/M)^{1/2}] \text{ g/cm}^2\text{-sec.}$$

Diffusion Coefficients for $SbCl_3$ and HCl. The diffusion coefficient for a species in a gas mixture can be estimated from the expression:

$$D_{12} = \frac{2.62 \times 10^{-3} [T^3 (M_1 + M_2)/2M_1 M_2]^{1/2}}{P \sigma_{12}^2} \text{ cm}^2 \text{sec}^{-1},$$

as derived from hard sphere (i.e., adiabatic elastic collisions) kinetic theory, e.g., see Hirschfelder et al. [p. 544 (16)]. The average collision diameter σ_{12} between unlike species (e.g. N_2 and $SbCl_3$) is estimated from the expression:

$$\sigma_{12} = (\sigma_1 + \sigma_2)/2, \text{ Å}$$

Values of the molecular diameters σ_1 and σ_2 are taken as:

$\sigma(SbCl_3) \sim 7.0$ Å

$\sigma(HCl) = 2.54$ Å, and

$\sigma(N_2) = 2.10$ Å,

using molecular parameters given by Sutton (17). Therefore, at 350 °C,

$$D(SbCl_3 \text{ in } N_2) = 0.28 \text{ cm}^2 \text{sec}^{-1},$$

and

$$D(HCl \text{ in } N_2) = 1.34 \text{ cm}^2 \text{sec}^{-1}.$$

Boundary Layer Thickness δ. An upper limit to the value of δ is given by the flame stand-off distance, which is typically observed to be of the order of 0.1 cm (21). Alternatively, if we assume that the common gas dynamic case of laminar flow over a flat plate applies here, then δ can be estimated from the expression:

$$\delta = 1.5 \text{ L } N_S^{-1/3} N_R^{-1/2},$$

where L is the length of travel by the gas over the plate,

$$N_S = \nu/D\rho \qquad \text{(Schmidt number)},$$

$$N_R = VL\rho/\nu \qquad \text{(Reynolds number), and}$$

where V is the bulk gas velocity
 ν is the coefficient of viscosity for the gas, and
 ρ is the gas density ($\rho = PM/RT$),
e.g., see Graham and Davis (18) and Turkdogan et al. (19). To a good approximation both ν and ρ can be taken as the established values for air i.e.,

$$\nu = 3.18\text{x}10^{-4} \text{ gm/cm-sec}$$

and

$$\rho = 5.6\text{x}10^{-4} \text{ gm/cm}^3 \text{ at } 350 \text{ °C}$$

[Handbook of Chemistry and Physics (20)]. The bulk gas velocity V is an adjustable quantity. For the present purpose we assume a value of:

$$V = 10 \text{ cm sec}^{-1},$$

which is similar to the experimental values of Hirano et al. (21) for a burning cardboard system. This choice of V is not particularly critical to the arguments which follow.

The effective sample length L is also an adjustable parameter. In more classical boundary layer problems, L is defined as the sample length and may be experimentally measured. However, for a propagating fire system some geometrical assumptions are required in determining L. In the absence of direct evidence, the following arguments are used to estimate likely values for L. Since the phenomenon of flame spread allows for a continual replenishment of fuel and heat across the surface, the convective movement of gas is maintained at the flame front and a small effective value for L can therefore be expected. Now a fire can be considered as an aggregation of individual flame elements. Each flame element will consist of a pre-flame, reaction zone and burnt gas region, as suggested in Figure 10. For purposes of flame propagation and inhibition, the region near and at the reaction zone will be of primary significance. This is the region where the concentration of propagating radicals, such as H, is in appreciable excess of the thermodynamic equilibrium value. It is known that for either diffusion or premixed fuel-rich CH_4-air flames the thickness of this region varies between about 0.1-0.4 cm. Likewise, the concentration of suspected inhibitor species such as HCl, HBr or SbO is predominant in this region, as indicated in Figure 10 (22). Thus as the flame front moves across the combustible surface the passage of an inhibitor, such as $SbCl_3$, through the boundary layer is likely to be most significant over a distance of about 0.4 cm, coinciding with the passage of the pre-flame and reaction zone regions. Given the non-ideal geometry of real fires, this distance could vary and an estimated factor of three uncertainty in L should be allowed for.

From the above considerations, a value of the boundary layer thickness δ is calculated as:

$$\delta = 0.24 \text{ cm for } V = 10 \text{ cm/sec}.$$

This distance is of a similar order of magnitude as the flame stand-off distance which tends to support our interpretation of L for flame-spread applications.

Flux Predictions. With these estimated data for D and δ, either the mass transport flux or the partial pressure differential across the boundary layer can be calculated from the flux relationship given above. The partial pressure differential for an inhibitor species, such as $SbCl_3$, can be estimated *a priori* provided there are no kinetic limitations to the movement of $SbCl_3$ and HCl through the polymer substrate to the vaporizing surface. This involves a determination of the value of $P(SbCl_3)$ at the surface from the reaction free energies given in Table I. The flux of retardant entering a flame front may be derived as follows. Consider a polymer substrate containing a halogen source equivalent to 6 mole % HCl, together with 2 mole % $[Sb_2O_3]$. We also assume that the partial pressure of H_2O in the atmosphere is fixed by the air humidity which is typically equivalent to $3x10^{-2}$atm. From the previously determined equilibrium constants for reaction (1), the amount of $SbCl_3$ present at the surface of the substrate can be calculated. Figure 11 shows the general relationship between $P(SbCl_3)$ and the initial unreacted P(HCl) for various typical pyrolysis temperatures. Thus for the assumed condition of P(HCl) total = $6x10^{-2}$ atm we have $P(SbCl_3) = 1.3x10^{-2}$ atm and $P(HCl) = 2.1x10^{-2}$ atm at 350 °C, and 65% of the HCl is converted to $SbCl_3$. The effect of an order of magnitude increase in $P(H_2O)$ on $P(SbCl_3)$ may also be noted in Figure 11. From the previous determinations of K_3, the production of SbOCl can be shown to be negligible for these conditions.

These partial pressures are assumed to represent the halogen retardant species concentrations at the substrate surface. The corresponding concentrations in the bulk gas can be expected to be significantly reduced by the predicted presence of a mass transport-limiting boundary layer. An upper effective limit for $P(SbCl_3)$ in the bulk gas can be estimated from the amount of $SbCl_3$ required in the flame for significant flame inhibition to occur. From our observations of burning velocity and H atom concentration reductions in the presence of $SbCl_3$, this pressure is known to

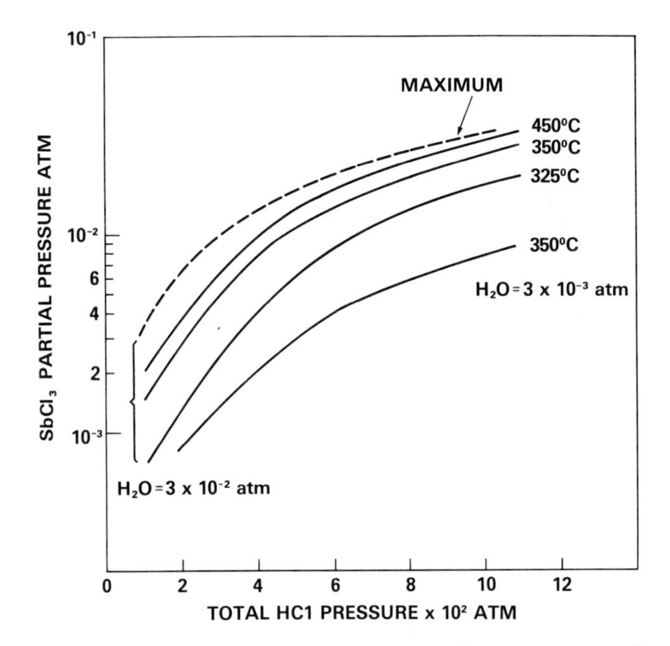

Figure 11. Dependence of $SbCl_3$ partial pressure on total HCl pressure and temperature for conditions of H_2O partial pressure equal to 3×10^{-2} or 3×10^{-3} atm, based on the K_1 values in Figure 4. Broken line indicates maximum possible $SbCl_3$ pressure, i.e., *complete conversion of HCl to $SbCl_3$. Note that SbOCl is a negligible species for these conditions.*

fall in the range of 10^{-4}-10^{-3} atm (see the following discussion). In practice, this is an upper limit since the flame removes $SbCl_3$ from the bulk gas (see Fig. 10). We can therefore reasonably assume the partial pressure of $SbCl_3$ in the bulk gas to be small compared with the corresponding pressure at the surface. From these estimates it follows that the flux of $SbCl_3$ is

$$J(SbCl_3) = 6.8 \times 10^{-5} \text{ g/cm}^2\text{sec } (=3.0\text{x}10^{-7} \text{ mole/cm}^2\text{sec}).$$

This value may be compared with the maximum Langmuir flux (i.e., no boundary layer) of 0.35 g/cm^2sec. Thus,

$$J/J(\text{max}) \sim 2\text{x}10^{-4}$$

and a significant reduction in flux over the maximum allowed by free molecular vaporization appears likely under actual fire conditions.

Similar considerations to those given above for $SbCl_3$ indicate the HCl flux to the flame front to be:

$$J(HCl) = 0.9\text{x}10^{-4} \text{ g/cm}^2\text{sec at 350 °C.}$$

At 450 °C the fluxes are calculated as:

$$J(SbCl_3) = 0.9\text{x}10^{-4} \text{ g/cm}^2\text{sec,}$$

and

$$J(HCl) = 0.4\text{x}10^{-4} \text{ g/cm}^2\text{sec.}$$

Thus the amount of $SbCl_3$ relative to HCl increases with increasing pyrolysis temperature. This effect is largely a result of the absence of stable antimony oxychloride solids at elevated temperatures (i.e., >420 °C).

Predicted Flame Inhibitor Species Concentrations. From a knowledge of the flux of inhibitor entering a flame front we can derive inhibitor species concentrations for a well defined fire system, such as that represented in Figure 10. For example, consider a hypothetical polyethylene-terephthalate substrate containing [Sb_2O_3] together with a halogenated hydrocarbon equivalent to 6 mole % HCl. The predominant gaseous fuel component in this system is acetaldehyde, CH_3CHO. A stoichiometric combustion reaction between CH_3CHO and air, i.e.:

$$CH_3CHO + 5/2O_2 = 2CO_2 + 2H_2O$$

consumes 0.078 mole fraction CH_3CHO, with respect to the fuel-air mixture. The N_2 component in the bulk gas mixture amounts to 0.725 mole fraction. This amount can also be considered as an upper limit to the N_2 concentration at the substrate surface. Based on experimental observations for analogous systems such as polymer candle burning, e.g., see Fenimore and Martin (23), we may consider 0.4 N_2 mole fraction as a likely lower limit surface concentration. If we assume O_2 to be of negligible concentration at the surface then the fuel component concentration should therefore be in the region of 0.27-0.6 mole fraction.

Consider now the effect of the diffusion boundary layer on the flux of CH_3CHO entering the flame front. At a temperature of 350 °C we calculate:

$$D(CH_3CHO) \text{ in } N_2 = 0.82 \text{ cm}^2 \text{ sec}^{-1} [\text{using } \sigma(CH_3CHO)\ 2.9\ \text{Å}].$$

If $P(CH_3CHO) = 0.26$ atm at the surface then

$$J(CH_3CHO) = 7.6\text{x}10^{-4} \text{ g/cm}^2\text{sec}.$$

Comparison of this flux with that for $SbCl_3$ gives a _molar_ flux ratio of:

$$J(SbCl_3)/J(CH_3CHO) = 1.68\text{x}10^{-2}$$

Now at the flame reaction zone, under the characteristic stoichiometric reaction condition of a diffusion flame, 0.078 mole fraction of CH_3CHO is required. Hence, from the above flux ratio, the amount of $SbCl_3$ within the flame is:

$$SbCl_3 \text{ in flame} = 1.2\text{x}10^{-3} \text{ mole fraction}.$$

From the observed burning velocity (2)--and H atom concentration--reductions (see Fig. 12 given later) in premixed flames this amount of $SbCl_3$ is more than sufficient to strongly inhibit a hydrocarbon -- or even H_2 -- fueled flame. Similar arguments to those above indicate:

$$HCl \text{ in flame} = 1.0\text{x}10^{-2} \text{ mole fraction}.$$

An alternative method of converting the $SbCl_3$ flux to a flame concentration is to consider the flux of air entering the flame. For example, consider a volume element determined by the boundary layer thickness δ(= 0.24 cm), the effective reaction zone thickness L(= 0.4 cm) and a unit width (see Fig. 10). Let the flow of air across the surface into this volume element be V(= 10 cm/sec). Then the molar air flux will be given by:

$$J(\text{air}) = V/V_m$$

i.e.

$$J(\text{air}) = 2.1\text{x}10^{-4} \text{ mole/cm}^2\text{sec at } 350\ °\text{C},$$

where V_m (=22.4x10^3 cm^3 at N.T.P.) is the molar volume of air. From before:

$$J(SbCl_3) = 3.0\text{x}10^{-7} \text{ mole/cm}^2\text{sec},$$

hence the concentration of $SbCl_3$ in a flame under these assumed conditions of air flux is determined as:

$$SbCl_3 \text{ in flame} = 1.3\text{x}10^{-3} \text{ mole fraction},$$

which is similar to the previous estimate.

We may recall that macroscopic flammability tests for substrates containing a similar chloride/antimony oxide content to that assumed here indicate an effective degree of flame retardancy, e.g. see the Figure 8 given by Pitts (13). Note also from Pitts that for substrate mole ratios of Sb:Cl > 1:3 very little gain in flame retardancy is achieved. Our data are in accord with this observation owing to the finite values of K_1 ensuring the presence of excess substrate [Sb_2O_3], even for stoichiometric reaction conditions.

Fate of $SbCl_3$ in the Pre-Flame Region

Once $SbCl_3$ enters a flame it encounters a far different chemical environment to that at the substrate surface. In particular, the diffusion of H_2O as a combustion product into the pre-flame region, and the increased temperature, will tend to favor the conversion of $SbCl_3$ to SbOCl via reaction (3). For instance, at some point in the pre-flame region we can expect the following conditions to occur:

$$T = 1200 \text{ K}$$

$$P(H_2O) = 10^{-1} \text{ atm}$$

$$P(SbCl_3) = 5\text{x}10^{-5} \text{ atm}$$

and

$$P(HCl) = 10^{-3} \text{ atm}.$$

From the thermodynamic data given in the Appendix, the free energy change for reaction (3) at 1200 K is 19.7 kcal/mol. Hence, we calculate:

$$P(SbOCl) = 1.3\text{x}10^{-3} \text{ atm}.$$

and SbOCl becomes the predominant antimony containing species at the expense of $SbCl_3$ for these conditions.

From premixed flame studies (3), we know that mass diffusion from the reaction zone provides a significant concentration of H-atoms in the pre-flame region (e.g. $\sim 10^{-3}$-10^{-4} atm). The exothermic bi-molecular reaction:

$$SbOCl + H \rightarrow HCl + SbO \qquad (F.1)$$

can therefore be expected to occur readily. As reaction (3) is considerably endothermic it will most likely be slow as compared with reaction F.1. It therefore follows that the steady-state concentration of SbOCl would be small and this accounts for its non-observation in our mass spectrometric flame sampling studies (22). On the other hand, both HCl and SbO were observed as major products of $SbCl_3$ decomposition in the pre-flame region (22). Thus SbOCl is most likely a significant intermediate species (but of low steady-state concentration) in the pre-flame conversion of $SbCl_3$ to the flame inhibiting moieties of SbO and HCl. These arguments do not preclude the presence of additional reactions such as:

$$SbCl_3 + H \rightarrow HCl + SbCl_2,$$

which were suggested earlier (22).

Action of $SbCl_3$ and HCl in Flame Inhibition

Thus far we have considered the chemistry of $SbCl_3$ from the point of its initial substrate production to its conversion to other species in the pre-flame region. These secondary species of SbO and, to a lesser extent, HCl are considered to be directly responsible for flame inhibition. As has been argued elsewhere (1,22), we consider that these species are catalytically involved in accelerating the recombination of flame radicals and particularly H atoms. For instance,

$$H + HCl = H_2 + Cl$$

$$Cl + RH \rightarrow HCl + R$$

$$Cl + HO_2 \rightarrow HCl + O_2$$

$$Cl + H + M \rightarrow HCl + M$$

where RH is a fuel species and M a third body such as N_2, and

$$SbO + H \rightarrow SbOH$$

$$SbOH + H \rightarrow SbO + H_2.$$

These processes are likely to be more rapid than the normal radical recombinations of:

$$H + OH + M \rightarrow H_2O + M$$

$$H + H + M \rightarrow H_2 + M.$$

Using the Li/Na technique for monitoring H atom concentrations we have demonstrated that the presence of antimony does indeed cause a reduction in the steady-state flame H atom concentration as shown, for example, in Figure 12. Further evidence for SbO as a significant flame species was also obtained by the observation of known emission spectra for SbO (System A at 506.2 nm)-as shown in the Figure 12. It is interesting to note that a similar degree of effectiveness in reducing H atom concentrations is shown by HCl at 30 times the concentration of $SbCl_3$. This result is similar to that found by burning velocity measurements (2). From the flux arguments given above an $HCl/SbCl_3$ flame concentration ratio of about 8 was indicated. Hence, for these conditions, the HCl contribution to flame inhibition would be negligible as compared with that provided by the antimony component of $SbCl_3$. Further flame measurements are in progress to establish the molecular details of the apparent catalytic role of antimony species in effecting radical recombination.

Conclusions

Our studies indicate that the effectiveness of antimony-halogen flame retardant systems depends on the following principal factors:

a. The degree of conversion of the substrate halogen component to $SbCl_3$ and SbOCl species,

b. The flux of these antimony halide species from the substrate surface through a diffusion boundary layer to the flame front,

c. The conversion in the flame of the species $SbCl_3$ and SbOCl to SbO and HCl,

d. The interaction of SbO and, to a lesser extent, HCl with H atoms with a resultant loss of chain branching and reduced burning velocity.

We have shown that the first factor (a) is strongly influenced by the formation of intermediate solid oxychloride phases. The production of the volatile species SbOCl can also be significant in the presence of H_2O vapor. As oxychlorides are generally more volatile species than the corresponding halides,we are considering their possible role as retardant species for other metal oxide systems such as [SnO_2], [Al_2O_3], [TiO_2] and [Cr_2O_3]. Here-to-fore, the less than favorable thermodynamics for the conversion of HCl (or HBr) to volatile metal chlorides such as $TiCl_4$ or $CrCl_3$ has limited consideration of such systems as [Sb_2O_3] substitutes.

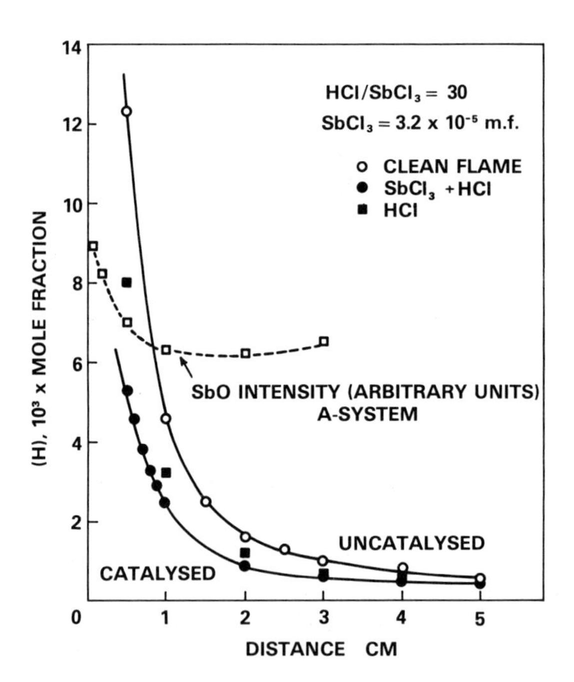

Figure 12. Concentration profiles for H atoms in the $H_2/O_2/N_2$ flame 3/1/3.7 (1650 K) showing catalytic effects of 3.2×10^{-5} mole fraction $SbCl_3$ and 9.6×10^{-4} mole fraction HCl on the loss of H atoms. Additives were introduced with the premixed combustion mixture nebulized aqueous droplets.

With regard to role of $SbCl_3$ or SbOCl in flame inhibition, it appears that the halogen moiety is of secondary importance; its primary function being to transport the metal component to the flame reaction zone.

Appendices: Basic Data

Estimated Thermodynamic Properties for the Species SbOCl. The species SbOCl has not been previously identified and hence no thermodynamic data exist. In view of its significance to the present work we have made the following estimations for the stability and entropy properties of SbOCl. These estimates are believed to be sufficiently reliable to allow calculations of its probable concentration in a "real fire" environment.

The heat of formation ΔH_f is estimated as follows. Consider the atomization energy for the process:

$$Cl - Sb = O \rightarrow Cl + Sb + O,$$

as ΔH atoms = D(Sb-Cl) + D(Sb=O). We assume D(Sb-Cl) to be the average bond energy in SbCl ; namely, 74.8 kcal/mol. Also D(Sb=O) is taken as the bond dissociation energy for the free diatomic species SbO. Literature values of D(SbO) vary from about 90 to 105 kcal/mol. From the known periodic trends for the similar species of GeO (157±3), AsO (114±3), SeO (100±15), SnO (126±2) and TeO (92.5±2) it appears that D(SbO) $\sim$ 105 kcal/mol is the most appropriate value. This result is supported by a similar value of 104 reported by Labsham as cited by the review of Brewer and Rosenblatt (24). Hence, the atomization energy for SbOCl is taken as 179.8±10 kcal/mol. This is equivalent to

$$\Delta H_{f298}(SbOCl) = - 28.7 \text{ kcal/mol}$$

which compares reasonably with the estimated value of - 25.5 kcal/mole given by Wagman et al (26).

The entropy of SbOCl can be satisfactorily determined from estimated geometrical and vibrational parameters using the well known treatment of statistical thermodynamics (26). Based on the known frequencies of vibration for $SbCl_3$ [see Nakamoto (27)] and SbO [see Brewer and Rosenblatt (24)] we estimate for SbOCl frequencies of 710±30 cm^{-1}, 360±20 cm^{-1} and 200±50 cm^{-1}. An O = Sb-Cl bond angle of 120° is assumed and internuclear distances of 2.32 Å and 1.85 Å for Sb-Cl and Sb = O, respectively, are used. It follows from these estimates and the statistical treatment that

$$S_{298}(SbOCl) = 71.6 \pm 0.8 \text{ cal/deg mol.}$$

Literature Thermodynamic Data. The literature thermodynamic data for the other species of interest may be obtained from the compilations of Wagman et al (25) and the JANAF Tables (28), with several exceptions. Pertinent data are summarized in Table A. It should be noted that ΔH_T and ΔS_T can be determined to a good approximation from these ΔH_{f298} and S_{298} data since, for the reactions of interest, the heat content and entropy changes are almost independent of temperature.

TABLE A

Summary of Basic Thermodynamic Data

Species	ΔH_f kcal/mol	$S°_{298}$ cal/deg mol	Comments
[Sb_4O_6]	-344.3	52.8	Senarmonite, i.e., cubic form.
[SbOCl]	- 89.4	37±5	$S°_{298}$ our estimate.
[$2SbOCl \cdot Sb_2O_3$]	-346.9 } -354.5	--	
HCl	- 22.06	44.64	
H_2O	- 57.8	45.1	
$SbCl_3$	- 75.0	80.7 or 82.1	Larger $S°_{298}$ value is from Wilmshurst (29).
SbOCl	- 28.7 or - 25.5	71.6 0.8	Larger ΔH_f value is our estimate.

Mass Spectral Fragmentation Data. In order to utilize mass spectral ion intensities for species partial pressure determinations it is necessary to establish the mass spectral fragmentation pattern for each species of interest. These are given as follows for 90 eV ionizing electron energy:

$SbCl_3$: $SbCl_3^+$ (22.6), $SbCl_2^+$ (60.5), $SbCl^+$ (7.2), Sb^+ (9.7)

SbOCl: $SbOCl^+$ (18.9), SbO^+ (19.9), $SbCl^+$ (20.2), Sb^+ (41.0)

Sb_4O_6: $Sb_4O_6^+$ (9.2), $Sb_3O_5^+$ (2.1), $Sb_3O_4^+$ (17.2), SbO^+ (71.5)

The results for $SbCl_3$ and Sb_4O_6 compare favorably with available literature data, i.e.:

$SbCl_3$ [Preiss (30)]
$SbCl_3^+$ (23.2), $SbCl_2^+$ (58.5), $SbCl^+$ (7.8), Sb^+ (10.5)

Sb_4O_6 [Boerboom et al (31)]
SbO^+ (68), $Sb_4O_6^+$ (13.6), $Sb_3O_4^+$ (6.8).

It is pertinent to note that, contrary to the observations of Boerboom et al (31) and Kazanas et al (32), the relative abundances for the mass spectral ions in the Sb_4O_6 system were temperature independent, as is to be expected if Sb_4O_6 is the sole neutral precursor.

Literature Cited

1. Hastie, J. W., J. Res. N.B.S. (1973) 77A, 733.
2. Lask, G. and Wagner, H. G., Eighth Symposium (International) on Combustion (1962) Williams and Wilkins Co., p. 432.
3. Hastie, J. W., Comb. Flame (1973) 21, 187.
4. Hastie, J. W., Int. J. Mass Spec. Ion Phys. (1975) 16, 89.
5. Grimley, R. T., Mass Spectrometry in "The Characterization of High Temperature Vapors", 195, Ed. J. L. Margrave, John Wiley and Sons, Inc., New York (1967).
6. Jungermann, E. and Plieth, K., Z. Physik. Chem. (1967) 53,1.
7. Drowart, J. and Goldfinger, P., Angew Chem. Int. Ed. (1967) 6, 581.
8. Gaydon, A. G., "Spectroscopy of Flames", Chapman and Hall, London (1957).
9. Bulewicz, E. M. and Sugden, T. M., Trans. Faraday Soc. (1956) 52, 1475.
10. Bulewicz, E. M., James C. G., and Sugden, T. M., Proc. Roy. Soc. (1956) A255, 89.
11. Mavrodineanu, R. and Boiteux, H., "Flame Spectroscopy", John Wiley and Sons, New York (1965).
12. Touval, I., "A comparison of antimony oxide and stannic oxide hydrate as flame retardant synergist" (1973) Polymer Conf. Series, University Detroit, May 22, 1973; see also J. Fire and Flammability (1972) 3, 130.
13. Pitts, J. J., "Inorganic flame retardants and their mode of action" p. 133 in "Flame Retardance of Polymeric Materials", Vol. I, Eds. W. C. Kuryla and A. J. Papa, Marcel Dekker, Inc., New York (1973).
14. Behrens, R. G. and Rosenblatt, G. M., J. Chem. Thermodynamics, (1973) 5, 173.
15. Jost, W., "Diffusion in Solids, Liquids, Gases", Academic Press, New York, (1960).
16. Hirschfelder, J. O., Curtiss, C. F., and Bird, R. B., "Molecular Theory of Gases and Liquids", John Wiley and Sons, New York (1954).

17. Sutton, L. E., "Tables of Interatomic Distances and Configuration in Molecules and Ions", Spec. Publ. No. 18, Chemical Society, Londong (1965).
18. Graham, H. C. and Davis, H. H., J. Amer. Ceram. Soc. (1971) 54, 88.
19. Turkdogan, E. T., Grieveson, P. and Darken, L. S., J. Phys. Chem. (1963) 67, 1647.
20. "Handbook of Chemistry and Physics", 55th Edition, CRC Press, New York (1974).
21. Hirano, T., Noreikis, S. E. and Waterman, T. E., Comb. Flame (1974), 22, 353; see also Interim Technical Report No. 2, Project J1139 "Measured velocity and temperature profiles of flames spreading over a thin combustible solid", IIT Research Institute, Chicago, Illinois, May (1973).
22. Hastie, J. W., Comb. Flame (1973) 21, 49.
23. Fenimore, C. P. and Martin, F. J., "The Mechanisms of Pyrolysis, Oxidation, and Burning of Organic Materials", Nat. Bur. Stand. (U.S.) Spec. Publ. 357 (June 1972).
24. Brewer, L. and Rosenblatt, G. M., "Advances in High Temperature Chemistry", (L. Eyring, Ed.) 2, 1 Academic Press, New York (1969).
25. Wagman, D. D., Evans, W. H., Parker, V. B., Halow, I., Bailey, S. M. and Schumm, P. H., "Selected Values of Chemical Thermodynamic Properties", N.B.S. Tech. Note 270-3 (U. S. Govt. Printing Office, Washington, D. C.)(1968).
26. Herzberg, G., "Infrared and Raman Spectra of Polyatomic Molecules", Van Nostrand, New York (1945).
27. Nakamoto, K., "Infrared Spectra of Inorganic and Coordination Compounds", John Wiley & Sons, Inc., New York (1963).
28. JANAF Tables, Joint Army Navy Air Force Thermochemical Tables, 2nd Ed. NSRDS-NBS 37, U. S. Govt. Printing Office, Washington, D. C. (1971).
29. Wilmshurst, J. K., J. Mol. Spec. (1960) 5, 343.
30. Preiss, H., Z. Anorg. Allg. Chem. (1972) 389, 280.
31. Boerboom, A. J. H., Reyn. H. W., Vugts, H. F. and Kistemaker, J., "Thermochemistry of antimony and antimony trioxide" p. 945 in Advances in Mass Spectrometry 3, Ed. W. L. Mead (1966).
32. Kanzanas, E. K., Chizhikov, D. M., Tsvetkov, Yu.V. and Ol'Shevskii, M. V., Russ. J. Phys. Chem. (1973) 47, 871.

DISCUSSION

A. W. BENBOW: Perhaps the major difficulty inherent in the study of the mechanism of the synergistic flame-retardant action of metal oxides (particularly antimony oxide) with halogen-containing compounds is that it is a multistage process. Dr. Hastie's work and recent work carried out at the City University, London, have concentrated on different aspects of the inhibition process and the results tend to be complementary. Following our experimental work in which largely unsuccessful attempts were made to incorporate a wide range of metal halides into various polymers, it became obvious that an important stage in the metal oxide/halogen compound interaction was the reaction between them in the condensed phase to produce volatile species such as those mentioned by Dr. Hastie. As an example of our studies, a halogenated compound (actually I.C.I. Cereclor 70, a chlorinated wax with 70% chlorine content) and antimony oxide were heated in air in a sensitive thermobalance, and the kinetics of the volatilization reaction were monitored. Variation of the ratio by weight of antimony to halogen heated together showed that the predominant volatilized species was the trihalide. This was confirmed by chemical analysis. Other halogen compounds, which volatilized in preference to decomposing in the condensed phase, required higher than theoretical ratios of halogen: antimony but the volatilized species was the trihalide in each case.

I was very interested to hear Dr. Hastie stressing the importance of antimony oxyhalides (e.g., SbOCl) in the flame. Such entities have previously been considered important in the condensed phase reactions between antimony oxide and halogen compounds. In fact the endothermic breakdown of antimony oxychlorides in the condensed phase was frequently postulated as an important part of their mode of action. However, our present work has shown that the reaction between antimony oxide and halogenated compounds is in fact moderately exothermic. Thus the involvement of oxyhalides in the condensed phase seems unimportant.

When attempts are made to use other metal oxides

as substitutes for antimony oxide in real flame-retardant compositions the results are generally very poor. The only exceptions are some of the hydrated metal oxides (aluminum, stannic, zinc) which are now being brought on to the market, and it seems that generally these metal oxides have a different mode of action from that of antimony oxide.

J. W. HASTIE: I agree with the general comments of Benbow with the following exception. Our results do show that oxychlorides are present as condensed phase intermediates. This does not preclude the possibility of an overall exothermic reaction (observed by Benbow and Cullis) between antimony oxide and the halogen source since the condensed oxychloride forming reactions are exothermic as shown in Table I of our paper.

5

Effect of CF_3Br on Counterflow Combustion of Liquid Fuel with Diluted Oxygen

K. SESHADRI and F. A. WILLIAMS

Department of Applied Mechanics and Engineering Sciences,
University of California, San Diego, La Jolla, Cal. 92037

To obtain basic information on mechanisms of fire suppression, well controlled experiments on flame inhibition and extinction are needed. Most such laboratory tests concern premixed flames and provide information relating to the effects of suppressants on burning velocity, flammability limits and premixed reaction-zone structure (1, 2). Yet fires generally are thought to be diffusion flames. A few basic studies (3-8) have been focused on the structure and extinction of diffusion flames subjected to inhibiting agents. All of these involve gaseous fuels instead of the more commonly encountered liquid fuels. Recently an experiment has been designed (9) for testing the effects of suppressants on diffusion flames of liquid fuels under carefully controlled conditions. The present work provides further quantitative data helping to validate the technique and exploits the method by applying it to a variety of different liquid fuels.

Counterflow and coflowing configurations are employed in the two principal competing approaches for studying diffusion flames. The former contains a stagnation point toward which fuel and oxidizer flow from opposite directions. For the latter, an example of which is provided by fuel issuing from a tube into a quiescent oxidizing atmosphere, the fuel and oxidizer streams flow in approximately the same direction, and a stagnation point need not be present, although often there is a stagnation line in the gas near the lip of the wall at which the two streams first meet. Both coflowing (3, 4, 8) and counterflow (5, 6, 7) configurations were used in earlier studies of effects of inhibitors on gaseous fuels. In coflowing systems residence times and observation volumes typically are larger, convection toward the flame zone is at a minimum throughout most of the field, but concern can arise over possible premixing near the cool wall where fuel and oxidizer gases first come into contact. In counterflow systems closer confinement usually is needed to assure stability of the flow and results in smaller observation volumes and shorter residence times, convection normal to the reaction zone always is present and affects concentration and temperature profiles, but the possibility of premixing other than that resulting directly from diffusion of reactants through the flame is eliminated completely. That both types of configurations are useful is demonstrated by the fact that conclusions drawn from careful work with both systems tend to agree. The present studies employ a counterflow configuration in which a flat laminar diffusion flame is established in a stagnation-point boundary layer by directing the oxidizing gas stream downward onto the surface of a burning liquid fuel.

In experiments with liquid fuels the suppressant is added to the oxidizing gas. Although there is greater freedom in working with gaseous fuels, in that the suppressant also can be added to the fuel, real-world situations generally are approximated better by addition to the oxidizer. Details of the apparatus employed in the present experiments may be obtained from Figure 1; a description is given by Kent and Williams (9). The pressure is atmospheric, and the gas consists of mixtures of O_2, N_2 and CF_3Br. The thin flame sits a few millimeters above the liquid surface, well on the oxidizer side of the stagnation point for all stoichiometries achievable with undiluted liquid fuels. Thus, convective velocities in the flame zone are directed toward the fuel in all tests. The present paper reports on measurements of temperatures within the liquid fuel, of gas-phase temperature profiles, of concentration profiles for major stable species, and of extinction conditions observed for methanol, benzene, iso-propyl ether, butyl-vinyl ether, methyl methacrylate, polymethyl methacrylate, iso-octane, kerosene and various normal alkanes. The latter results provide overall kinetic information of a quantitative character in the absence of CF_3Br but as yet enable only qualitative deductions to be drawn concerning kinetics in the presence of the chemical inhibitor.

Literature for Gaseous Fuels

All of the existing evidence suggests that the gas-phase combustion processes are similar for gaseous and liquid fuels. Therefore earlier results on chemical inhibition of diffusion flames for gaseous fuels are relevant to the present investigation.

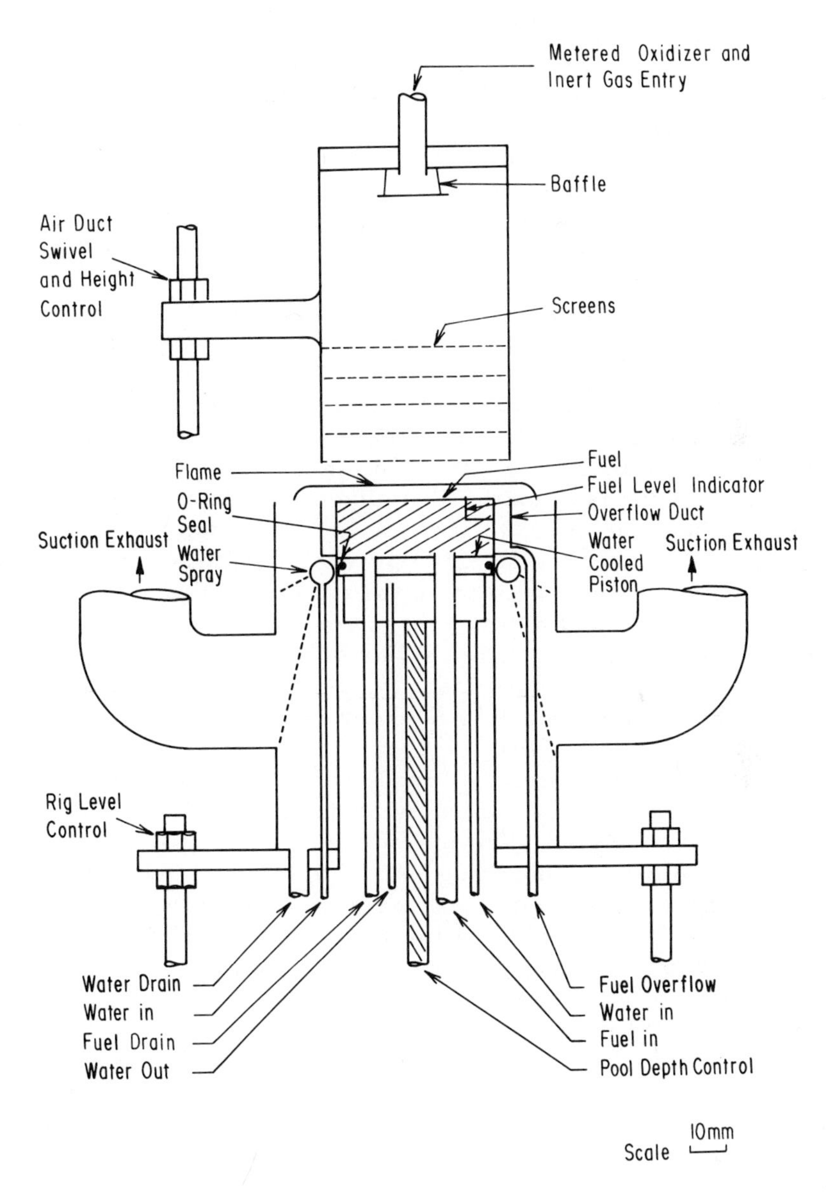

Figure 1. Schematic of apparatus

Coflowing Streams. In the early works of Simmons and Wolfhard (3) and Creitz (4) results may be found for gaseous hydrocarbons as well as for carbon monoxide and hydrogen, with the inhibitors CH_3Br, CF_3Br and Br_2. Flame extinction occurred when the added inhibitor concentration in the fuel or air stream was high enough to lift the flame off the burner rim. It was found that the suppressants acted more efficiently as extinguishers when added to the air side of the flame rather than to the fuel side; this effect was explored in detail by Creitz (4) for CH_3Br and CF_3Br. The result is understandable from the fact that the reaction zone in diffusion flames seeks out a position where the fuel-oxidizer ratio is approximately stoichiometric. Since there is more oxygen than fuel in a stoichiometric mixture, there is more suppressant in the flame zone when it is added to the oxidizer.

Simmons and Wolfhard (3) made spectroscopic examinations of the flames and found that the flame structure was considerably altered by addition of the halide inhibitors. Halons increased carbon luminosity on the fuel side of the flame. In the case of methyl bromide added to air, an extra reaction zone was formed on the air side, in which premixed combustion of the inhibitor with air occurred when the mixture encountered sufficiently high temperatures. These results and the observation that smaller volumes of halides than of nitrogen needed to be added to produce extinguishment (3, 4) led to the belief that the effect of the halide inhibitors is chemical in nature and not merely thermal quenching.

More recently, Creitz (8) made composition and temperature measurements in a coflowing propane-air diffusion flame inhibited with CF_3Br. The measurements showed that carbon, hydrogen and carbon monoxide were the only fuel species on the air side of the yellow zone, indicated that the oxygen concentration vanished before reaching the yellow zone, and demonstrated that the concentration of CF_3Br, which was added to the air, dropped to zero well away from the blue flame reaction zone. It was concluded that decomposition of CF_3Br occurred by chemical means rather than thermally, since the measured temperature when CF_3Br disappeared was only about $1080^{\circ}K$. It was inferred that inhibiting effects were due to the decomposition products rather than the intact molecule.

Counterflow Streams. Friedman and Levy (5) used opposed-jet methane-air diffusion flames to test the effects of alkali metal vapor and organic halide inhibitors. Extinction occured in the center of the flame sheet upon application of the inhibitor, as is usual for opposed-jet flames. In our experiment, on the contrary extinction occurs suddenly and completely over the entire surface of the liquid pool; the difference may be attributed to the fact that afterburning is quenched by a water spray in our apparatus, while longer residence times at elevated temperatures are available at the sides in typical opposed-jet tests.

Friedman and Levy (5) give curves of inhibitor volume concentrations at varying flow rates, showing that on this basis, when added to the fuel stream CF_3Br and CH_3Br are more effective than $CH_3C\ell$ and $CC\ell_4$ and considerably more effective than N_2. Ibiricu and Gaydon (6) made spectrographic studies and temperature measurements of inhibited hydrocarbon and hydrogen flames. Their results show that the halide inhibitors generally decrease the emission of OH and HCO and increase the emission of C_2. Carbon luminosity increases and temperature decreases with increasing inhibitor concentration. Milne, Green and Benson (7) measured the effectiveness of various halide inhibitors and powders on hydrocarbon fuels. Their curves of gas velocity versus mole fraction of inhibitor required to extinguish the flame, show that CF_3Br is more effective than CH_3Br and confirm that the inhibitors are more effective when added to the air side.

The Experiment and its Instrumentation

The apparatus employed in the present tests differed from that described previously (9) in only one respect. The fuel feed system was replaced by an arrangement similar to the Bajpai feed device (10, 11). This simplified the operation, eliminating the need to run the peristaltic pump and heat exchanger during the tests; the pump was run only initially to fill the Bajpai leveling reservoir. This modification caused no observable differences in experimental results.

Temperature Measurement. Liquid temperatures were measured with chromel-alumel thermocouples and gas temperatures with coated Pt-Pt 10% Rh thermocouples (12). Wire diameters of 0.076 mm were employed for the thermocouples used in the present tests. Within the liquid, this provided excellent spatial resolution, and measurements with different lead configurations demonstrated that conductive losses were negligible. Within the gas, conductive losses were made negligible by aligning the leads parallel to the flame sheet so that they extended along an isotherm for a length of approximately 5 mm.

Although slight misalignment has no effect, errors up to 200°C can occur if the leads are oriented normal to the flame sheet.

The noncatalytic coating (12) can be important in the diffusion flame; a bare thermocouple indicated a flame temperature about 200°C higher than a similar coated thermocouple. A correction for radiant loss of energy also is important in the flame. The two-thermocouple method was employed, with a linear extrapolation to zero wire diameter, for calibrating thermocouples of the size employed. Radiation corrections did not exceed 150°C. Estimated accuracies of final corrected temperatures are ± 50°C in the flame.

Composition Measurement. Concentrations of gaseous species were measured by gas chromatographic analysis of samples withdrawn through fine quartz probes. Data were obtained for O_2, N_2, H_2O, CO_2, CO, H_2, CH_4, C_2H_2, C_2H_4, C_2H_6, C_3H_6, CF_3Br and the fuel. The chromatograph contained two 6.4 mm diameter columns, each 305 cm long, one a Porapak Q and the other a molecular sieve 5A. A valve allowed the molecular sieve to be either isolated or connected in series downstream from the Porapak column. The oven was operated at a constant temperature of 80°C, and the carrier gas was helium flowing at a rate of 80 ml/min. A thermal conductivity cell served as the detector.

Calibration of the instrument was performed by mixing gases with nitrogen in a 1 liter pressure cylinder. The compositions of the mixtures were determined from partial pressures which were read on a mercury manometer. After allowing the gases time to mix thoroughly in the cylinder, the mixture was introduced via the sampling valve to the 3 ml sampling loop of the gas chromatograph. The sampling loop, as well as the sampling and column switching valves, are located in the gas-chromatograph oven and are at oven temperature to prevent condensation of vapors in the sample. The areas of the chromatograms were approximated by the product of the peak height and width at half height.

A schematic flow diagram of the sampling system is shown in Figure 2. Samples are withdrawn from the flame with a quartz microprobe of the type described by Fristrom and Westenberg (13). The probe has a sonic orifice with a throat diameter of about 60 microns. A flexible teflon tube connects the probe to a stainless-steel tube of larger diameter, which then leads to the oven-mounted sampling valve and the sampling loop of the gas chromatograph. The complete sampling line, except

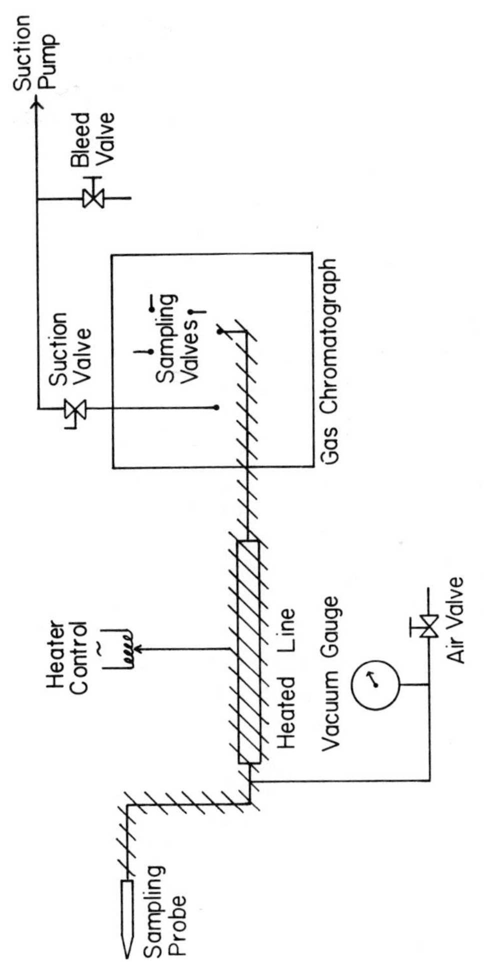

Figure 2. Schematic of sampling and chromatographic systems

for the probe itself, is wrapped in heating tape to avoid condensation of the sample on the walls. If condensation occurs in the probe, it is easily seen and may be dispelled by temporarily heating the probe externally with a bunsen flame. Downstream of the gas chromatograph, the sampling line leads to the suction pump.

Samples were withdrawn from the flame at a line pressure of 63.5 cm Hg below atmospheric. A sampling time of about 4 minutes is required with a 60 micron probe. During this time, the sample flushes through the entire flow line, including the sampling loop of the gas chromatograph. This procedure leads to good repeatability for H_2O measurement. Water tends to adsorb on the walls of the sampling line, and the system must be "conditioned" in this way to obtain reliable results.

After the sampling system has been completely flushed, the suction valve, downstream of the sampling loop, is closed and the line is quickly brought to atmospheric pressure by opening the air valve shown in Figure 2. The tube volume ahead of the sampling valve is large enough to prevent air from reaching the sampling loop when pressurization takes place. When the system has reached atmospheric pressure, the sampling valve is switched to sweep the sample onto the columns.

To analyze the sample, the columns initially are adjusted so that the molecular sieve is in series with the Porapak Q. During the first sixty seconds, the species H_2, O_2, N_2, CO, and CH_4 elute from the Porapak Q and are swept onto the molecular sieve. At 60 seconds after the sample is introduced, the molecular sieve is isolated. The species CO_2, C_2H_2, C_2H_4, C_2H_6, H_2O and C_3H_6 elute from the Porapak Q column in that order and go to the detector. Then at about seven minutes after the sample was introduced, the molecular sieve column is again switched on line, and the species H_2, O_2, N_2, CH_4, and CO are eluted in that order. At about fourteen minutes after the sample is introduced, the molecular sieve column is switched out and the Porapak Q column is reverse flushed. This causes the heavy component, the fuel, to be eluted, which completes the analysis. Accuracies of composition measurements are estimated to be better than $\pm 10\%$.

Thermal Environment of the Liquid

Much work has shown that the thermal environment of the liquid is important in extinguishment of burning pools (14, 15). Since the objective of the present study is to extract information concerning gas-phase kinetics, consideration must be given to possible thermal influences from the condensed phase. Analysis of the data is based on a parallel theoretical effort (16) in which the influence of heating and vaporization of the fuel is taken into account. However, the theory ignores lateral liquid motion and heat loss from the liquid to its surroundings. To test the accuracy of these approximations, temperature measurements within the liquid were made for representative fuels.

Horizontal profiles of temperatures at various depths within the liquid are shown in Figure 3 for methyl methacrylate. Corresponding vertical temperature profiles along the centerline and at the wall of the cup are shown in Figure 4. It is seen that the maximum liquid temperature is achieved at the top near the rim of the cup and the minimum on the bottom at the center. The temperatures are consistent with the fact that the cup is cooled with water from the bottom. The thermal conductivity of the metal walls exceeds that of the liquid, thereby accounting for the shallower gradient along the wall. This causes heat to be transferred from the liquid to the side wall near the top of the cup and from the side wall to the liquid near the bottom (Figure 3).

Observations of the motion of dust on the surface of the liquid indicate a radially outward flow at velocities of a few cm/s, small compared with gas velocities. This outward flow and the temperature profiles suggest the existence of a weak toroidal vortex in the liquid. The vortex may be driven by shear at the gas-liquid interface due to the gas flow, by flow from the nearly centrally located fuel inlet in the bottom of the cup (a very small effect) or by buoyancy forces due to the nonuniform temperature of the liquid. With respect to the last effect, the vertical stratification always is stable, but the lower temperature along the upper wall, in comparison with the temperatures of the nearby liquid, can generate horizontal pressure gradients that tend to produce motion in the observed direction.

The temperature just below the liquid-gas interface is seen in Figure 3 to peak near the side wall. Continued heating of transversely convected liquid may contribute to this, but the principal cause is believed to be that due to increased gas flow velocities produced partially by exhaust suction, the flame is slightly closer to the liquid near the periphery. This increases the local heat transfer from the gas to the liquid. The interpretation is strengthened by Figure 5, which shows that moving the flame closer to the liquid surface at the stagnation point by increasing the mass fraction of oxygen in the gas stream, has the effect of reducing the difference between near-surface liquid temperatures at the center and on the periphery.

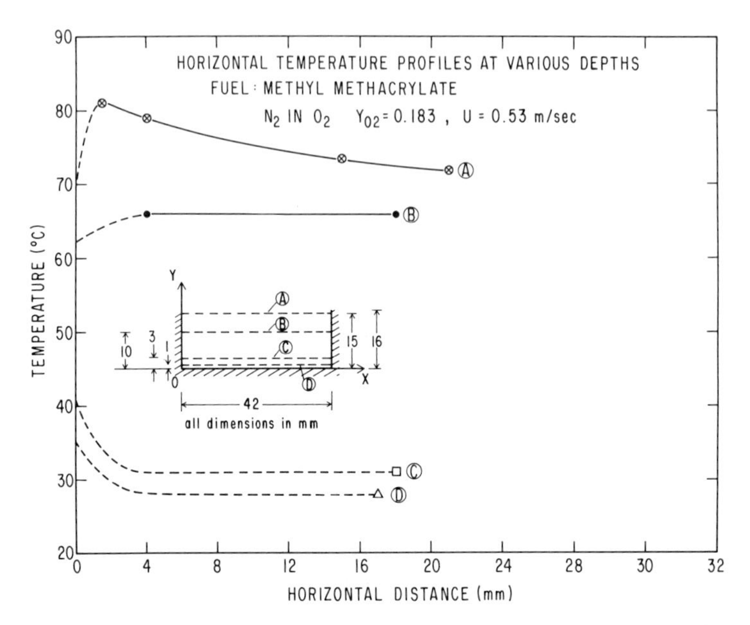

Figure 3. Horizontal profiles of liquid temperature at various depths for methyl methacrylate burning in an O_2–N_2 mixture having Y_{O2} = 0.183 with U = 0.53 m/s

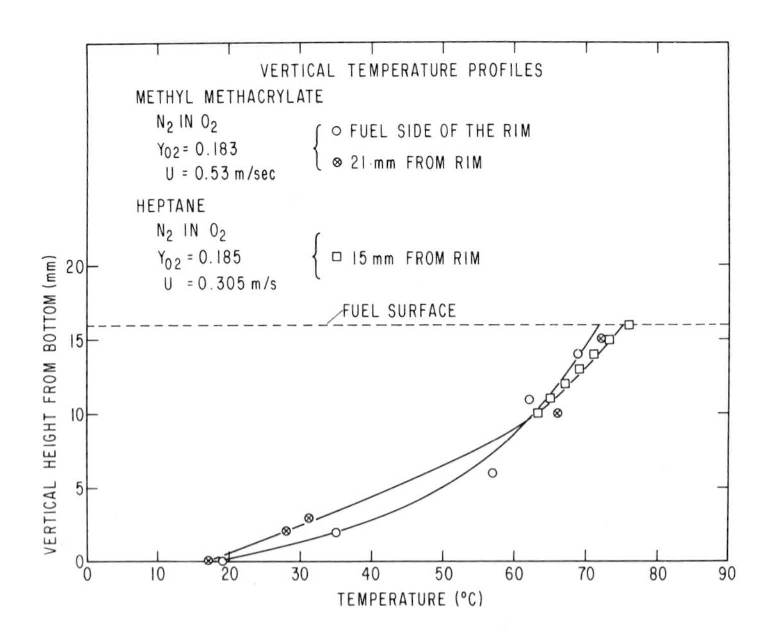

Figure 4. Vertical profiles of liquid temperature for methyl methacrylate and heptane burning in O_2–N_2 mixtures

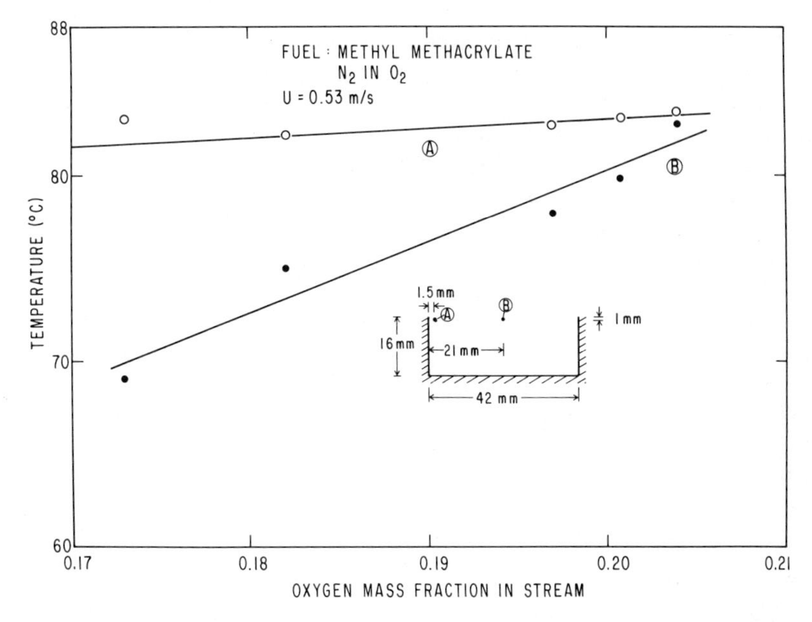

Figure 5. Dependence of liquid temperatures on oxygen content of gas stream at U = 0.53 m/s for O_2–N_2 mixtures

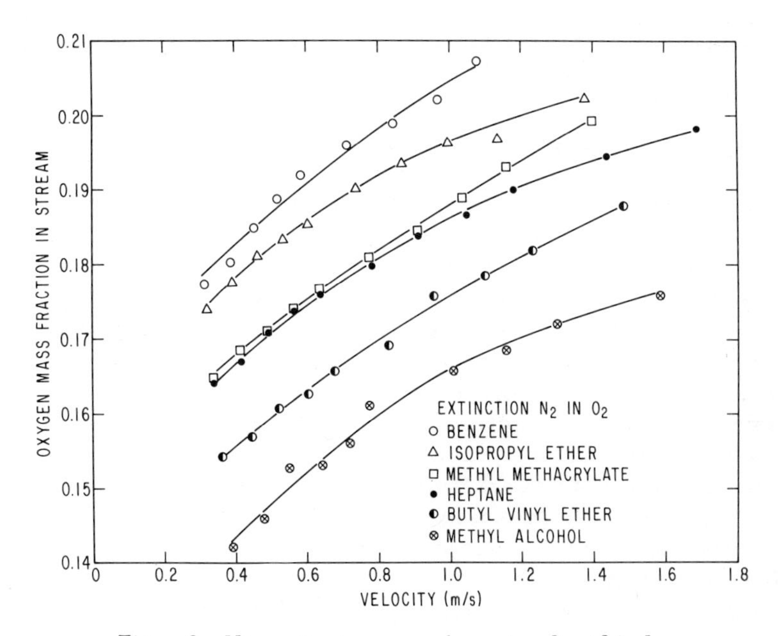

Figure 6. New extinction curves for various liquid fuels

The measured temperature gradients were used to estimate the heat loss from the liquid to its surroundings. Values less than 0.5 cal/s were obtained; this is less than 5% of the heat flow required to vaporize the fuel. Errors of this magnitude, even if systematically dependent on conditions of gas flow, are too small to influence calculated kinetic parameters significantly. Therefore it was concluded that variations in the thermal environment of the liquid affect deduced gas-phase kinetic quantities negligibly in the present experiments.

An additional observation was made to strengthen this conclusion and to indicate further that the liquid motion *per se* was unimportant. Extinction curves for heptane in O_2 - N_2 mixtures were generated for conditions under which the cup was filled with steel wool and screening to a height 1 mm below the liquid-gas surface. Within the accuracy of rotameter measurement, extinction conditions were identical with those obtained in the absence of this flow-inhibiting, conduction-enhancing modification.

Results for Liquid Fuels in Mixtures of Nitrogen with Oxygen

Since the effect of nitrogen addition is known to be purely thermal, an understanding of the influence of nitrogen must be developed before a comparably thorough understanding of effects of chemical inhibitors can be obtained. Experiments with nitrogen addition help to validate the approach. A number of results with nitrogen, beyond those given earlier (9), have recently been obtained and are reported here.

Nitrogen Quenching. The simplest measurements to obtain are the velocity of the oxidizing stream at extinction as a function of the composition of the stream. New data of this type, generated for various fuels, are given in Figure 6. This data, as well as data obtained earlier for other fuels, are summarized in Figure 7. The measurements are highly reproducible on these graphs; systematic differences in the third digit for the extinguishment value of Y_{O2} at a given gas flow rate were observed for different samples of methyl methacrylate and of heptane. The oxygen mass fraction was selected as the vertical coordinate because in the earlier work (9) a correlation was obtained on this basis for extinguishment with the different inert diluents N_2 and CO_2. It is clear from Figures 6 and 7 that data for different fuels do not correlate on such a graph. Moreover, systematic variations with either stoichiometry or expected reactivity are not evident. Similar graphs were prepared using mole fractions and also using stoichiometrically adjusted quantities (measuring molar and mass compositions in the flame zone), but no better correlations were obtained.

Since the effects of nitrogen are thermal, it is more logical to seek a correlation on the basis of the adiabatic flame temperature. Equilibrium dissociation influences the adiabatic flame temperature appreciably only at the higher calculated temperatures (above 2000°K), but regardless of whether dissociation is taken into account, the dependence of the adiabatic flame temperature on the extinction velocity is found to be qualitatively similar to the dependence of oxygen mass fraction shown in Figures 6 and 7; the correlation is no better. The difficulty is that the extinction velocity alone is not a suitable parameter for characterizing differences in the structure of the stagnation-point boundary layer for different fuels.

The theory (16) on which analysis of the extinction experiments is based has been summarized earlier (9). The flame-sheet approximation is employed in lowest order to describe the stagnation-point flow field. The effect of finite-rate chemistry is then incorporated as a singular perturbation. The complex chemical kinetics in the flame are approximated as a one-step Arrhenius process, of first order with respect to both fuel and oxidizer. The justification for this approximation, gross from the viewpoint of underlying chemistry, is that extinction tests can provide at most two parameters, an overall activation energy and an overall pre-exponential rate factor. Clearly a formidable task will remain in relating these overall rate parameters to chemical mechanisms. However, the gross description itself can be useful, even in the absence of further fundamental elucidation. The rate parameters obtained will be specified to diffusion flames and as such should be applicable better than available overall premixed rate parameters to diffusion flames in other configurations.

The theory (16) predicts that UFM_F/T^2 (see Nomenclature) at extinction serves to extract the overall chemical kinetics, being proportional to the pre-exponential factor (cm^3/mole sec) times the Arrhenius factor. The factor F accounts for the influences of the stagnation-point flow field with vaporization. It is calculated from computer solutions for the flow in the flame-sheet approximation, and it depends on physico-chemical properties of the fuel. The definition of F employed here is that of Krishnamurthy and Williams (16) and differs slightly from that of Kent and Williams (9). Fuels having the same pre-exponential factor and activation energy will follow the same extinction curve in a plot of adiabatic flame temperature as a function of UFM_F/T^2. Since actual flame temperatures at extinction are so low that equilibrium dissociation is negligible, from the viewpoint of overall energetics it is proper to exclude dissociation when calculating the adiabatic flame temperature for use in connection with the theory. The resulting plot is shown in Figure 8. From this graph, it finally becomes possible to identify trends in chemical reactivity.

Contrary to Figures 6 and 7, it is seen that with the exception of the line for isopropyl ether, all lines in Figure 8 have approximately the same slope. This turns out to imply roughly equal overall activation energies for these fuels. The higher curves imply extinguishment at higher adiabatic flame temperatures and therefore correspond to less reactive fuels. It is seen that lines for the normal alkanes and also for benzene nearly coincide, suggesting that these species have approximately equal reactivities. Kerosene and particularly iso-octane extinguish at higher flame temperatures, implying that these fuels are somewhat less reactive. The oxygenated fuels are seen to be appreciably more reactive as might be expected. The variation in overall activation energy is much greater for the oxygenated fuels, and because of this the differences in their reactivities are less than Figure 8 suggests; the cube of the activation energy belongs in the numerator inside the logarithm if reactivities of fuels with different activation energies are to be compared (16). This correction for activation energy shows that butyl vinyl ether is less reactive and isopropyl ether more reactive than Figure 8 would indicate. The difference between the overall reaction rates at any given temperature for the least and most reactive of all of the fuels is not quite one order of magnitude.

It is tempting to seek correlations with the carbon to hydrogen ratio C/H of the fuel. If this is done by plotting the adiabatic flame temperature at extinction as a function of C/H for a fixed value of U, then a very rough correlation is obtained, in which the extinction flame temperature increases with C/H. If the better approach of keeping UFM_F/T^2 (instead of U) fixed is employed, then there is no correlation for the complete set of fuels, although correlations of subclasses (e.g., oxygenated fuels and normal alkanes) appear, in which the extinction flame temperature decreases with increasing C/H. From the fundamental viewpoint, C/H correlations are to be expected only for subclasses, if at all. As yet, too few fuels have been tested to ascertain whether such correlations actually exist.

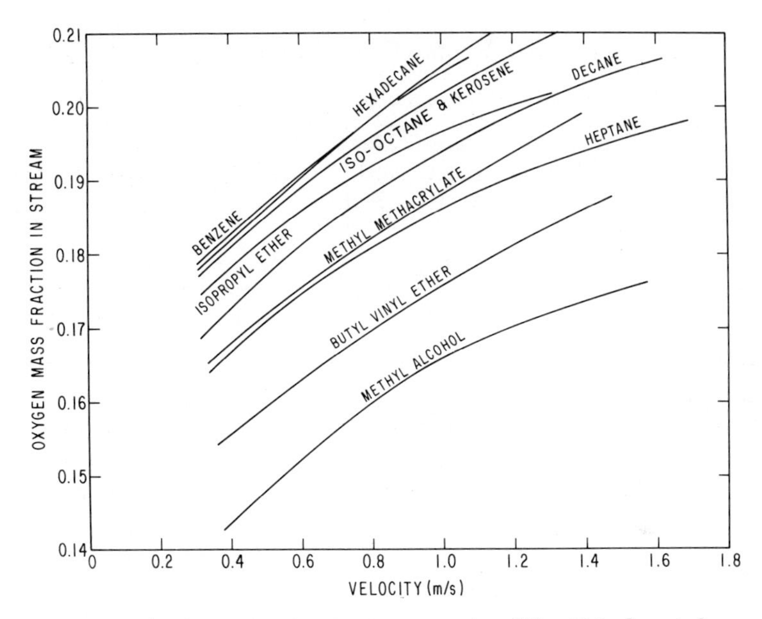

Figure 7. Summary of extinction curves for all liquid fuels tested

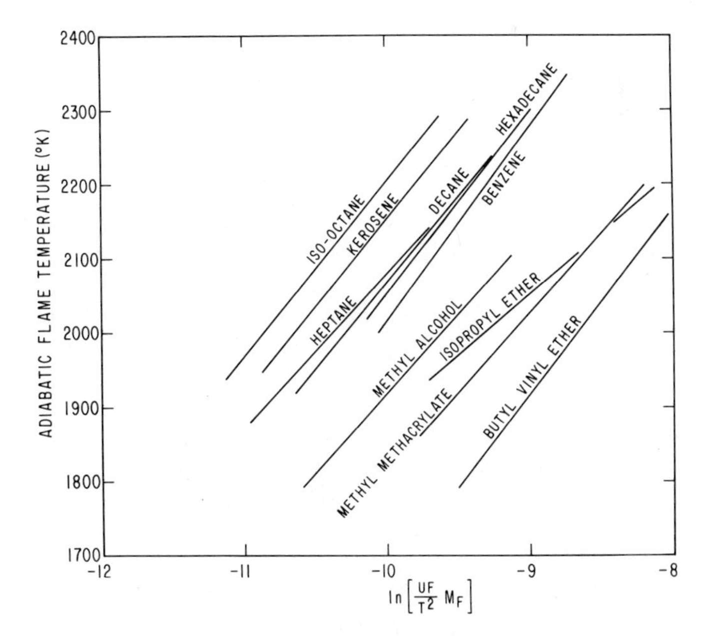

Figure 8. Dependence of adiabatic flame temperature on UFM_F/T^2 for various liquid fuels

Table I Thermal Properties of Fuels (17, 18, 19, 20)

Fuel	Heat of Formation (gas) at 25°C kcal/mole	Latent Heat at Normal Boiling Point cal/gm	Boiling Point °C	Liquid Heat Capacity at Boiling Point cal/gm°C
Benzene	19.82	94	80.1	0.456
Iso-propyl Ether	68.40 (Bond Energy Calc.)	80 (Same as Di-propyl Ether)	68	0.55 (Same as Di-propyl Ether)
Butyl Vinyl Ether	33.28 (Bond Energy Calc.)	74 (Same as Di-propyl Ether)	93	0.58 (Same as Di-propyl Ether)
Heptane	44.89	75	98.4	0.602
Methyl Methacry-late	81.46	85	101	0.46

Rate constants are derived from Arrhenius plots of the type shown in Figure 9. From the theoretical development (16) it is known that calculated rather than measured flame temperatures are to be used in the computations. The measured flame temperatures, shown in Figure 9, obtained under conditions near extinction are 450°C to 500°C below the adiabatic values, in agreement with earlier observations for heptane (9); the difference is due mainly to incomplete reaction.

The physical properties of fuels listed in Table I were employed in calculating the rate constants; when properties were not available in tables, those of similar compounds were used, or bond-energy calculations were made.

Derived rate constants are listed in Table II. The main source of error in the calculation of the activation energy is the

Table II Overall Rate Constants

Fuel	Chemical Formula	Activation Energy (kcal/mole)	Pre-Exponential Factor (cm^3/mole sec)
benzene	C_6H_6	36	6.0×10^{14}
isopropyl ether	$[(CH_3)_2CH]_2O$	50	1.2×10^{17}
butyl-vinyl ether	$CH_3(CH_2)_3OCH:CH_2$	31	5.0×10^{14}
methyl methacrylate	$C_5H_8O_2$	37	2.4×10^{15}
polymethyl methacrylate	$(C_5H_8O_2)_n$	42	7.6×10^{15}
methanol	CH_3OH	40	4.6×10^{15}
heptane	C_7H_{16}	38	2.2×10^{15}
decane	$C_{10}H_{22}$	37	1.6×10^{15}
hexadecane	$C_{16}H_{34}$	35	8.2×10^{14}
iso-octane	C_8H_{18}	35	4.5×10^{14}
kerosene	$C_{10}H_{20}$ (approx.)	35	5.4×10^{14}

determination of the value of F. This error was estimated not to exceed ± 2 kcal/mole. Other sources of error, such as uncertainties in the data listed in Table I, and experimental inaccuracies, were estimated to produce uncertainties not exceeding ± 1 kcal/mole.

The most striking aspect of Table II is the similarity in overall rates of the fuels tested. Certain systematic trends may be seen, such as a tendency for the overall activation energy to decrease with increasing chain length for the alkanes. Many of the differences lie within experimental error. However, the overall activation energy for isopropyl ether quite definitely is different from those of the other fuels.

Flame Structure. It has been indicated (9) that for alkanes the diffusion flames in the present apparatus are pale blue, there being insufficient residence time on the fuel side for formation of a luminous carbon-containing zone. For the fuels tested here, the only exception to this result occurred with benzene, which exhibited a broad orange zone beneath the blue flame. With increasing dilution the orange zone diminished in size and disappeared entirely just before extinction. Rapid buildup of carbon deposits at the rim of the cup limited testing times with benzene.

Additional calculations were made for heptane using temperature and composition profiles previously reported (9). The measured composition profiles were employed to calculate a local adiabatic temperature. The resulting temperature profile, shown in Figure 10, agrees well with the measured temperature profile. The entire discrepancy can be attributed to the uncertainty in the absolute value of the vertical position; the zero points for temperature and concentration may differ by a fraction of a millimeter. The agreement implies the absence of an appreciable difference between the coefficient of diffusion for heat and the average coefficient of diffusion for chemical species. It also supports the validity of the temperature and composition measurements.

If residence time were sufficiently large, complete equilibrium would be expected to exist at each point in the diffusion flame. The fact that measured flame temperatures lie appreciably below equilibrium adiabatic values (Figure 9) implies that complete equilibrium does not exist. Nevertheless, it is of interest to test whether equilibrium may occur for certain reactions. The temperature and composition profiles for the heptane flame (9) were employed for this purpose. The equilibrium constant K_p was evaluated separately from the measured temperature and from the measured concentrations for the overall reactions $H_2 + \frac{1}{2} O_2 \rightleftarrows H_2O$, $CO + \frac{1}{2} O_2 \rightleftarrows CO_2$ and $CO_2 + H_2 \rightleftarrows CO + H_2O$. Throughout the flame, the K_p evaluated from temperature was much larger than that evaluated from concentrations for the first two reactions. The K_p's for the third reaction are shown in Figure 11. The agreement is reasonably good in the central region where the composition measurements are most accurate. The agreement might be quite generally valid in

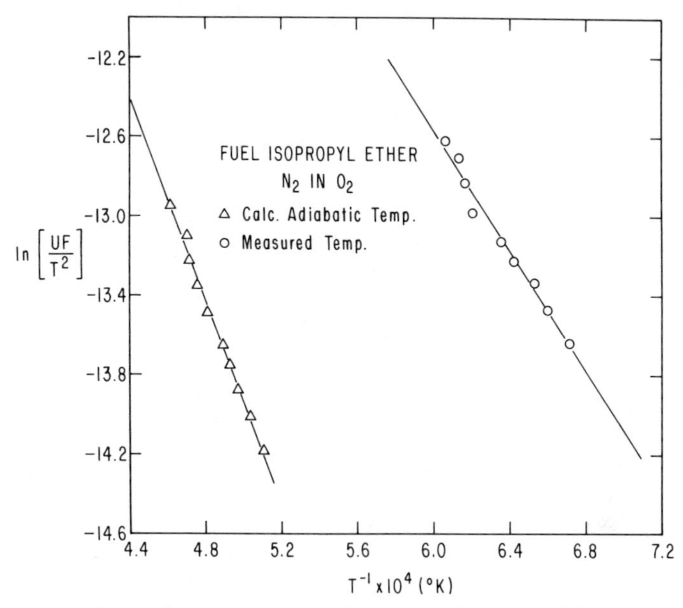

Figure 9. Arrhenius plot for adiabatic and measured flame temperatures of isopropyl ether

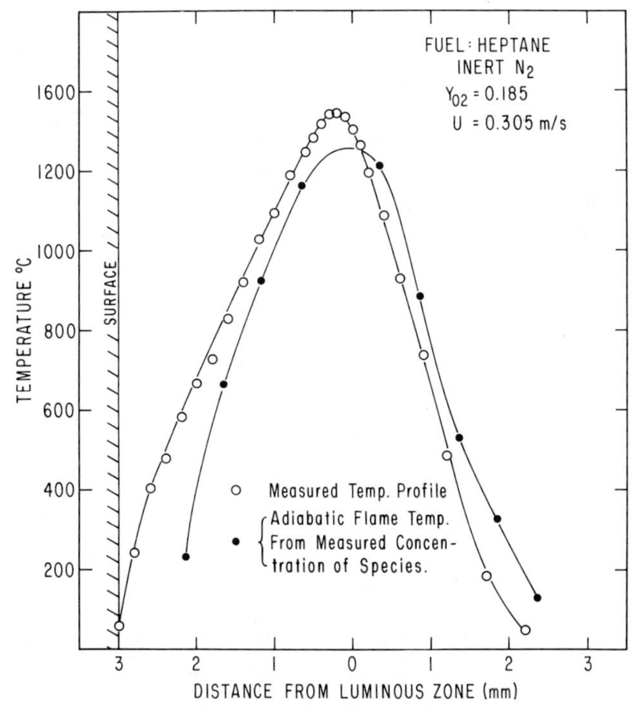

Figure 10. Measured temperature profile vs. *that calculated from measured local composition on the basis of adiabaticity for heptane*

hydrocarbon-air diffusion flames.

A profile of the C/H ratio was calculated from the measured concentration profiles (9) for the heptane flame. The result, shown in Figure 12, indicates that the ratio tends to peak in the center of the flame. A result of this type is acceptable if it is assumed that H_2 is produced near the center of the flame and diffuses preferentially in both directions because of its higher diffusion coefficient. If the alternative view is taken that the variation arises from inaccuracies in composition measurements, then corrected profiles, shown in Figure 12, can be generated by assuming a constant C/H ratio, the corrections being made in the species whose measurements were deemed least accurate. The correction greatly increases water concentration in the center of the flame; the total hydrogen element concentration there must be increased by about a factor of two. Since it is unlikely that the experimental error can be so large, the results tend to support the occurrence of preferential H_2 diffusion.

Extinguishment of Polymethyl Methacrylate

There has been continuing interest (21, 22) in the gas-phase chemical kinetics for combustion of polymethyl methacrylate, which serves as a model solid fuel. Extinction measurements have been made in oxygen-enriched air by use of a combustion tunnel (22). Since there is evidence that the polymer unzips to the monomer in gasification, the gas-phase kinetics may be expected to be the same as for methyl methacrylate. One of the reasons for including methyl methacrylate in the present experiments was to test this hypothesis.

It was found to be possible to test the polymer itself in the present apparatus. Cylinders 5cm in diameter were cut into slices 2cm thick, and each slice was placed on top of the cup and ignited. The outlet of the duct for delivering the oxidizing gas mixture was positioned 1cm above the burning polymer surface, and extinction conditions were measured in the same manner as for liquid fuels. The flame was similar in appearance to those of the liquid fuels, except that around the sides of the polymer a yellow zone existed. Since the oxidizing gas is more dilute than air, the new extinction results complement those obtained earlier (22).

Extinction curves for the polymer and its monomer are given in Figure 13. The polymer data at the higher flow velocities are poorly reproducible because in the altered experimental configuration the flame is less stable. The points shown correspond to the minimum oxygen content for extinction; often

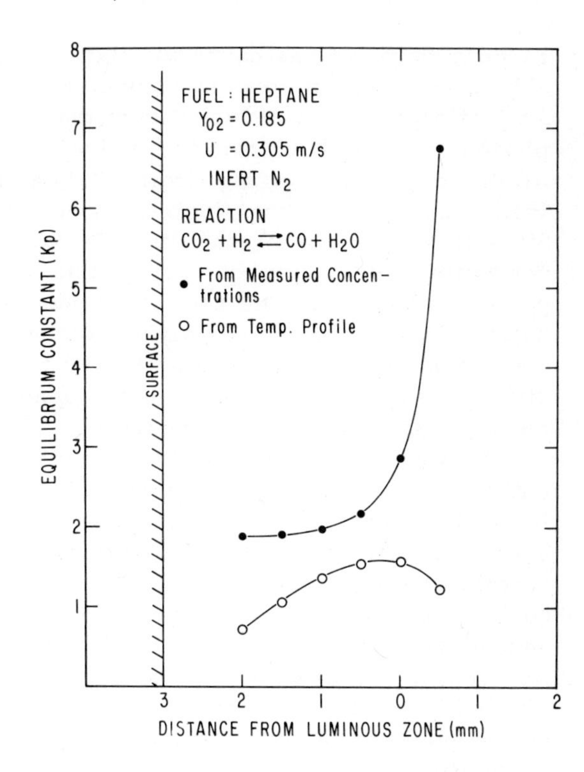

Figure 11. Equilibrium constant for: $CO_2 + H_2 \rightleftharpoons CO + H_2O$, calculated separately from measured temperatures and measured compositions, for heptane

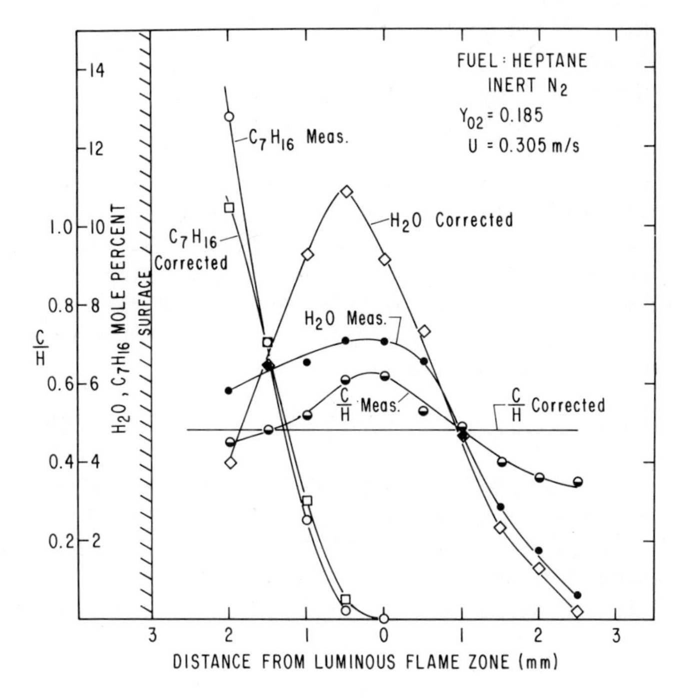

Figure 12. Profiles of the C/H ratio, fuel, and water for heptane at Y_{O_2} = 0.185, U = 0.305 m/s

random disturbances cause the flame to extinguish at considerably higher oxygen concentrations, at the highest velocities. Activation energies deduced from the nitrogen-quench data plotted in Figure 13 for the polymer vary with oxygen mass fraction, ranging from 42 kcal/mole at the lower velocities to 55 kcal/mole at the higher velocities. Since the high-velocity results are inaccurate, the low-velocity activation energy has been adopted as most reliable and is reported in Table II. The rate constants shown in Table II for the polymer and the monomer differ somewhat, but the accuracy of the experiment is not quite sufficient to establish whether the difference is real. The polymer combustion might involve monomer combustion in the gas phase.

It may be noted that activation energies reported earlier for enriched atmospheres (21, 22) were in the vicinity of 20 kcal/mole. In the calculations that produced this value, dissociation was neglected even though, contrary to the present experiments, actual flame temperatures at extinction were sufficiently high for equilibrium dissociation to be important energetically. This procedure was followed because there exists no theory for extracting accurate overall activation energies from experiments in which dissociation occurs. However, approximate corrections can be made in an *ad hoc* manner by including dissociation in the calculation of the adiabatic flame temperature and by correcting the overall heat release for dissociation in deriving values of F. Application of these corrections to the data obtained in enriched atmospheres produces activation energies ranging from 42 kcal/mole in air to 48 kcal/mole at $Y_{O2} = 0.53$. Although these corrected results are in reasonable agreement with results obtained from the new data reported herein, it should be emphasized that the correction procedures are open to question. There is need for further theoretical work directed toward accounting for effects of dissociation. Nevertheless, it may be significant that all of the calculations for the polymer seem to show a tendency for the effective overall activation energy to increase with increasing oxygen content.

Results for Liquid Fuels in CF_3Br - Air Mixtures

Quenching by CF_3Br. Various liquid fuels burning in air were extinguished by adding CF_3Br to the air stream. The conventional plot of the volume percent of CF_3Br in the oxidizing gas, required for extinguishment, as a function of the velocity of the oxidizing gas, is shown in Figure 14. At the lower

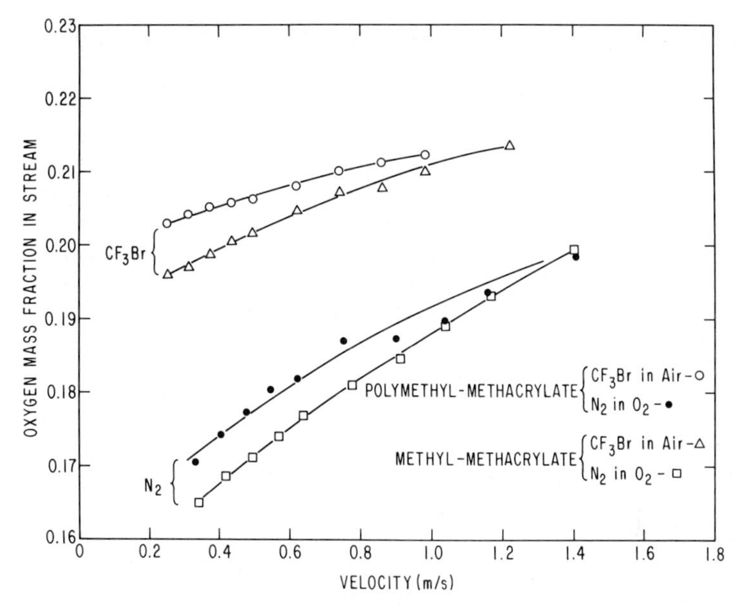

Figure 13. Extinction curves for poly (methyl methacrylate) and for its monomer, burning in air with added N_2 and CF_3Br

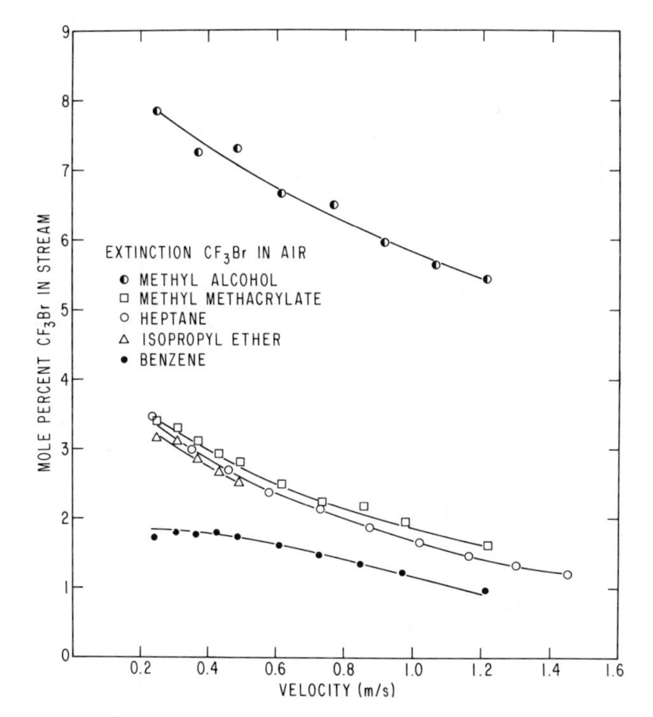

Figure 14. Vol % CF_3Br in air required to quench various fuels

velocities, the concentrations needed for extinguishment are quite similar to those reported by many other investigators, for methanol and heptane. Data for isopropyl ether and for methyl methacrylate have not been reported in the past; here they are found to nearly coincide with the data for heptane. The results for isopropyl ether are quite comparable with earlier results for diethyl ether. The CF_3Br volume percentage required for extinguishment of benzene was found to be less than the lowest value (2.9%) reported earlier, although the general trend toward comparatively easy extinguishment of benzene is verified. It is understandable that lower inhibitor concentrations are needed at higher air velocities because as the velocity increases the residence time in the reaction zone decreases.

It has been emphasized (9) that quenching efficiencies on a mass basis can differ from those on a volume basis. Mass-based extinction curves, similar to Figure 6, are shown in Figure 15 for extinguishment with CF_3Br. Although relative rankings of various extinguishants differ on mass and volume bases (9), it is seen that the dependence on the type of fuel is unchanged.

For nitrogen quenching, plots like Figure 15 were turned into plots like Figure 9 for extracting overall kinetic data. This has not yet been attempted with CF_3Br because of uncertainty in the proper theoretical adiabatic flame temperature to adopt. If CF_3Br is treated as an inert, then low flame temperatures are obtained, but if equilibrium formation of species such as HF and HBr is included, then the flame temperatures are high. Exploratory calculations of the sensitivity of derived overall kinetic parameters to different equilibrium assumptions would be of interest.

Differences in inhibitor effectiveness with alkanes and methanol, for example, often are attributed to differences in stoichiometry. Since the flame sits at the stoichiometric position, proportionally less inhibitor penetrates to the flame for methanol. To see if this effect can explain observed differences in effectiveness for different fuels, an extinction plot using the stoichiometrically adjusted mole percentage was prepared (Figure 16). The adjusted percentage is defined as the molar ratio of inhibitor to oxygen in the gas stream, divided by one plus this quantity plus the stoichiometric molar ratio of fuel to oxidizer. Since the methanol curve lies closer to that for heptane in Figure 16 than in Figure 15, it appears that there does exist an effect of the accessibility of the reaction zone. However, it is clear that there exist additional fuel-specific influences. In particular,

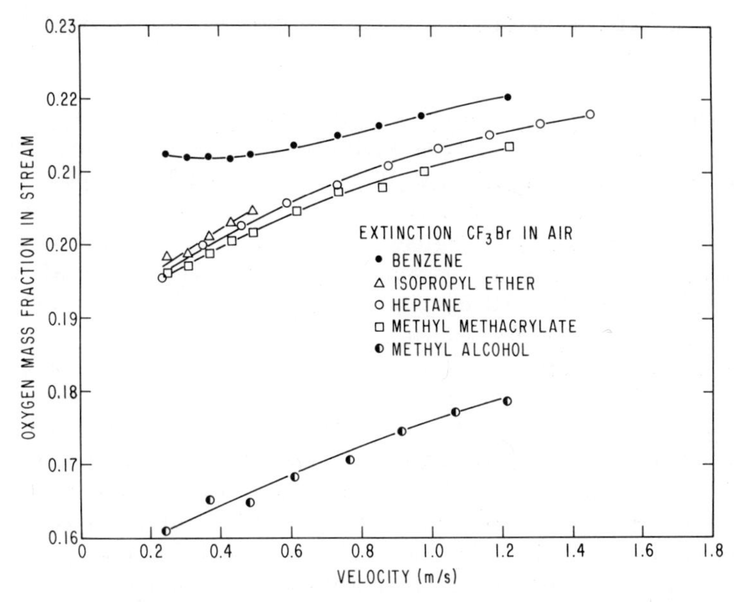

Figure 15. Mass-based extinction curves of CF_3Br in air for various fuels

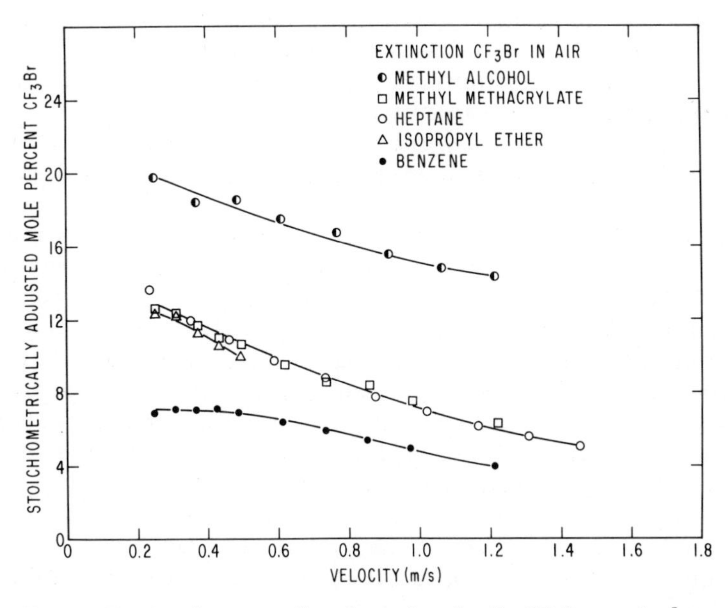

Figure 16. Stoichiometrically adjusted mole % CF_3Br required to quench various fuels burning in air

relative positions of fuels remain the same in all plots yet constructed.

Effect of Varying Relative Proportions of N_2, O_2 and CF_3Br. A graph for heptane extinction has been given previously (9) showing that although CF_3Br is less effective in pure oxygen than in air, chemical inhibition remains an important aspect of its mode of operation. Figure 17 shows that the same phenomenon occurs for benzene. It appears that curves of the shape shown in Figure 17 will be representative of all hydrocarbon fuels. In O_2 - CF_3Br mixtures, extinction occurs at very high flame temperatures.

Flame Structure. Addition of CF_3Br to the air side of the flame produces an orange carbon zone on the fuel side. It also changes the color of the flame from blue to blue-green. These observations tend to agree with the results reported by Simmons and Wolfhard (3) and by Ibiricu and Gaydon (6) for gaseous fuels, the green being attributable to enhanced C_2 emission. Clearly, the influence of the inhibitor penetrates well into the fuel side of the flame. The augmentation in carbon production was so great that coating of probes by soot prevented accurate temperature or composition measurements from being made on the fuel side of the center of the blue-green flame.

Composition profiles on the air side of a heptane flame with added CF_3Br have been reported earlier (9). A temperature profile has now been obtained for the same conditions, and the complete results are shown in Figure 18. These results should be compared with corresponding results of Creitz (8); there are both similarities and differences, the differences probably being attributable to longer residence times in the flame studied by Creitz.

The only fuel species other than carbon that Creitz found on the air side of the yellow zone were CO and H_2. In Figure 18, some C_2H_2 and CH_4 also are present, but in lower concentrations than CO and H_2, and also in slightly lower concentrations than found (9) in the corresponding uninhibited flame. It may be assumed that the longer residence times in the flame of Creitz allowed time for C_2H_2 and CH_4 to be converted to carbon, H_2 and CO before emerging from the yellow zone and that addition of CF_3Br enhances fuel pyrolysis.

Contrary to the results of Creitz, the oxygen concentration is appreciable at the edge of the yellow zone in Figure 18; it certainly cannot be stated that fuel pyrolysis occurs without

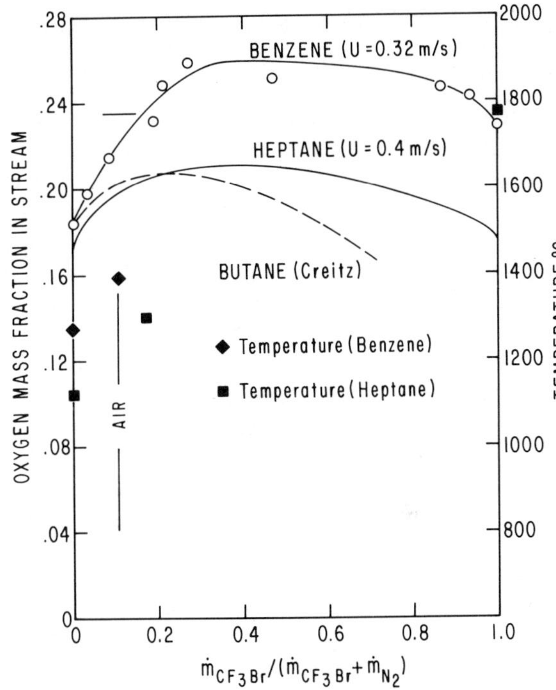

Figure 17. Effects of varying CF_3Br–N_2 ratio on oxygen content and temperature at extinction for benzene

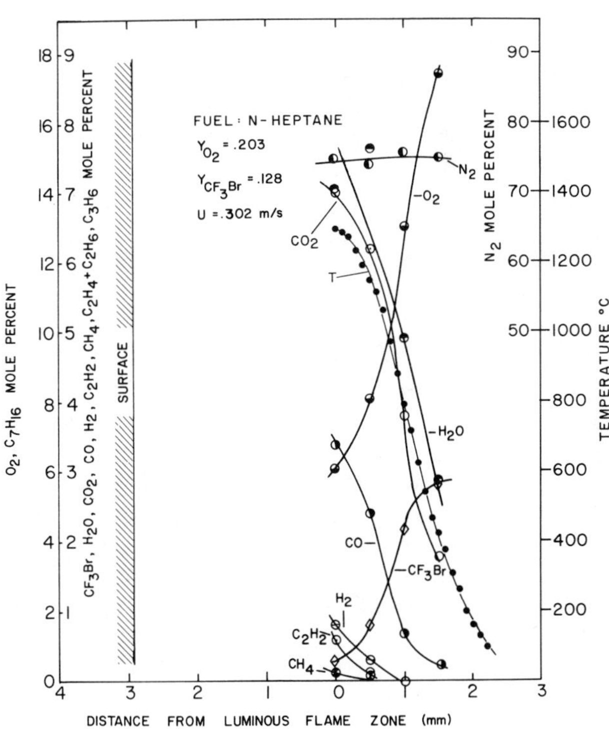

Figure 18. Composition and temperature profiles on the air side of a heptane flame with added CF_3Br

oxygen in the present experiment. The tendency toward a higher oxygen concentration in the flame again may be attributed at least partially to the shorter residence time. However, an additional effect observed in the present program is a tendency for the oxygen concentration in the flame to be higher with CF_3Br addition than in the corresponding flame without CF_3Br (9). In the center of the blue zone of our uninhibited flames, the oxygen mole fraction was found in repeated tests to range from 1.5% to 4%, while two measurements of inhibited flames gave values of 6% and 9%. Although there is uncertainty in these values, mainly because of uncertainty in measurement of absolute positions, the flame-zone oxygen contents with CF_3Br definitely tend to be two to three times as large as in the absence of the inhibitor. Chemical mechanisms proposed for the action of the inhibitor will have to try to explain this observation.

Fielding, Woods and Johnson (23) made composition measurements in the effluent gases of a furnace in which carbon was burned in an air stream with added CF_3Br. They demonstrated that the inhibitor decreased the CO_2 concentration and increased the CO concentration by about a factor of two. No such changes were found in the present study; addition of CF_3Br had very little effect on the CO or CO_2 profiles. The difference suggests that even in the sooty, inhibited, diffusion flame, carbon combustion is not the major source of CO.

Finally, the profile of CF_3Br itself should be considered. Unlike the inert N_2, the inhibitor drops very nearly to zero concentration at the center of the flame, as if it were an air-side reactant. In fact, the apparent sink into which CF_3Br diffuses seems to be slightly on the air side of the O_2 sink, the flame center. This certainly implies production of dissociation or reaction products of CF_3Br and, in agreement with the conclusion of Creitz (8), implies that the intact CF_3Br molecule is not responsible for the chemical inhibition effect.

That not CF_3Br but rather its products cause the enhanced fuel pyrolysis and carbon production, is consistent with the observation of Simmons and Wolfhard (3) that addition of Br_2 produced approximately the same intensification of the carbon zone as addition of CH_3Br. Bromine atoms or HBr are among the candidates for the pyrolysis-augmenting agents, but apparently no kinetic mechanism for the process has been proposed. Enhancement of oxygen leakage, followed by oxygen-catalyzed pyrolysis, is another possibility which is consistent with our observations.

Although CF_3Br disappears more rapidly than O_2, it is

not consumed nearly as early as in the flame of Creitz (8). From the temperature profile in Figure 18 it is seen that at the Creitz disappearance temperature of $1080^{\circ}K$, more than half of the original CF_3Br remains in the counterflow flame. It is unlikely that this difference is effected by use of heptane rather than propane as the fuel. Instead it may be attributable to differences between diffusivities for heat and for CF_3Br, with the influence of the relative immobility of the larger molecule magnified at the longer residence time in the coflowing system.

The mechanism by which CF_3Br is attacked remain unclear. Thermal decomposition seems unlikely at these temperatures. The attack by oxygen atoms, proposed by Frankiewicz, Williams and Gann (24), is an attractive possibility, especially since the inhibitor tends to be consumed on the air side of the flame. On the other hand, in Figure 18 the relative positions of H_2 and CF_3Br profiles are somewhat reminiscent of a diffusion flame in which these two species react rapidly. Unfortunately, the H_2 profile is not accurate enough to warrant drawing conclusions. However, since H_2 diffuses easily (recall C/H profile in Figure 12), and since H atoms diffuse even more quickly than H_2, the popular idea that H atoms attack CF_3Br cannot be excluded on the basis of our data, and probably not even on the basis of the data of Creitz. There is a great deal of chemistry occurring within the concentration profiles shown in Figure 18.

Concluding Remarks

There is need for a considerable amount of additional work before the methods employed herein can achieve clarification of reaction rates and mechanisms for chemically inhibited diffusion flames, comparable to the clarification that has been obtained in the absence of chemical inhibition. The approach to determining overal reaction rates, discussed in connection with nitrogen quenching, should be explored both theoretically and experimentally for flames to which chemical suppressants are applied. In addition, measurements of concentration and temperature profiles, of the type discussed in the preceding section, should be repeated to test reproducibilities and should be extended to other compositions and velocities of the oxidizing gas and to other fuels, so that systematic trends can be identified. Knowledge of such trends may be quite helpful in shedding light on the inhibition mechanisms of chemical fire suppressants.

Acknowledgement

The contributions of J. H. Kent and of L. Krishnamurthy to these studies are gratefully acknowledged. The work was supported partially by the Naval Research Laboratory, U. S. Navy and partially by the National Science Foundation through the project on Research Applied to National Needs, Grant No. GI-36528X.

Nomenclature

F	Stagnation-point flow parameter
M_F	Molecular weight of fuel
T	Theoretical adiabatic flame temperature
U	Gas velocity in the approach stream
Y_{O2}	Mass fraction of oxygen in the approach stream

Literature Cited

1. Fristrom, R. M., "Combustion Suppression", Fire Research Abstracts and Reviews (1967), 9, 125-152.
2. McHale, E. T., "Survey of Vapor Phase Chemical Agents for Combustion Suppression", Fire Research Abstracts and Reviews (1969), 11, 90-104.
3. Simmons, R. F. and Wolfhard, H. G., "The Influence of Methyl Bromide on Flames. Part 2 - Diffusion Flames", Transactions of the Faraday Society (1956), 52, 53-59.
4. Creitz, E. C., "Inhibition of Diffusion Flames by Methyl Bromide and Trifluoromethyl Bromide Applied to the Fuel and Oxygen Sides of the Reaction Zone", Journal of Research of the National Bureau of Standards-A. Physics and Chemistry (1961), 65A, 389-396.
5. Friedman, R. and Levy, J. B., "Inhibition of Opposed-jet Methane-air Diffusion Flames, The Effects of Alkali Metal Vapors and Organic Halides", Combustion and Flame (1963), 7, 195-201.
6. Ibiricu, M. M. and Gaydon, A. G., "Spectroscopic Studies of the Effect of Inhibitors on Counterflow Diffusion Flames", Combustion and Flame (1964), 8, 51-62.
7. Milne, T. A., Green, C. L. and Benson, D. K., "The Use of the Counterflow Diffusion Flame in Studies of Inhibition Effectiveness of Gaseous and Powdered Agents", Combustion and Flame (1970), 15, 255-263.

8. Creitz, E. C., "Gas Chromatographic Determination of Composition Profiles of Stable Species Around a Propane Diffusion Flame", Journal of Chromatographic Science (1972), 10, 168-173.
9. Kent, J. H. and Williams, F. A., "Extinction of Laminar Diffusion Flames for Liquid Fuels", Fifteenth Symposium (International) on Combustion, The Combustion Institute, Pittsburgh (1975), 315-325.
10. Bajpai, S. N., "An Investigation of the Extinction of Diffusion Flames by Halons", Factory Mutual Research Corp., Rept. No. 22391.2, Nov. 1973.
11. Kubin, R. F. and Gordon, A. S., "Inhibition of Methanol Pool Fires", presented at Clay Preston Butler Meeting on Experimental Methods in Fire Research, Stanford Research Institute, May 1974.
12. Kent, J. H., "A Noncatalytic Coating for Platinum-Rhodium Thermocouples", Combustion and Flame (1970), 14, 279-281.
13. Fristrom, R. M. and Westenberg, A. A., "Flame Structure", McGraw-Hill, New York (1965), 197.
14. Rashbash, D. J., "Extinction of Fires by Water Sprays", Fire Research Abstracts and Reviews (1962), 4, 28-53.
15. Roberts, A. F., "Extinction Phenomena in Liquids", Fifteenth Symposium (International) on Combustion, The Combustion Institute, Pittsburgh (1975).
16. Krishnamurthy, L. and Williams, F. A., "Asymptotic Theory of Diffusion-Flame Extinction in the Stagnation-Point Boundary Layer", unpublished.
17. Gallant, R. W., Physical Properties of Hydrocarbons, Volume 1, Gulf Publishing Company, Houston (1968).
18. Gallant, R. W., Physical Properties of Hydrocarbons, Volume 2, Gulf Publishing Company, Houston (1970).
19. Rossini, F. D. et. al., Selected Values of Properties of Hydrocarbons, NBS Circular C461, National Bureau of Standards, Washington (1947).
20. Stull, D. R. et. al., JANAF Thermochemical Tables, U. S. Department of Commerce, National Bureau of Standards, Washington (1971).
21. Krishnamurthy, L. and Williams, F. A., "Laminar Combustion of Polymethylmethacrylate in O_2/N_2 Mixtures", Fourteenth Symposium (International) on Combustion, The Combustion Institute, Pittsburgh (1973), 1151-1164.

22. Krishnamurthy, L., "Diffusion-Flame Extinction in the Stagnation-Point Boundary Layer of PMMA in O_2/N_2 Mixtures", Combustion Science and Technology, to appear.

23. Fielding, G. H., Woods, F. J. and Johnson, J. E., "Halon 1301: Mechanism of Failure to Extinguish Deep-Seated Fires", Journal of Fire and Flammability (1975), 6, 37-43.

24. Frankiewicz, T. C., Williams, F. W. and Gann, R. G., "Rate Constant for $O(^3P)$ + CF_3Br from 800° to 1200^oK", Journal of Chemical Physics (1974), 61, 402-406.

DISCUSSION

A. S. GORDON: We have found that hydrogen-air diffusion flames partially inhibited by CF_3Br are very luminous. At least part of the luminosity must be from carbon particles, since we see black particles circulatory inside the flame cone. We have also noted that HBr added to a hydrogen flame causes luminosity, so that part of the luminosity from the CF_3Br flame arises from the HBr.

F. A. WILLIAMS: That is very interesting. I do not understand the luminosity in the case of HBr and hydrogen, but I can believe that carbon would be formed by addition of CF_3Br to the hydrogen flame. In our experiments, all of which involved carbon-containing fuels, I feel strongly that the greatly enhanced luminosity is attributable mainly to carbon that originated in the fuel itself. The suppressant enhances fuel pyrolysis. This agrees with observations of Simmons and Wolfhard who found that addition of Br_2 to hydrocarbon-air diffusion flames increased carbon production.

L. D. SAVAGE: In premixed flame sheets of methane, we see no formation of carbonacious matter when CF_3Br is added, however the spectral changes from the bromine compounds give a resulting color similar to that from carbon particulate emission.

F. A. WILLIAMS: In our diffusion flames there is definite evidence for enhanced production of carbonaceous material by addition of CF_3Br. For example, with benzene, our fuel that exhibited the greatest propensity for soot production, in the absence of CF_3Br, we could burn for ten minutes before carbonaceous deposits that built up on the side of the cup produced flow nonuniformities, but when CF_3Br was added we were lucky to have two minutes running time before soot deposition disrupted the flow. The magnitude of this effect appears to be much too great to be attributed only to the carbon contained in the CF_3Br. In lean premixed flames carbon formation should not occur even in the presence of CF_3Br, but in sufficiently rich flames sooting might be influenced by the additive if it can participate directly in the pyrolysis mechanism. One of the possible explanations for our observations is that the inhibitor enhances oxygen leakage through the diffusion flame, which in turn promotes oxygen-catalyzed pyrolysis; if this is the operative mechanism, then in premixed flames CF_3Br need not increase sooting at all.

J. DEHN: Have you used HBr, and how do HBr, CF_3Br and N_2 compare in efficiency?

F. A. WILLIAMS: We have not used HBr at all. On a molar basis, CF_3Br is more efficient than N_2. The same is usually true on a mass basis, but to a lesser extent. We have emphasized (9) that relative efficiencies on mass molar bases can differ and that efficiencies of CF_3Br are lower in pure oxygen than in air because of exothermicity.

R. G. GANN: In your graph of concentration profiles, you show no fluorinated or brominated species, although we might expect them since the CF_3Br is virtually totally consumed. COF_2, HBr and HF are difficult to sample, but CH_3Br has been observed to appear just as CF_3Br in declining. Would you comment on the non-appearance of halogenated species other than the parent CF_3Br.

F. A. WILLIAMS: Our system is designed to measure only major stable species. Its sensitivity as employed is such that species present in concentrations less than approximately one half mole percent would not be detected. Although exhaustive searches for additional species were not performed, no unidentified peaks were apparent in the chromatograms. I think that only HBr and HF, which condensation problems prevented us from measuring, would be present in high enough concentrations to be seen at our level of sensitivity.

6

Halomethane and Nitrogen Quenching of Hydrogen and Hydrocarbon Diffusion and Premixed Flames

R. F. KUBIN, R. H. KNIPE, and ALVIN S. GORDON

Naval Weapons Center, China Lake, Cal. 93555

It has been shown in previous work (1-4) that flame inhibitors may function by chemical as well as physical inhibition. Often the efficiency of inhibitors has been determined by the amounts of inhibitor required to produce a given percentage reduction in the flame speed of a premixed flame (5,6) or by the observed flame speed reduction for a given constant inhibitor percentage (1). However, it should be noted that the ability to reduce the flame speed may not bear any relationship to quenching efficiency. For example, low concentrations of hydrocarbons substantially reduce the speed of a hydrogen flame. Also, it has been reported that inhibitors are more effective in rich rather than lean premixed hydrogen flames (1). The ambiguities introduced by such methods are eliminated by considering the amount of inhibitor required to quench.

For diffusion flames few inhibitors have been studied. Quenching concentrations have been reported for CF_3Br, CH_3Br and N_2 for several hydrocarbons, hydrogen and carbon monoxide flames (7). In more recent studies the opposed jet diffusion flame has been employed to quantify inhibitor effectiveness (8,9,10).

In this work we have compared premixed and diffusion flame results for various inhibitors in a bunsen flame configuration to try to gain insight into the combustion and inhibition process. This paper contrasts the effect of CF_3Br with those of CF_4 and N_2 on both premixed and diffusion flames. Data are also presented on CF_2Br_2, HBr, and He inhibition.

Experimental

The burner used for most of the work was a 10.6 mm ID quartz burner approximately 50 cm long. The remainder of the apparatus consisted of separate air and fuel systems. The air and fuel systems each contained an appropriate number of gas metering devices and a gas sampler for stream analysis.

In the premixed flame configuration the metered air and fuel streams were led to a mixing tee before going to the burner. The burner was surrounded with an approximately 15 cm diameter fine mesh screen to minimize drafts.

For the diffusion flame configuration the fuel was metered to the burner and the air metered to a 30 mm ID quartz chimney surrounding the burner, subsequently termed the standard configuration. The chimney was sealed at the bottom and extended approximately 10 cm above the top of the burner to minimize secondary air effects. Additionally, to test the effects of burner and chimney size, a 18.9 mm ID quartz burner, subsequently termed the large burner, and a 65.2 mm ID quartz chimney, subsequently termed the large chimney, were substituted in some of the experiments.

The metering devices were Fischer Porter Tri Flat Precision Rotameters calibrated against a wet test meter. The reproducibility of gas calibrations was better than 2% for those rotameters with a nominal diameter greater than 1/16-inch. For the 1/16-inch rotometer the reproducibility was 3-5% which was dependent on the flow rate, the slower the flow, the poorer the reproducibility. Because of the critical nature of the gas mixing to the experiment, air, fuel, and inhibitor gas mixtures were sampled and analyzed by mass spectrometry and gas chromatography. These analyses showed that to better than the reproducibility of the experiments (5-8%), the rotometers were stable analytical devices. Because of its corrosiveness and solubility in water, a calculated calibration was used for HBr flows.

Observation of the quench points was visual except for those hydrogen flames where the inhibitor imparted no color to the flame. For these essentially invisible flames a simple biased PbS IR detector with maximum sensitivity at 2.5μ was used. The d.c. signal intensity was monitored by an oscilloscope using a differential amplifier. At the quench point, the signal dropped sharply to essentially zero. For some diffusion flames, the flame did not actually quench but would rise from a few mm to 1 cm or more off the burner and stabilize. The point of lift off was reproducible and was taken as the quench point since it was felt that if the chimney did not extend beyond the burner the flame would have blown off. This definition of quench was also employed by Creitz (7).

The quench points for both premixed and diffusion flames were observed to be dependent on the gas flow rates. In order to correlate the results for differing flow rates, we have chosen to use the volumetric flow rate divided by the cross-sectional area of the flow channel to parameterize the flow. For the premixed flames and the fuel stream of the diffusion flames, the flow channel is the burner tube. For the oxidizer stream of the diffusion flames, the flow channel is the annular region between the burner tube and the chimney. This flow parameter is designated by $\bar{v}$.

For fully developed laminar flow where the velocity distribution can be scaled by the channel dimensions, the velocity distribution corresponding to a given $\bar{v}$ is independent of the nature of the gas. We recognize that in the critical region near the burner port the large temperature and concentration gradients will influence the velocity distribution in a manner which cannot be described by the parameter.

Air was dried and compressed into cylinders and the oxygen used was aviators grade. The hydrogen was industrial grade, methane ultra high purity, butane industrial grade, acetylene technical grade and carbon monoxide Matheson purity grade. The reagent gases were used as available and not further purified except for CO and C_2H_2. The acetylene was further purified by scrubbing with concentrated sulfuric acid and drying over calcium sulfate; the acetone being removed quantitatively and the loss of acetylene small. The carbon monoxide was run through the same system since it removed the iron pentacarbonyl contaminant always present. Mass spectrometric analysis showed no acetone or other contaminant present in the C_2H_2 and no iron pentacarbonyl in the CO after treatment. For the inhibitors, CF_3Br was approximately 98% pure, the remainder showing up as unidentified peak on the gas chromatograph, the CF_2Br_2 and CF_4 were both found to be mass spectrometrically pure and N_2 was technical grade.

All experiments were performed at ambient pressure, approximately 700 torr.

Results

Premixed Flames. A summary of our results of quenching of some premixed flames is shown in Figure 1. To minimize the effect of oxygen in the air external to the flame, the flames were made up to be slightly on the lean side of stoichiometric. All fuel-air mixes were such that the adiabatic flame temperature was kept constant at 2214 ± 14°K. It is seen that all the hydrocarbons and CO exhibit a characteristic flow dependence for quenching by halomethanes; faster flows require less inhibitor to quench the flame. In contrast, hydrogen flames require more CF_3Br and CF_2Br_2 with increasing flow. When nitrogen is the quencher, hydrogen shows little flow dependence. In the case of CO, at the flow rates used in this work, the pure gas would not stabilize and burn in air until 7% hydrogen was added to the fuel.

CF_2Br_2 tends to be less than twice as effective as CF_3Br per mole under our test conditions. For methane and butane at the lowest flows, quenching tends to correlate with the number of bromine atoms but the relationship becomes poorer at the higher flow rates. This relationship is in qualitative agreement with the work of Rosser, et al, (6), based on flame speed reduction measurements for near stoichiometric methane-air flames.

The effect of fuel/air ratio on the amount of CF_3Br required to quench premixed butane or hydrogen flames at a constant v of 75 cm/sec is illustrated in Figure 2. Over the range from slightly fuel rich to lean mixtures, the correlation between the quencher/

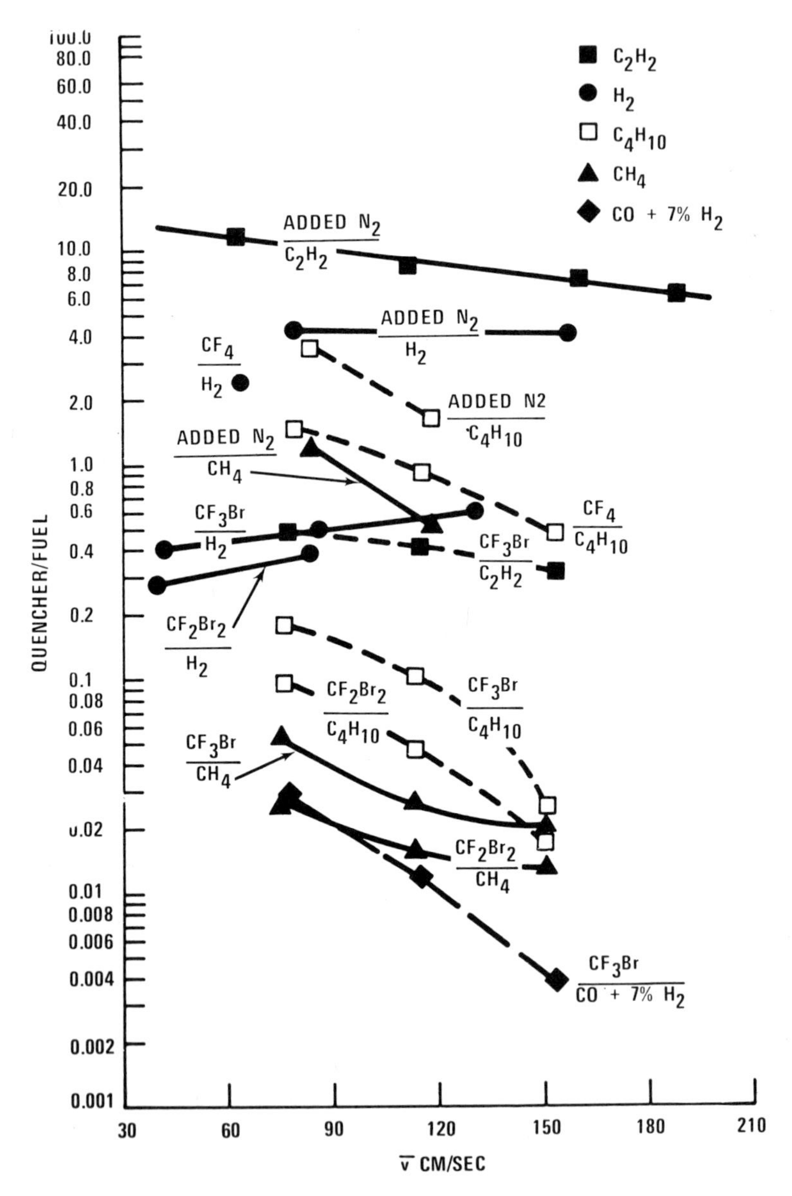

Figure 1. Quencher/fuel ratios at quench for slightly lean premixed flames as a function of flow parameter $\overline{v}$. (Added N_2 is nitrogen in excess of that present in the air premixed in the fuel stream.)

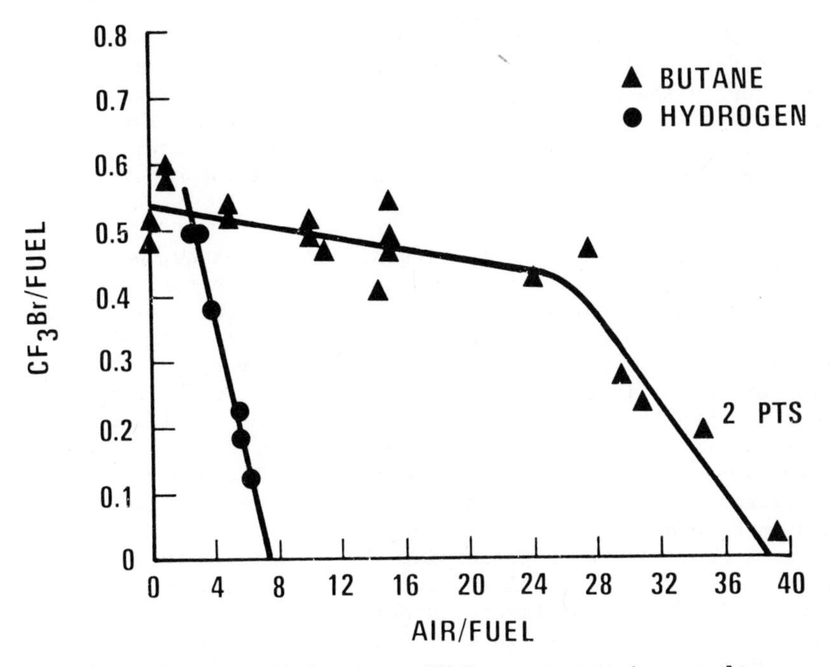

Figure 2. Effect of air/fuel ratio on CF_3Br requirement for quenching premixed butane and hydrogen flames with $\bar{v} = 75$ cm/sec

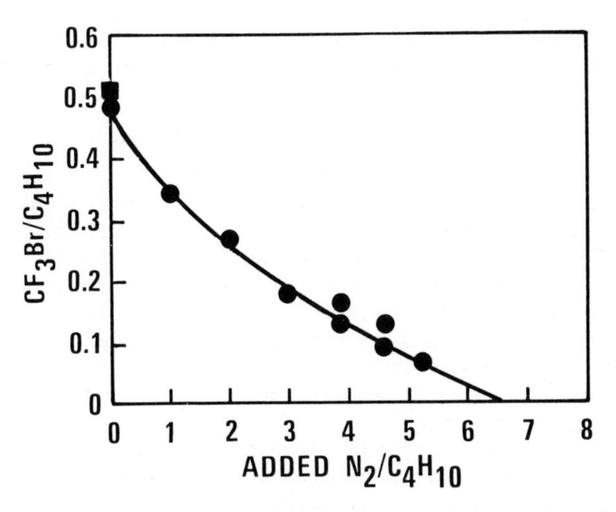

Figure 3. CF_3Br/butane ratio at quench vs. *added N_2/butane in rich flames (v = 75 cm/sec)*

fuel ratio at quench and the air/fuel ratio is approximately linear. For rich butane/air mixtures, the quencher/fuel ratio shows little dependence on the air/fuel ratio. In this air/fuel regime flames with various (added N_2)/butane ratios were studied. The effect of added nitrogen on the CF_3Br/butane ratio at quench is illustrated in Figure 3 where an approximately linear relationship may be noted.

Diffusion Flames. In studying diffusion flames, we have investigated the effects of quenchers in both the fuel stream and in the oxidizer stream. As a consequence, there are two flow parameters which are under experimental control: that of the oxidizer stream $\bar{v}_{ox}$ and that of the fuel stream $\bar{v}_{fuel}$. The qualitative effect of independent variation of these parameters on the amount of quencher required to quench the flame is illustrated in Figures 4 and 5 for butane and in Table I for methane. The figures suggest that the effect of having different flow parameters for the two streams is not a great deal larger than the reproducibility of the quench points (∿8%) over our experimental range.

As a convenience in presenting the results, much of the data were taken where the two flow parameters were equal at the quench point within their experimental uncertainty. This entailed an iterative technique where initial fuel and oxidizer flows were established and quencher added to one of the flows until the flame quenched. Based on the disparity of the two flows, new initial flows were established and a new quench point was determined. This was continued until the disparity of the flows was negligible. The quench point was found to be insensitive to the disparity in the flow parameters near equality.

The effect of nitrogen concentration in the oxidizer on the amount of fuel stream nitrogen required for quench is illustrated in Figure 6. The approximately linear relationships exhibited suggest that there is an equivalence between nitrogen in the oxidizer stream and in the fuel stream such that at quench

$$(N_2/\text{fuel})_{\text{fuel stream}} + q_f(N_2/O_2)_{\text{oxidizer stream}} \simeq D_f \qquad (1)$$

where q_f and D_f are parameters dependent on the particular fuel.

The data in Figure 6 show that when atmospheric air is used as the oxidizer, the nitrogen/fuel ratio necessary to quench is approximately the same for hydrogen and butane flames. Figure 7 shows much the same effect for other quenchers considered in this study. In each case, the relative quencher requirement for the hydrogen and butane flames depends on the flow parameter $\bar{v}$. The trend of the data with $\bar{v}$ is similar to that observed for the premixed flames (Figure 1) as are the relative magnitudes of the quencher/fuel ratios at quench.

In Figure 8, both the fuel stream and the air plus quencher stream have the same flow parameter $\bar{v}$. Since nitrogen is always present in the air stream, the effect of additional nitrogen

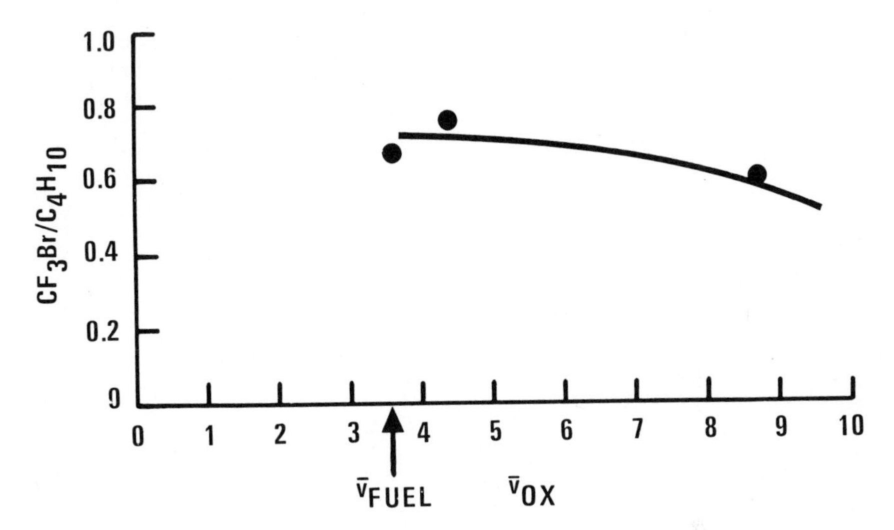

Figure 4. Effect of oxidizer stream flow parameter v_{ox} *on* CF_3Br *addition to fuel stream to quench a butane diffusion flame* ($\bar{v}_{fuel} = 3.6$)

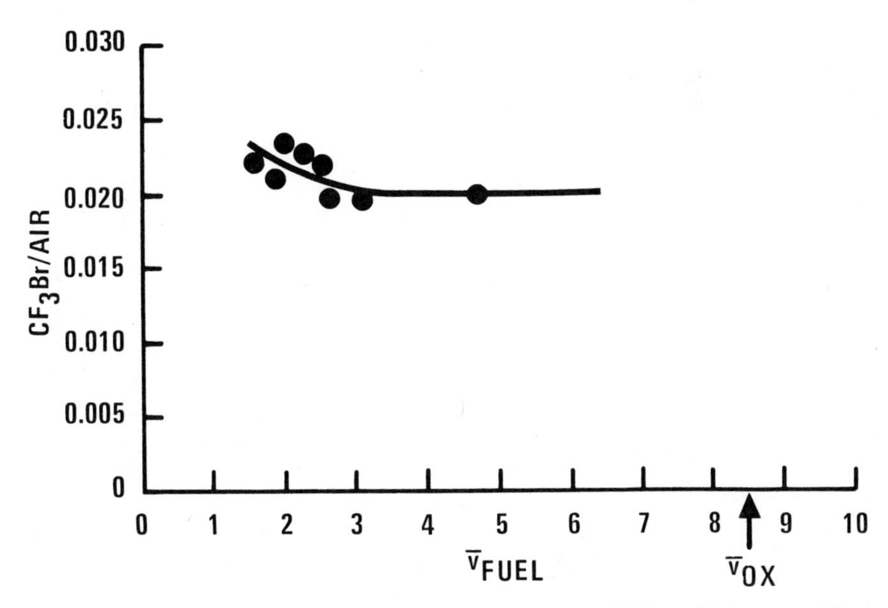

Figure 5. Effect of fuel stream flow parameter $\bar{v}_{fuel}$ *on* CF_3Br *addition to oxidizer stream to quench a butane diffusion flame* ($v_{ox} = 8.8$)

TABLE I. QUENCHER REQUIREMENTS FOR SOME DIFFUSION FLAMES.

			FUEL STREAM ADDITION			OXIDIZER STREAM ADDITION			
Fuel	$\bar{V}_{Fuel}$ [a]	$\bar{V}_{Ox}$	CF_3Br / Fuel	CF_2Br_2 / Fuel	Added N_2 / Fuel	CF_3Br / Air	Added N_2 / Air	(CF_3Br/Fuel) / (CF_3Br/Air)	(Added N_2/Fuel) / (Added N_2/Air)
CH_4	3.4	4.9	0.34			0.019	0.36	18	
CH_4	6.9	9.8	0.33	0.22	2.9	0.016	0.26	21	11
CH_4	10.3	14.7	0.30	0.21	2.2	0.013	0.21	22	10
CH_4	13.8	19.6	0.29		1.9		0.19		10
CH_4	3.4	9.8	0.30			0.016	0.26	19	
CH_4	6.9	4.9	0.39			0.020	0.36	20	
CH_4	10.3	28.2	0.24	0.16					
CH_4	13.8	37.7	0.22	0.16					
C_2H_2	4.2	10.2	0.74		13	0.092	0.98	8.0	13
C_2H_2	6.3	15.2	0.76			0.088	1.02	8.6	
C_2H_2	8.4	20.4	0.75			0.085	1.01	8.8	
CO +2% H_2	18.2	8.2	0.056			0.038		1.5	
CO +2% H_2	27.3	12.2	0.056			0.038		1.5	
CO +2% H_2	36.5	16.3	0.058			0.037		1.6	
CO +1% H_2	36.5	16.3				0.017			

[a] $\bar{V}_{Fuel}$ and $\bar{V}_{Ox}$ are the values of the flow parameters for the fuel and air streams determined prior to addition of the quencher.

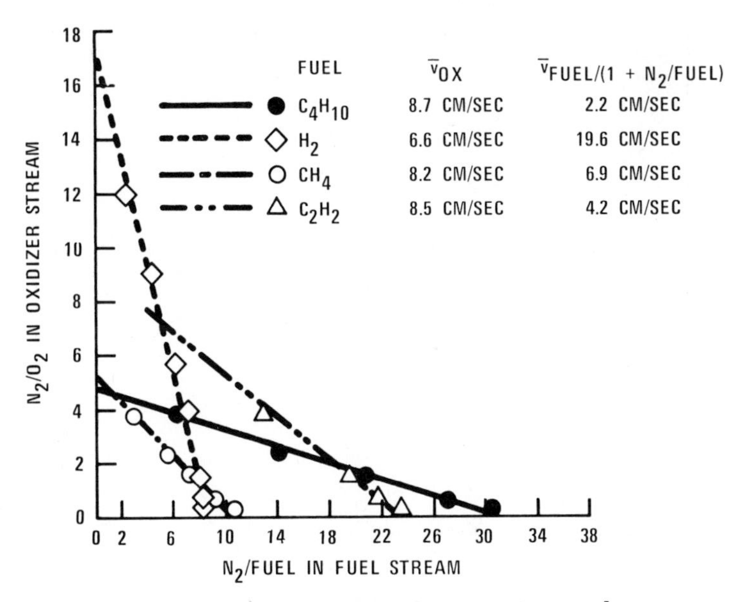

Figure 6. Quench point relationship between N_2 in oxidizer stream and N_2 in fuel stream for diffusion flames of several fuels. (v_{ox} and v_{fuel} are flow parameters for oxidizer stream and fuel stream.)

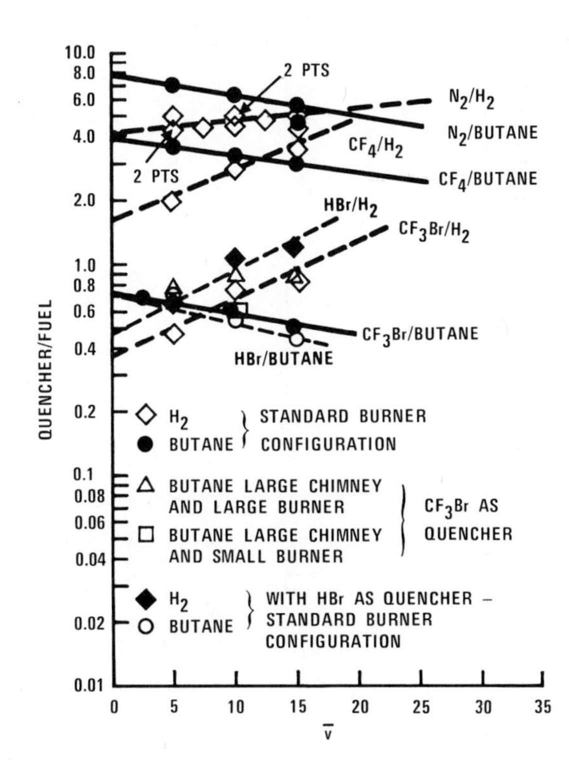

Figure 7. Quencher/fuel ratios for hydrogen and butane diffusion flames vs. flow parameter. Flow parameter for the fuel plus quencher stream is equal to that of the air stream at quench. (Reproducibility of data shown by the many points defining trend for nitrogen quenching of hydrogen.)

TABLE II. RELATIVE QUENCHER CONCENTRATION REQUIREMENTS AT QUENCH.

Fuel ($\bar{V}$)[a]	N_2/CF_4	N_2/HBr	N_2/CF_3Br	CF_4/CF_3Br
		OXIDIZER STREAM		
H_2 (5)	2.6	9.7	11.9	4.5
H_2 (10)	2.7	12.4	11.4	4.3
H_2 (15)	2.6	11.4	10.6	4.1
C_4H_{10} (5)	2.6	21.1	13.8	5.2
C_4H_{10} (10)	2.2	11.2	14.5	6.5
C_4H_{10} (15)	2.5	6.8	13.2	5.3
		FUEL STREAM		
H_2 (5)	2.3	6.6	9.6	4.2
H_2 (10)	1.7	4.4	6.2	3.8
H_2 (15)	1.8	4.2	6.0	3.4
C_4H_{10} (5)	1.9	8.8	10.6	5.5
C_4H_{10} (10)	1.9	11.3	10.3	5.4
C_4H_{10} (15)	1.8	12.3	10.6	5.9

[a] Both the fuel stream and the oxidizer stream have the same flow parameter $\bar{V}$.

rather than total nitrogen is compared with the addition of the other quenchers. This is consistent with the striking equivalence of the incremental addition of nitrogen and CF_3Br in a rich butane flame exhibited in Figure 3. The flow dependence exhibited by these data is somewhat different from that shown in Figure 7 for the same systems. The significance of the difference is open to question since in general with the same $\bar{v}$ the velocity distribution is different in a tube and an annular channel. In spite of this deficiency, it is evident that, independent of $\bar{v}$, that N_2, CF_4, CF_3Br and HBr show approximately the same relative efficiency for both fuels independent of whether they are added to the fuel stream or to the oxidizer stream (see Table II). It is particularly significant that the relative efficiencies of N_2 and CF_3Br are similar in magnitude to that exhibited by the average slope of the data for the premixed flame plotted in Figure 3 ($\sim$14).

Based on the approximately linear correlations shown in Figure 6, it is of interest to compare the quencher requirements for fuel stream addition and for air stream addition. The apparent flow dependence is different for the two cases, and this comparison reflects that dependence (see Figure 9). It is significant that within the uncertainty associated with the flow dependence, the air stream to fuel stream comparison correlates well with the air/fuel ratios for stoichiometric reaction to H_2O and CO_2. This correlation is also illustrated in Figure 10, where the range of the (quencher/fuel)//(quencher/air) values for hydrogen and butane as well as the data from Table I, where the flows were not as closely similar, are plotted as a function of the stoichiometric O_2/fuel ratio.

Simmons and Wolfhard (11) considered the possibility of this type of stoichiometric correlation for the quenching of a number of different hydrocarbon diffusion flames and the hydrogen diffusion flame by CH_3Br. In these cases, they showed that there was no evidence of such a correlation. It must be recognized however that CH_3Br is itself a fuel (7). At the level of addition required to quench (quencher/fuel ranged from 0.9 for CH_4 to 3 for H_2) the additional oxygen requirement would significantly alter the stoichiometry. Thus, at quench any such correlation is strongly dependent on the extent to which the CH_3Br is participating as a fuel molecule.

Further evidence of the correlation with stoichiometric O_2/fuel ratio is shown in Figure 7, where the lines were constructed such that the value of q_f in Eq. (1) was the stoichiometric O_2/fuel ratio and D_f was the mean value of $(N_2/\text{fuel}) + q_f(N_2/O_2)^2$ for the measured points. The values of D_f thus obtained for H_2, C_2H_2, CH_4 and C_4H_{10} were 8.7 $\pm$ 1, 23 $\pm$ 1, 10.5 $\pm$ 0.7, and 30 $\pm$ 1 where the quoted uncertainties represent the extremes of the range of the individual values. Apparently similar conclusions were drawn by Wolfhard but no data have been published (11).

On the basis of the above correlations we are led to generalize Eq. (1) and express the quench condition by

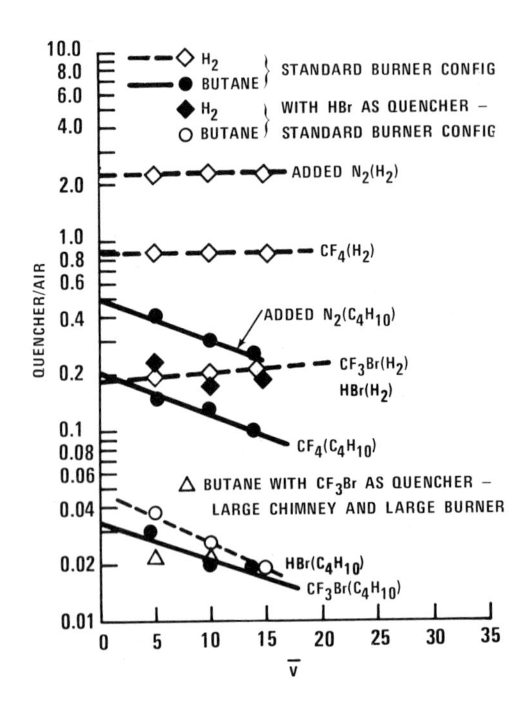

Figure 8. Quencher/air ratios for hydrogen and butane diffusion flames vs. flow parameter. Flow parameter for fuel stream is equal to that of the air plus quencher stream.

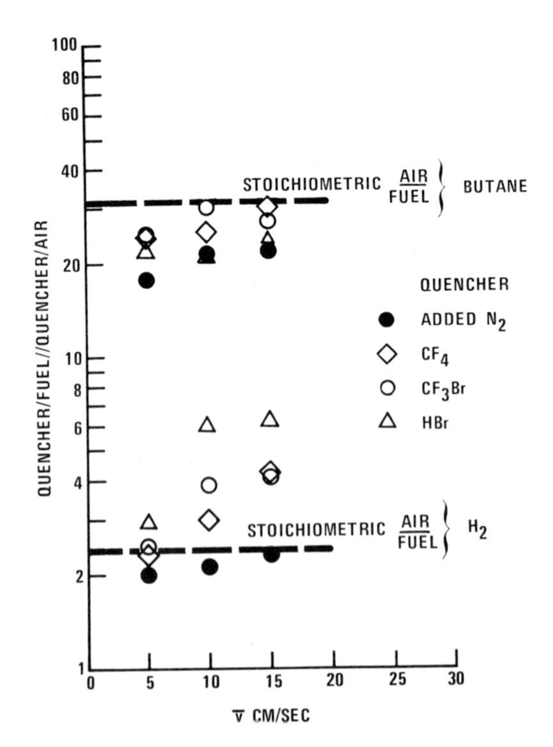

Figure 9. Ratio of quencher/fuel ratios from data in figure 8 to the quencher/air ratios from data in Figure 9 vs. flow parameter v

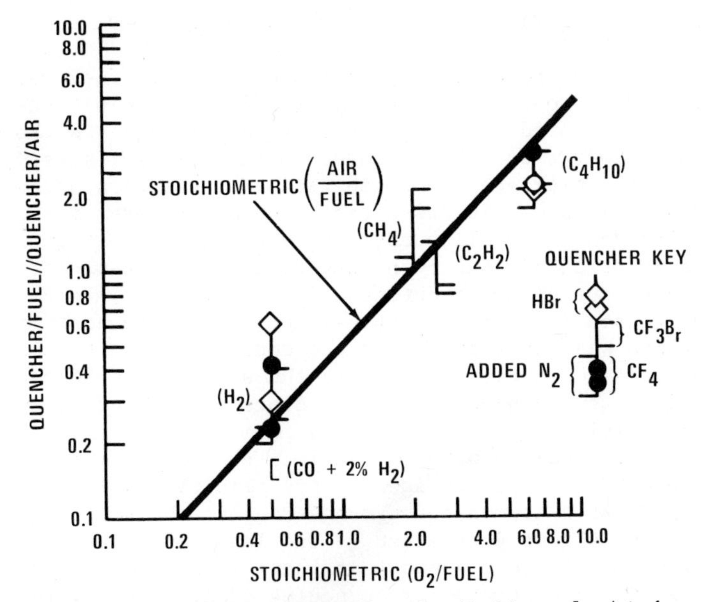

Figure 10. Range of values of quencher/fuel/quencher/air from diffusion flame data of Figures 8 and 9 and Table II vs. stoichiometric O_2/fuel ratio. (Stoichiometric air/fuel ratio is plotted on the same scale.)

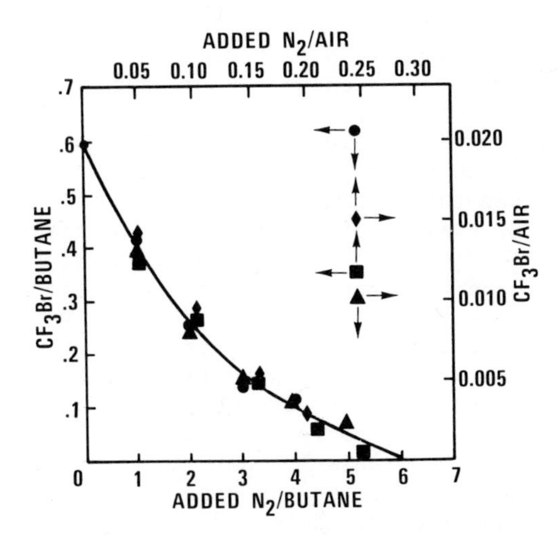

Figure 11. Effect of nitrogen addition on CF_3Br requirement for quenching butane diffusion flames ($\overline{v}_{fuel} = \overline{v}_{ox} = 10$ cm/sec)

$$\varepsilon(Q/F)_F + (N_2/F)_F + q_f(N_2/O_2)_O + q_f\varepsilon(Q/O_2)_O \simeq D_f \qquad (2)$$

In this latter equation, $(Q/F)_F$ is the quencher to fuel ratio for quencher Q added to the fuel stream, $(N_2/F)_F$ is the nitrogen to fuel ratio for nitrogen added to the fuel stream, $(Q/O_2)_O$ is the quencher to oxygen ratio for quencher Q added to the oxidizer stream, $(N_2/O_2)_O$ is the nitrogen to oxygen ratio for nitrogen added to the oxidizer stream, and ε is the effectiveness parameter for Q relative to nitrogen where all quantities are evaluated at quench.

This relationship has been investigated in detail for nitrogen and CF_3Br quenching of butane diffusion flames. In Figure 11, the relative scaling for fuel stream and air stream addition of the quenchers was determined from the empirical result taken from Figure 9 for $\bar{v}$=10 cm/sec rather than the stoichiometric equivalence ratio. As a consequence, the points corresponding to single component addition both to the fuel stream and to the air stream are plotted as a common point. The fact that the remainder of the points where both quenching agents were simultaneously added tend to scatter about a common curve suggests that the scaling parameters are only slightly affected by the level of addition. On the other hand, Figure 9 shows that there is a significant effect of $\bar{v}$ as well as the particular quenching agent on the value of q_f.

The curve drawn through the data in Figure 11 illustrates the large decrease in relative effectiveness of CF_3Br as a quenching agent as the amount of nitrogen is decreased.

A less complete but similar study of the addition of nitrogen and CF_3Br to the hydrogen diffusion flame is illustrated in Figure 12 where the relative scaling of the fuel stream and air stream data was again based on the empirical values from Figure 9. The gross behavior of the data is similar to that for butane cited above. However, the consistent difference between the results for fuel stream and air stream addition of the quenching agents suggests that not only is there a difference in the scaling parameters q_f dependent on the quencher but also that the dependence of the efficiency parameter ε on nitrogen level is somewhat different for fuel side and oxidizer side addition. These effects are, however, small compared to the gross change in the magnitude of these parameters over the total range of the data.

As an extreme example of the reduction of efficiency of CF_3Br at low nitrogen levels, we have quenched the hydrogen-oxygen diffusion flame both by fuel stream and oxidizer stream addition where no nitrogen was present. These limiting values were CF_3Br/H_2=1.8 and CF_3Br/O_2=2.7 where the flow parameters of both streams was kept at $\bar{v}$=10 cm/sec. By way of comparison, the corresponding values when air is the oxidizer are CF_3Br/H_2=0.77 and CF_3Br/O_2=0.95 at $\bar{v}$=10 cm/sec. The complete range of CF_3Br efficiencies relative to nitrogen is shown in Figure 13 where the relative scaling of the fuel stream and oxidizer stream data is such that the points without nitrogen addition are forced to coincide.

Relative to the bromine atom effect, it is of interest that

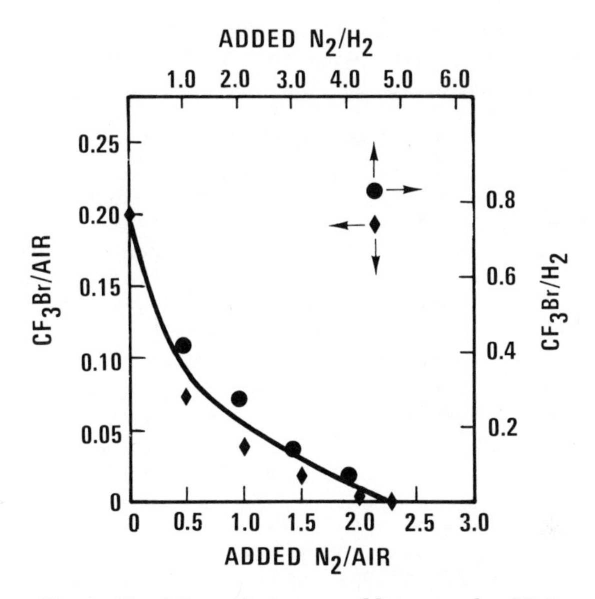

Figure 12. Effect of nitrogen addition on the CF_3Br requirement for quenching hydrogen diffusion flames ($v_{ox} = \overline{v}_{fuel} = 10$ cm/sec)

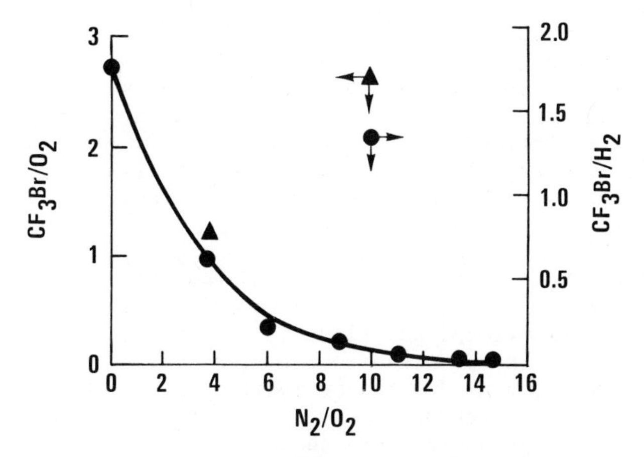

Figure 13. Full range of CF_3Br requirements for quenching hydrogen diffusion flames as a function of nitrogen content of oxidizer stream ($\overline{v}_{ox} = \overline{v}_{fuel} = 10$ cm/sec)

HBr has approximately the same quenching ability per mole as CF_3Br (see Figures 7 and 8). The data for fuel stream quenching of methane quoted in Table I show that, per mole, CF_2Br_2 is approximately 1.5 times as effective as CF_3Br.

The data for the CO flames quoted in Table I are of some interest relative to the effect of hydrogen in stabilizing the CO flame. It is evident from the nitrogen quench data in Figure 7 that CO is behaving as a fuel in the CO-H_2 flame, since the mix has about ten times as much CO as the nitrogen necessary to quench a hydrogen diffusion flame. However, the experimental concentrations of CF_3Br necessary to quench the CO-H_2 diffusion flame are roughly proportional to the hydrogen content of the fuel, and are consistent with the hypothesis that CF_3Br is reacting with those flame species which are common to the H_2-O_2 and the hydrogen stabilized CO-O_2 flames.

Discussion

A phenomonological description of the conditions for the stability of a burner flame was given by Lewis and von Elbe (13). According to this description, the stability is controlled by the relationship between the gas flow speed and the flame propagation speed adjacent to the surface of the burner. Due to the quenching action of the surface on the radical species responsible for flame propagation, the propagation speed increases rapidly with distance normal to the burner surface from zero to a value characteristic of the gas mixture. In contrast, the gas flow speed which is also zero at the surface increases more gradually. A stable flame results when the two speeds become equal at some value less than the homogeneous flame propagation speed. As the gas flow to the burner is increased a stable flame becomes unstable. The critical gas flow occurs when the slopes of the two speeds are equal at the point of intersection and with larger gas flows the flame blows off. This description has been generalized in terms of the transport properties of the gas and reaction rates to a broadly applicable theory of flames through the use of the Damköhler number (14).

Lewis and von Elbe demonstrated the consistency of their description for laminar natural gas (principally methane)-air flames. Wohl, et al, investigated butane-air flames from the same point of view (15). For their experimental conditions employing large flow rates, the latter group observed that blow off was replaced by lift off where, after leaving the immediate vicinity of the burner surface, the flame was subsequently stabilized at a greater distance from the surface by turbulent effects. In both of the above investigations, the flame increased in stability as the fuel/air ratios were increased to the rich side of stoichiometric. This was interpreted as due to the modification of the gas composition in the critical region at the base of the flame by admixture of air from the surrounding atmosphere. For butane, it was shown that the blow off/lift off limit as a function of percentage butane increased monotonically to a limit characteristic of the diffusion flame. This is qualitatively similar to our observations for the quenching of butane flames with CF_3Br as plotted in Figure 2.

Rather than inducing blow off by increasing the gas flow rate, we induce blow off by reducing the flame speed in that critical region adjacent to the burner lip as a consequence of adding quenching agents. In the diffusion flame studies reported here, the flow parameters are at least an order of magnitude smaller than those required for lift off in the absence of the quenching agents. For the premixed flame studies this is not the case. In fact, the data of Wohl, et al, (15) suggests that the upper limit of flow parameter for the butane data plotted in Figure 1 is close to the blow off limit.

With the exception of the data for hydrogen flames, the trends of the data with flow parameter are in qualitative accord with the simple stability criterion proposed by Lewis and von Elbe (13). It must be recognized that the velocity distribution adjacent to the burner surface cannot be described in terms of a single variable, e.g., the flow parameter, and the same is true for the gas composition which controls the flame propagation velocity. In fact the locus of final flame attachment to the burner surface at quench may in general be different from that of the uninhibited flame. This was very evident in the case of quenching butane flames by fuel side addition of CF_3Br. The uninhibited flame was attached near the air side edge of the burner lip. Near quench, attachment was $\sim$1 mm below the lip on the air side of the burner tube. We do not believe that the anomolous dependence of hydrogen flame quenching on flow parameter negates the basic description of flame stability, but rather reflects the details of the velocity and concentration distributions over the burner surface for high mobility fuel under flow conditions which are far from uninhibited blow off or lift off.

The possibility that the anomalous flow dependence of the hydrogen quench points might be associated with the high thermal conductivity of hydrogen was considered. If this were true, it should be possible to alter the flow dependence of the butane quench points by using helium as the quenching agent. Some helium quench points were determined for butane diffusion flames for flow parameters in the range 5 to 15 cm/sec. The results showed that helium was very nearly equivalent to nitrogen as a quenching agent and the dependence of the quench point on $\bar{v}$ was essentially the same as that exhibited by the nitrogen quench points in Figures 7 and 8. On this basis, it is concluded that thermal conductivity of the gas does not play an important role in determining the flow dependence of the quench point.

Those data in Figures 7 and 8 which show the effect of the large chimney and large burner on the CF_3Br quenching of the butane diffusion flame suggest that the flow dependence of the quench points is largely controlled by the detailed fluid dynamics in the neighborhood of the burner lip. In the case of the large chimney and large burner where the apparent flow dependence is much less than for the standard configuration, it would be anticipated that for fully developed laminar flow, the flow velocity would change more slowly with the distance from the burner surface than would be the case for the standard configuration. Considering the complexity of a detailed description of the conditions at the base of the flame, the overall consistency of the results for different fuels and different quenching conditions with the correlation given by Eq. (1) and (2) is all the more remarkable. The left-hand side of Eq. (1) is the total moles of nitrogen per mole of fuel that would be obtained as a result of simply mixing appropriate aliquots from the fuel and oxidizer streams to obtain a stoichiometric fuel/oxygen mix. The left-hand side of Eq. (2) is the nitrogen equivalent of the total quencher that would be present in a similarly constructed stoichiometric mix. While it is obvious that these equations are not quantitative for the representation of the data that have been presented, they provide a useful semi-quantitative tool for discussing the effects of the quenchers. For example, it is readily apparent that, relative to hydrogen, butane diffusion flames require much less oxidizer stream quencher as a consequence of its greater oxygen demand. It is also apparent that the apparently trivial oxidizer stream addition of CF_3Br ($\sim$2% in the flow region of this data) is the order of one molecule of CF_3Br for every two molecules of butane, when considered relative to the fuel molecules which enter into the reaction.

In order to discuss the chemistry of burner flame quenching, it would be desirable to know the composition of the gas in that critical volume at the base of the flame which determines the stability of the flame relative to blow off. On the strength of the correlations demonstrated in this paper, we propose that for qualitative considerations this composition can be represented by a stoichiometric mix constructed by mixing appropriate aliquots of the oxidizer and fuel streams.

In the case of diffusion flames, we have probed the dark region of a methane flame, between the flame and the port and find that there are all mixes of fuel and air present. Since mixtures in the neighborhood of stoichiometric have the greatest flame velocity, it is this mix which will control the flame attachment (all other mixes would be in the blow off regime). Interestingly, this concept can account for the great resistance of diffusion flames to blow off, relative to stoichiometric premixed fuel-air flames. In the critical region of flame attachment, the diffusion flame stability is controlled by a premixed stoichiometric flame, whereas the stoichiometric premixed flame is weakened by the admixture of outside air into this region.

The quenching of a burner flame by nitrogen is due to physical interactions in the sense that nitrogen is not chemically reacting with any constituent of the flame. If we consider the base of the flame, there is a bulk flow of gas into the flame zone. With air as the oxidizer, for all of the flames we have considered, nitrogen is the principle constituent of the gas that is flowing into the flame zone. By increasing the nitrogen in this flow over its normal air value relative to oxygen either by addition from the fuel side or from the air side in a diffusion flame, we are in effect decreasing the fuel and oxygen flow per unit volume of total flow and thus decreasing the heat release available per mole of nitrogen in that flow. This implies an overall reduction in the temperature of the flame. In addition, the reduced concentration of fuel and oxidizer will result in the reduction of the rates of reaction and hence, the rate at which heat is released. Therefore, the net result of nitrogen addition is to reduce the flame temperature, the temperature gradient into the unburned gas, and the reaction rate in the unburned gas. Taken together these effects result in a reduction in the flame propagation speed. Thus, relative to blow off, for a sufficient addition of nitrogen to the supply gases, the flame propagation speed is reduced to the point where it is no longer sufficient to keep up with the flow of nitrogen and reactants into the flame zone. In a quiescent fuel-oxygen mix, the addition of nitrogen would eventually give the critical situation where the heat loss from the unburned mix would be greater than the heat release rate from the chemical reaction.

While quenchers such as CF_3Br, CF_2Br_2 and HBr will participate in a physical way much the same as nitrogen, the fact that they are effective at an order of magnitude lower concentration than nitrogen suggests that their primary interaction is chemical. Since CF_3Br reacts to form other molecules in the flame, it is reasonable to inquire as to whether the heat of reaction is endothermic enough to substantially reduce the flame temperature. Actually, to form CO, HF and HBr, the reaction is somewhat exothermic. Flame temperature measurements have shown that CF_3Br addition actually increases the flame temperature (10). Thus, it must be concluded that the chemical interaction must be such that it results in a substantial reduction in the rates of those reactions which are involved in the uninhibited flame.

The case of quenching by CF_4 is somewhat ambiguous, its heat capacity is 2 or 3 times that of nitrogen depending on temperature; however, if it were to react to CO and HF considerable heat would be produced. We have observed that the hydrogen flame which is normally non-luminous becomes highly luminous upon addition of amounts of CF_3Br which are small compared to the amounts necessary to quench the flame. The luminosity, in part, is due to carbon, since carbon particles can be seen inside the flame cone. In contrast there is no significant visible luminosity induced in the hydrogen flame by CF_4 even with the relatively large amounts required to quench. The initial chemical process would be fluorine atom abstraction in the case of CF_4 and bromine atom abstraction in the case of CF_3Br to yield CF_3 where the CF_3 radical is the source of the carbon which is responsible for at least part of the flame luminosity. We, therefore, conclude that there is little if any chemical attack on the CF_4 molecule in the hydrogen flame. This is in accord with the much greater bond strength for F relative to Br (125 kcal vs 68 kcal). On this basis, we conclude the CF_4 is primarily a physical quencher and that its greater effect relative to nitrogen is a consequence of its greater heat capacity.

There is a body of evidence which suggests that in the flame zone of a hydrocarbon-air flame, the predominant chemistry is that of the hydrogen-air flame. The hydrogen present in the reaction zone of the hydrocarbon-air flame is produced by the pyrolysis of the hydrocarbon in the lower temperature regions adjacent to the flame. Thus, if one understands the quenching chemistry of the hydrogen-air flame, one can extend this understanding to hydrocarbon-air flames in general. Fristrom and Sawyer (16) have analyzed chemical quenching relative to hydrogen-oxygen flame chemistry. From an investigation of the influence of HBr on the partial equilibrium concentrations of the various radicals, they conclude the quenching action must be kinetic in nature. They also conclude that the quenching action must occur in the low temperature region of the flame volume, rather than in the high temperature region. An extension of these considerations has proven fruitful in defining the nature of the kinetic interaction.

The following reactions are postulated to be the important steps in developing the radicals which lead to flame propagation:

$$H + O_2 \rightarrow OH + O \quad (3)$$

$$O + H_2 \rightarrow OH + H \quad (4)$$

$$OH + H_2 \rightarrow H_2O + H \quad (5)$$

Without the interference of other processes, once initiated, these reactions would proceed to a partial equilibrium limit at a temperature determined by the initial temperature and composition. At any point in this process,

$$\frac{d[H_2O]}{dt} = \frac{d[H]}{dt} + \frac{d[O]}{dt} \quad (6)$$

where this last relation would continue to hold even if the process

$$OH + OH \rightarrow H_2O + O$$

and its reverse were to play a role. Therefore, if the initial radical concentrations necessary to start the reaction system (3), (4) and (5) were neglected, at partial equilibrium $[H_2O] = [H] + [O]$. We have computed the partial equilibrium concentration constrained by this condition using JANAF thermochemical data (17). Starting with a stoichiometric H_2/O_2 mixture at 298°K, partial equilibrium would be achieved at a temperature of ∿650°K with ∿2/3 of the hydrogen present as water and the remainder essentially present as hydrogen atoms. In the presence of diluent gases the temperature would be lower. Thus, other processes must play an important role to obtain the high temperature and water yields characteristic of the hydrogen air flames. These processes are the termolecular radical recombination reactions, the most important of which are:

$$H + H + M \rightarrow H_2 + M \quad (7)$$

$$H + OH + M \rightarrow H_2O + M \quad (8)$$

$$H + O + M \rightarrow OH + M \quad (9)$$

$$O + O + M \rightarrow O_2 + M \quad (10)$$

At 600°K, the relative partial equilibrium concentrations were computed to be

$$\frac{[H]}{[H_2O]} \simeq 0.99, \ \frac{[O]}{[H_2O]} \simeq 0.01, \ \frac{[H_2]}{[H_2O]} = 0.05, \ \frac{[O_2]}{[H_2O]} = 0.26$$

where $[OH]/[O] \sim 0.02$.

It is clear that hydrogen atoms are the dominant radicals. With these concentrations, we compare the relative probability of a hydrogen atom reacting by reaction (3) or the reverse of reactions (2) and (5) with that for reacting via reactions (7), (8) or (9), using rate constants tabulated by Kondratiev (18). The result is a factor of $\sim 10^3$ in favor of reacting via the termolecular reactions. Therefore, long before partial equilibrium is achieved, the termolecular reactions will enter into the reaction scheme in an important way. The effect of the termolecular reactions is twofold: 1) relation (6) becomes

$$\frac{d[H_2O]}{dt} > \frac{d[H]}{dt} + \frac{d[O]}{dt}$$

thus allowing an increased yield of water from the overall reaction system, and 2) the enthalpy that goes into radical production via reactions (3), (4) and (5) is returned to the system as heat through the recombination of the radicals, thus allowing high temperature to be achieved. While the relative activation energies of reaction (3), (4) and (5) suggest that hydrogen atoms will tend to build up in the system relative to O and OH, this will probably not be as extreme as suggested by the partial equilibrium results quoted above as a consequence of the termolecular reactions.

Thus relative to the propagation of the flame into unburned gases which are supplied to the reaction zone both the bimolecular reactions and the termolecular reactions play important roles: the bimolecular reaction through the production of radicals and water and the termolecular reactions through the production of heat which in turn leads to an acceleration of the bimolecular reactions leading to H_2O.

If bromine is introduced into the system either as bromine molecules, HBr or as part of an organic compound, four additional reactions need be considered*

$$H + Br + M \rightarrow HBr + M \tag{11}$$

$$H + HBr \rightarrow H_2 + Br \tag{12}$$

$$Br + Br + M \rightarrow Br_2 + M \tag{13}$$

$$H + Br_2 \rightarrow HBr + Br \tag{14}$$

The net result of these reactions is to increase the competition for hydrogen atoms by processes other than reaction (3) with an accompanying reduction in the rate of radical production by the bimolecular processes (3), (4) and (5). With sufficient addition of bromine to the system, the burning speed decreases because the flame propagation depends on the reaction rate of the mix and this is decreased by addition of bromine. The flame eventually blows off as described previously. For a quiescent flammable mix, the effect of bromine on the spread of flame is similar. Eventually with bromine addition, the reaction rate is inhibited to the point where heat leak out of the reaction volume is greater than the rate of heat generation and the flame does not propagate.

To the extent that the above is a valid description of the quenching effect resulting from the addition of bromine to the system, it implies that quenching depends on the number of bromine atoms introduced, independent of the bromine compound (bromine, organic bromides, or hydrogen bromide). Semiquantitative evidence that this is the case also has been reported by Rosser, et al (6).

The increased effectiveness of bromine addition with increased nitrogen can also be understood in terms of the temperature reduction associated with nitrogen dilution. The rate of reaction (3) decreases exponentially with the reduction in temperature thus increasing the effectiveness of reactions (11) thru (14) relative to reaction (3) since the former reactions have a lower energy of activation than the latter.

Creitz has previously proposed that the role of bromine as a quencher was principally to recombine oxygen atoms (19). Based on the analysis presented above, we feel that the hydrogen atom recombination reactions are most probably dominant in the critical region of the flame associated with its propagation into the unburned gases.

On the basis of the above mechanism, it is now possible to identify those unique properties which distinguish bromine containing compounds as good quenching agents. Where bromine is introduced into the flame as a hydrocarbon or fluorocarbon substituent

*For an organic bromide, the initial reactions are: $H + RBr \rightarrow R + HBr$ and pyrolysis, $RBr \rightarrow R + Br$.

or as Br_2, the strongest bond that can be made with the species present in the flame region is the HBr bond. Thus, to the extent that bond strengths are reflected in the rates of the abstraction and pyrolysis processes by which Br is introduced in the system, the kinetic drive in the flame region is to form HBr. The HBr bond is in turn weaker than the other possible H atom bonds in the system. As a consequence the kinetic drive is to liberate the bromine from HBr. This leads to the sequence of reactions (11) thru (14) where the bromine atoms are recycled with an attendant increase in the rate of hydrogen atom recombination in competition with radical production reactions (3) thru (5).

Summary and Conclusion

It has been shown that relative to the fuel, the quencher requirements are similar for diffusion flames and premixed flames. For those quenchers and fuels which we have studied, the relative effectiveness of the quenchers is approximately the same for all of the fuels where air is the oxidizer. In particular CF_3Br is about 10 times as effective as nitrogen on a per mole basis, and twice as effective on a weight basis. Because of the large cost difference between these agents, liquid nitrogen may in fact be the quencher of choice in specialized applications.

In our studies, a comparison of the quencher requirements for fuel stream introduction and oxidizer stream introduction into diffusion flames, has shown that they are qualitatively related to the stoichiometry of the uninhibited flame reaction. This suggests that for burner flames quenching is dominated by a reduction of the ability of the flame to propagate into an essentially stoichiometric mixture of fuel and oxidizer which is being convectively supplied to the base of the flame. From this point of view it is seen that the concentration of bromine containing species required for quench is of the same order of magnitude as the other reactive species in the flame region. It also follows that premixed flames are easier to quench than diffusion flames, because of the dilution of the mixture at the base of the flame by entrainment of gas from the surrounding environment.

The flow dependence of the quencher requirement for most flames studied was qualitatively consistent with a blow off description of quenching. Hydrogen flames, both premixed and diffusion flames, exhibited an anomalous flow dependence. Experiments with helium designed to delineate the cause of the anomaly established that the anomaly was not due to thermal conductivity. We concluded that the probable source was related to the detailed fluid dynamics in that region at the base of the flame which is associated with large gradients of temperature, gas velocity, and composition.

Of the quenchers studied, the large amounts of N_2 and CF_4 required for quench are consistent with the characterization of these agents as physical quenchers. The bromine compounds, CF_3Br, CF_2Br_2 and HBr behave as chemical quenchers where their relative effectiveness is determined primarily by the number of bromine atoms they carry into the flame.

For CF_3Br quenching of butane and hydrogen diffusion flames, it was clearly demonstrated that the effectiveness of CF_3Br as a quencher relative to nitrogen decreased markedly as the nitrogen content of the flame system was reduced.

In the case of hydrogen diffusion flames in the absence of nitrogen, the level of CF_3Br addition required for quench was sufficiently large that an appreciable dilution would result. This suggests that for some fires, the application of nitrogen and CF_3Br concurrently could greatly enhance the effectiveness of CF_3Br, and minimize the toxic and corrosive products.

An overall consistent picture was obtained from a consideration of the dominant chemical processes which relate to the ability of the reaction to propagate into the unburned gas. On this basis, it was inferred that the dominant radical species are hydrogen atoms which are produced, concurrently with water, by a bimolecular sequence of reactions where the slow step is $H + O_2 \rightarrow OH + O$. The reaction heat is liberated primarily by a set of termolecular recombination reaction, principally $H + H + M \rightarrow H_2 + M$, which competes with the hydrogen atom generating reactions. Both of these sets of processes are essential for flame propagation. The effect of physical quenchers is to dilute the reaction species to the point where the rates of reaction and, as a consequence, the temperature is reduced to the point where propagation cannot continue. The chemical quenchers introduce bromine into the system which catalytically recombine the hydrogen atoms via a sequence of reactions. With sufficient bromine addition, the recombination reactions become dominant over the radical, and water producing bimolecular reactions. As a consequence the hydrogen atom concentration is reduced to a point where the rate of the reaction and as a consequence the temperature is reduced so that flame propagation ceases.

Acknowledgements

The authors are indebted to Messers. J. H. Johnson and J. Calderwood who obtained many of the quench points which we report, to Dr. W. H. Thielbahr for illuminating discussions of the fluid dynamics at the base of the flame, and to Professors F. A. Williams and R. C. Corlett for valuable comments and suggestions on many aspects of this work.

Literature Cited

1. Miller, D. R., R. L. Evers, and G. B. Skinner; Combust. Flame 7 137 (1963).
2. Fristrom, R. M.; Fire Research Abstr. Re . 9 125 (1967).
3. Wilson, W. E., Jr., J. T. O'Donovan, and R. M. Fristrom; Twelfth Symposium (International) on Combustion, p. 929, 1968.

4. Simmons, R. F. and H. G. Wolfhard; Trans. Faraday Soc. 51 1211 (1955).
5. Lask, G. and H. Gg. Wagner; Eighth Symposium (International) on Combustion, p. 432, Williams and Wilkins, Baltimore, 1962.
6. Rosser, W. A., H. Wise, and J. Miller; Seventh Symposium (International) on Combustion, p. 175, Buttersworth, London (1959).
7. Creitz, E. C.; J. Res. Nat. Bur. Stand. 65A 389 (1961).
8. Friedman, R. and J. B. Levy; Combust. Flame 7 195 (1963).
9. Ibiricu, M. M. and A. G. Gaydon; Combust. Flame 8 51 (1964).
10. Kent, J. H. and F. A. Williams; Fifteenth Symposium (International) on Combustion, p. 315, Combustion Institute, 1975.
11. Simmons, R. F. and H. G. Wolfhard; Trans. Faraday Soc. 52 53 (1956).
12. Simmons, R. F. and H. G. Wolfhard; Combust. Flame 1 155 (1957).
13. Lewis, B. and G. von Elbe; J. Chem. Phys. 11 75 (1942).
14. Williams, F. A.; Combustion Theory; p. 131; Addison-Wesley Co., 1965.
15. Wohl, K., N. M. Kapp, and C. Gazley; Sixth Symposium (International) on Combustion, p. 3-21; The Williams and Wilkins Co., Baltimore, 1949.
16. Fristrom, R. and R. Sawyer, Flame Inhibition Chemistry, Thirty-Seventh AGARD Symposium, The Hague, 12 May 1971.
17. Stull, D. R. and H. Prophet, Project Directors; JANAF Thermochemical Tables, 2nd Ed., U.S. Dept. of Commerce, NSRDS-NBS-37, June 1971.
18. Kondratiev, V. N.; Rate Constants of Gas Phase Reactions; (in English Translation) U.S. Dept. of Commerce COM-72-10014, 1972.
19. Creitz, E. C.; J. Res. Nat. Bur. Stand. 74A 521 (1970).

7

The Effect of Halogens on the Blowout Characteristics of an Opposed Jet Stabilized Flame

R. W. SCHEFER, N. J. BROWN, and R. F. SAWYER

Department of Mechanical Engineering, University of California, Berkeley, Cal. 94720

Flame inhibitors are broadly classified as being physical or chemical. Physical inhibitors (e.g., N_2) are believed to act as diluents which lower flame temperatures. This results in slower reaction rates which in turn reduce flame propagation velocity. In contrast, chemical inhibitors (e.g., HBr) directly participate in the chemistry of the system and are believed to reduce reaction rates via a reduction in reactive radical concentrations.

There were two principal reasons for the present study of flame inhibition. First, the detailed mechanism of flame inhibition is not well understood. Even for those reactions which are thought to be important, rate constants are not known accurately over the temperature range of interest in practical combustion systems. Second, few quantitative measures of inhibitor effectiveness exist and relatively few measurements have been made on different inhibitors in identical combustion environments.

This investigation can be divided into two parts. In the first part, experimental measurements of the extinction characteristics of inhibited propane-air mixtures as a function of inhibitor type and inhibitor concentration were made. The inhibitors investigated were CH_3Br, HBr, HCl, N_2, and Ar. Measurements were done in an opposed reacting jet (ORJ). This system is characterized by intense turbulent mixing so that stability characteristics are strongly influenced by chemical kinetics as opposed to mixing phenomena. Part two is a computational study of inhibition in a well stirred reactor. The well stirred reactor represents the limiting case of a turbulent flow system in which mixing is assumed to occur instantaneously, making the system chemically limited. Inhibitor effectiveness based on an inhibition parameter proposed by Fristrom and Sawyer (1) is evaluated for both the opposed reacting jet and the well stirred reactor and the results are compared.

Work supported by NSF-RANN under Grant Number GI-43.

Experimental Apparatus and Procedure

Figure 1 shows the approximate flowfield in an opposed reacting jet. The opposed reacting jet used in the present study consists of a main premixed stream of fuel and air and a smaller stream of premixed air and inhibitor which is injected at a high velocity along the centerline and in a direction opposite to the main stream. The result is the formation of a stagnation region and zone of strongly recirculating combustion gases. A stable flamefront forms at the upstream end of the recirculation zone which spreads outward into the main flow.

A schematic diagram of the experimental apparatus used is shown in Figure 2. Air was supplied by the house air system. Main stream air was metered upstream of the combustor using a standard 1.52 cm diameter orifice. The pressure was reduced to approximately atmospheric prior to the combustor inlet. Relative humidity of the reactants was measured at each data point and was found to remain relatively constant at 20%. Propane used was Matheson commercial grade 99.9% pure with impurities consisting mainly of iso-Butane (0.35%) and n-Butane (0.05%). The propane was metered through a calibrated rotameter and then mixed with the main stream air in a venturi section. The flow then passes down a 1.0 meter long straightening section to assure a fully developed velocity profile and uniform composition. Measurements were made of the radial velocity profile at the test section inlet using a stagnation pressure probe in conjunction with a static pressure tap in the wall. The velocity profile indicated turbulent flow existed under the entire range of experimental conditions. Composition measurements using a quartz sampling probe and a mass spectrometer for analysis of O_2, N_2 and C_3H_8 were made. The mixture was ignited by a spark plug located just upstream of the combustor test section.

Jet air and inhibitor were metered separately through rotameters and mixed in a tee before passing into the jet injector. The HCl and CH_3Br had a minimum purity of 99.5%, with trace amounts of hydrocarbons and water. The HBr had a minimum purity of 99.8% with a maximum of 0.2% HCl. Inhibitor concentration in the jet stream was varied up to a maximum of 5% by volume of the total jet flow. No inhibitor was added to the main stream. The inhibitor was added in such a way as to replace air and not O_2 or N_2 separately.

The combustor test section consists of a 58 mm I.D. x 356 mm O.D. high temperature vycor liner which was wrapped in asbestos lagging, making the combustor more nearly adiabatic. Temperatures were measured at several locations inside the lagging to determine lagging effectiveness. The resulting temperature profiles indicated a heat loss of less than 0.3% of the total energy release due to combustion. The stainless steel jet injector extended 6.35 cm into the test section. It had a 6.35 mm O.D. and a

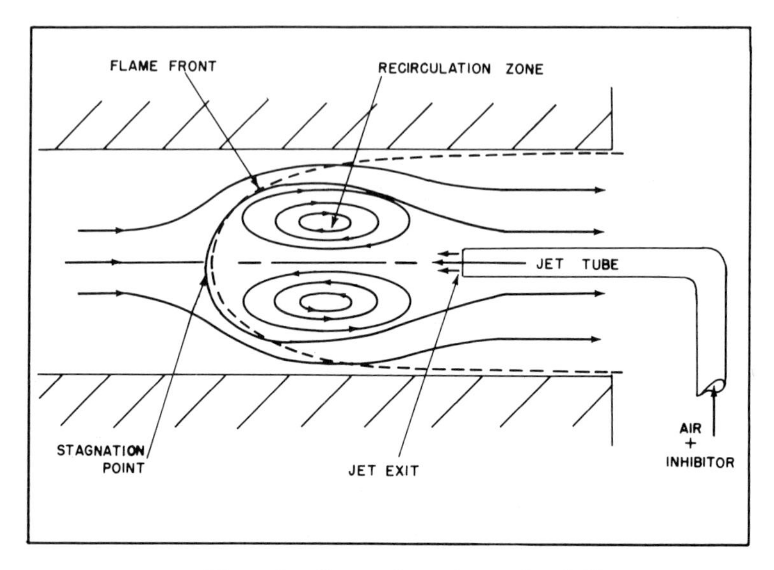

Figure 1. Opposed reacting jet flowfield

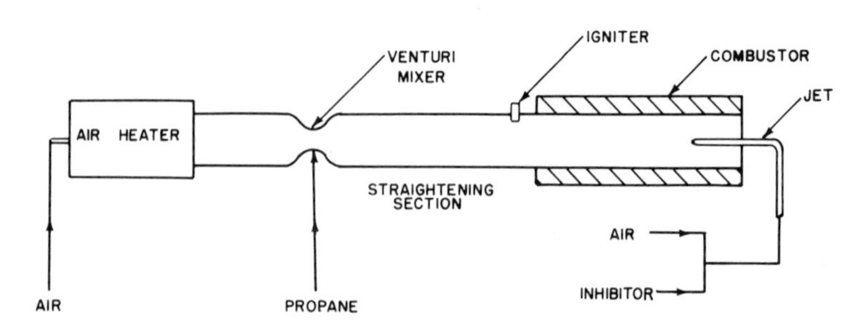

Figure 2. Experimental apparatus

3.97 mm I.D. except at the exit where the inner diameter converged to 1.60 mm. Water cooling of the jet walls allowed operation at high temperatures.

For all of the work presented here the total jet mass flow rate (air + inhibitor) was maintained constant at .25 gm/sec, corresponding to a jet exit velocity of~96.0 m/sec. To obtain extinction points for a given inhibitor concentration either the fuel flow could be decreased while maintaining a constant main stream air flow, or the air flow could be increased with a constant fuel flow. The former method of decreasing the fuel flow was found to provide the best reproducibility and was used in all cases. Typical main stream mass flow rates were in the range of 18.6 gm/sec to 32.2 gm/sec.

Computational Study of Inhibition

Introduction. Computational studies of inhibition were conducted for the inhibitors Ar, N_2, HCl and HBr. The case of inhibition by CH_3Br was not investigated since thermal data for CH_3Br was unavailable. A well stirred reactor model was used for this part of the study because it most closely simulates the experimental studies.

Although the ultimate goal of companion experimental and computational studies is quantitative agreement, qualitative agreement between the two is useful and often contributes to a general understanding. Here, there were two difficulties which made comparison at best qualitative: 1) possible uncertainties in the validity of the assumptions used to describe the fluid mechanics of the experimental system and 2) uncertainty in the kinetic data used in the computational studies. Hence the computational studies were directed toward assessing the sensitivity of various reaction steps in the proposed mechanism using the experimental results as guidelines.

Rate coefficients required for inhibition modeling have not been measured in the appropriate temperature range so that kinetic data can only be inferred from extrapolation of low or high temperature results; hence it is impossible to absolutely quantify error bounds on the rate coefficients. To assess the effects of various reactions in the assumed inhibition mechanism, the rate coefficients were varied by factors assumed to be extreme bounds on their uncertainties.

The remainder of this section contains a description of the well-stirred reactor formalism and a discussion of the mechanisms and rate data used.

Well-stirred reactor. The idealized well stirred reactor is a constant volume steady flow reactor in which mixing is assumed to occur instantaneously between cold incoming reactants and the reacting gases existing in the reactor. Thus the composition within the reactor is homogeneous and the process is kinetically

3) Equation of State:

$$\rho = \frac{p}{RT\sigma_m} \tag{7}$$

σ_m is the inverse of the mixture molecular weight and ρ is the density. Given the pressure p and rate of heat loss from the system $\dot{Q}$, equations (1) and (2) form a set of I+1 equations with I+1 unknowns σ_i, i=1,...,I and either $\dot{m}/V$ or T.

A computer program developed by Pratt and Bowman (3) and based on a Newton-Raphson iteration technique was used to solve the above set of equations.

Kinetic Mechanism. The chemistry of the present system can be divided into three parts; the oxidation of propane to the intermediate combustion products H_2 and CO, a mechanism for H_2 and CO combustion, and a mechanism which accounts for the effect of chemical inhibitor addition.

Due to the complexity of hydrocarbon oxidation (except perhaps for methane), little is known in the way of a detailed reaction mechanism. Thus, past studies have almost exclusively relied on some type of a global or quasi-global mechanism to describe hydrocarbon combustion. Two approximations for the mechanism have been widely used. The first (mechanism 1) and simplest assumes infinitely fast oxidation of the hydrocarbon to H_2 and CO (4). The second (mechanism 2), perhaps a somewhat better approach, is the use of a finite rate global expression for the oxidation of the hydrocarbon based on overall experimental observations in a particular system. Edelman and Fortune (5) derived such an expression for the oxidation of a parrafin hydrocarbon based on shock tube ignition delay times. For propane, the overall reaction can be written

$$C_3H_8 + \frac{3}{2} O_2 \rightarrow 3\ CO + 4\ H_2$$

and the global rate is given by

$$\dot{R} = \frac{5.52 \times 10^8}{p^{.825}} (C_3H_8)^{1/2}(O_2)T \exp(-\frac{12,200}{T}) \frac{moles}{cm^3 sec} \tag{8}$$

Caution should be excerised whenever a global reaction rate derived for one combustion configuration is used for another. Such an approximation entirely neglects any reactions between hydrocarbon intermediates. In two such different systems as a shock tube, where low radical concentrations are present during the initial stages of hydrocarbon oxidation, and a strongly backmixed system such as a well stirred reactor where high radical concentrations are present throughout the combustion process, it is probable that both the important reaction paths and the rate controlled.

The governing equations, using the notation of Jones and Prothero (2), are as follows:

1) Conservation of Energy:

$$\sum_{i=1}^{I} \sigma_i^* h_i - \sum_{i=1}^{I} \sigma_i h_i = \dot{Q}/\dot{m} \qquad (1)$$

σ_i^* and σ_i are the concentrations of the ith species (mole/gm) at the reactor inlet and exit respectively, h_i is the species enthalpy, $\dot{Q}$ is the rate of heat loss from the system and $\dot{m}$ is the total mass flow rate through the reactor.

2) Conservation of Species:

$$\frac{\dot{m}}{V}(\sigma_i^* - \sigma_i) = \sum_{j=1}^{J} (\alpha'_{ij} - \alpha''_{ij})(R_j - R_{-j}), \; i = 1,2 \ldots I. \qquad (2)$$

α'_{ij} and α''_{ij} are the stoichiometric coefficients for the ith species in the jth reaction as defined by the general reaction for the species S_i:

$$\sum_{i=1}^{I} \alpha'_{ij} S_i + \bar{\alpha}_j M = \sum_{i=1}^{I} \alpha''_{ij} S_i + \bar{\alpha}_j M, \quad j = 1,2 \ldots J. \qquad (3)$$

Here $\bar{\alpha}_j$ denotes the stoichiometric coefficient of the third body M and J is the total number of reactions considered in the system. R_j and R_{-j}, the forward and backward reaction rates of the jth reaction, are

$$R_j = k_j (\rho\sigma_m)^{\bar{\alpha}_j} \prod_{i=1}^{I} (\rho\sigma_i)^{\alpha'_{ij}}, \; j = 1,2,\ldots,J \qquad (4)$$

and

$$R_{-j} = k_{-j} (\rho\sigma_m)^{\bar{\alpha}_j} \prod_{i=1}^{I} (\rho\sigma_i)^{\alpha''_{ij}}, \; j = 1,2,\ldots,J; \qquad (5)$$

k_j is an Arrhenius rate coefficient of the form

$$k_j = 10^{B_j} T^{N_j} \exp(-E_j/RT).$$

Reverse rate coefficients were assumed to be related to the forward rate constants through the equilibrium constant K_{c_j}

$$k_{-j} = \frac{k_j}{K_{c_j}} \qquad (6)$$

will be different.

Bearing in mind the above limitations, it was decided as a comparative measure to use both an infinitely fast rate and the global reaction rate given by equation (8) for propane oxidation.

The reaction mechanisms and rate constants for H_2 and CO combustion are relatively well known. These are shown in Table I. The mechanism for H_2 consists of four propagation reactions and four recombination/dissociation reactions. In addition, three reactions involving HO_2 (reactions 9-11) were found to be important. Three reactions (12-14) were used to describe CO kinetics.

Fristrom and Sawyer (1) proposed a model for flame inhibition based largely on the experimental results of Wilson, et al. (11) The major assumption of their inhibition model was that the shuffling reaction

$$H + HX \rightleftharpoons H_2 + X$$

is in competition with the branching reaction

$$H + O_2 \rightleftharpoons OH + O \ .$$

For the case of $H_2 + O_2$ flames, inhibition would then occur through the replacement of the chain initiator (H atom) with an inhibitor atom (Br or Cl). A second radical-molecule shuffling reaction

$$HX + X \rightleftharpoons X_2 + H$$

was included since it is competitive with the abstraction reaction. Although dissociation reactions are likely to be slow in the temperature range 1000°K to 2000°K, it was felt the reverse recombination reactions might be important. Thus

$$HX + M \rightleftharpoons H + X + M$$

and

$$X_2 + M \rightleftharpoons X + X + M$$

were included to assess their realtive importance in the inhibition mechanism. The kinetic data associated with the four halogen reactions is shown in Table II. The data are believed to represent the best available data for these reactions and their selection was based on a survey of halogen kinetics by Brown (19).

Experimental Results

The first step in determining the effect of inhibitor addition on the stability characteristics of the ORJ was to obtain extinction curves. As discussed previously, extinction

Table I

C-H-O-N Chemical Kinetic Reaction Mechanism

	Reaction		
1.	$OH + H_2 \rightleftharpoons H + H_2O$	$10^{13.342}exp(-5140/RT)$	6
2.	$H + O_2 \rightleftharpoons OH + O$	$10^{14.342}exp(-16780/RT)$	6
3.	$O + H_2O \rightleftharpoons OH + OH$	$10^{13.833}exp(-18350/RT)$	6
4.	$O + H_2 \rightleftharpoons OH + H$	$10^{10.255}T'exp(-8900/RT)$	6
5.	$H + H + M \rightleftharpoons H_2 + M$	$10^{17.806}T^{-1}$	6
6.	$O + O + M \rightleftharpoons O_2 + M$	$10^{18.139}T^{-1}exp(338/RT)$	7
7.	$O + H + M \rightleftharpoons OH + M$	$10^{15.86}$	7
8.	$H + OH + M \rightleftharpoons H_2O + M$	$10^{16.788}exp(500/RT)$	8
9.	$H + HO_2 \rightleftharpoons OH + OH$	$10^{14.398}exp(-1890/RT)$	6
10.	$H + O_2 + M \rightleftharpoons HO_2 + M$	$10^{15.176}exp(1000/RT)$	6
11.	$H_2 + O_2 \rightleftharpoons H + HO_2$	$10^{13.740}exp(-57790/RT)$	6
12.	$CO + OH \rightleftharpoons CO_2 + H$	$10^{11.748}exp(-1080/RT)$	9
13.	$CO + O + M \rightleftharpoons CO_2 + M$	$10^{14.00}exp(-2500/RT)$	9
14.	$CO_2 + O \rightleftharpoons CO + O_2$	$10^{13.279}exp(-54130/RT)$	9
15.	$N + NO \rightleftharpoons N_2 + O$	$10^{13.204}$	10
16.	$N + O_2 \rightleftharpoons O + NO$	$10^{9.806}T'exp(-6256/RT)$	10

*units - k: $(cm^3 mole^{-1})^{n-1} sec^{-1}$ where n is reaction order;
R: cal $mole^{-1} °K^{-1}$; T: °K .

Table II

HBr - HCl Reaction Mechanism

	Reaction		
17.	$H_2 + Cl \rightleftharpoons HCl + H$	$10^{13.532}\exp(-4480/RT)$	12
18.	$H + Cl_2 \rightleftharpoons HCl + Cl$	$10^{14.57}\exp(-1800/RT)$	13
19.	$HCl + M \rightleftharpoons H + Cl + M$	$10^{13.623}\exp(-81000/RT)$	14
20.	$Cl_2 + M \rightleftharpoons Cl + Cl + M$	$10^{13.94}\exp(-48494/RT)$	15
21.	$H_2 + Br \rightleftharpoons HBr + H$	$10^{14.43}\exp(-19700/RT)$	16
22.	$H + Br_2 \rightleftharpoons HBr + Br$	$10^{14.833}\exp(-2200/RT)$	16
23.	$HBr + M \rightleftharpoons H + Br + M$	$10^{21.78}T^{-2}\exp(-88000/RT)$	17
24.	$Br_2 + M \rightleftharpoons Br + Br + M$	$10^{11.33}T^{0.5}\exp(-31300/RT)$	18

*units - k: $cm^3 mole^{-1} sec^{-1}$;

R: cal $mole^{-1} {}^\circ K^{-1}$; T: °K

points for a given inhibitor concentration were determined by maintaining a constant flow velocity and decreasing the fuel flow until stable combustion could not be maintained and extinction occurred. These measurements yielded extinction curves which are plots of main stream velocity at blowout as a function of main stream equivalence ratio. A curve was obtained for each concentration of a particular inhibitor. Typical extinction curves for an ORJ system are parabolic in shape with the maximum occurring at an equivalence ratio of approximately unity.

Figure 3 shows extinction curves for the various inhibitors. Fuel supply limitations made it possible to obtain only the lean portion of these curves. The region to the right of the curves represents stable combustion, and the region to the left represents a region in which stable combustion cannot be maintained. The effect of inhibitor addition is to shift the extinction curve to higher equivalence ratios (relative to the curve for an air jet). This corresponds to a reduction in stability limits for a given main stream velocity. Table III shows the reduction in the lean stability limit obtained for the inhibited mixtures of Figure 3. This reduction is indicated by the percent increase in equivalence ratio for the inhibited mixture relative to the uninhibited case.

Table III

Inhibitor	Mole Percent in Jet Stream	Percent Increase in Main Stream Equivalence Ratio
N_2	3	0.45
HCl	3	1.40
CH_3Br	1.5	6.80
HBr	1.5	6.80
CH_3Br	3	8.30
HBr	3	9.10

N_2 is least effective and HCl is slightly more effective while CH_3Br and HBr are the most effective.

Figure 3 gives a qualitative idea of the effect of inhibitor addition on the opposed reacting jet, however, it is important to quantify these results and have a measure of inhibitor effectiveness. An inhibition parameter which is independent of the experimental apparatus has been proposed by Fristrom and Sawyer (1) (see Appendix 1) as a quantitative means of evaluating the effectiveness of inhibitors. It is defined as

$$\phi_v = \frac{(O_2)}{(I)} \frac{\delta v}{v} \qquad (9)$$

where $\frac{\delta v}{v}$ is the fractional change in propagation velocity per unit inhibitor normalized to the oxygen consumption. If attention is confined to the effect of added inhibitor on the amount of additional fuel necessary to sustain a stable flame at a constant extinction velocity, then the ratio of inhibition parameters for two different inhibitors A and B becomes

$$\frac{\phi_{v_A}}{\phi_{v_B}} = \frac{\frac{(O_2)_A}{(A)}}{\frac{(O_2)_B}{(B)}} \qquad (10a)$$

since

$$\left.\frac{\delta v}{v}\right|_A = \left.\frac{\delta v}{v}\right|_B \qquad (10b)$$

for constant extinction velocity. In a premixed lean system where nearly all the fuel is consumed the O_2 consumption approximately equals the initial fuel concentration, and equation (10a) can be approximated as:

$$\frac{\phi_{v_A}}{\phi_{v_B}} \simeq \frac{\frac{(\text{fuel})_A}{(A)}}{\frac{(\text{fuel})_B}{(B)}} \quad . \qquad (11)$$

To use the relationship expressed by equation (11) for the ORJ, the concentration terms appearing in equation (11) must be related to conditions existing in the ORJ. Past investigations of the ORJ have indicated that flame stabilization is determined by conditions in a small critical zone located just behind the nose of the flame in the stagnation region (20). Turbulent mixing in this region is sufficiently intense that the zone acts as a homogeneous reactor. Thus the concentrations of fuel and inhibitor to be used in equation (11) are those that enter this zone.

Based on sodium tracer studies, Schaeffer (20) concluded the total mass flow into the critical zone could be divided into three parts: the primary and jet stream contributions, and the products of combustion entrained by the jet stream. No fuel exists in the jet stream, and under fuel lean conditions very little unburned fuel should exist in the entrained products of combustion. Thus the only fuel entering the critical zone is due to the main stream contribution. Bellamy (21) derived an empirical correlation for the ratio of main stream to jet stream contribution to the critical zone for an ORJ based on the fact that the equivalence ratio in the critical zone at the peak in extinction curves is unity. This has been verified by spectroscopic measurements (22). The resulting equation is

$$\frac{\dot{m}_M}{\dot{m}_j} = 9.85\left(\frac{M_J}{M_M}\right)^{0.37} \qquad (12)$$

where M_M and M_J are the rates of momentum transport (gm cm /sec^2) at the main stream inlet and jet stream exit respectively.

In the present model it was assumed that all of the jet stream enters the critical zone. This assumption has been used quite successfully in the past to model the extinction characteristics of an opposed reacting jet (20), (21). Bellamy further assumed that the contribution of the entrained combustion products was of the same order as the jet stream. Since the entrained products contain very little unburned fuel (fuel lean system) and no inhibitor, the net effect of the entrained gases is to act as a diluent in the critical zone. The amount of jet entrained gases depends primarily on the jet velocity, which is maintained constant in the present experiment. Thus the dilution effect would be the same for all extinction conditions. Since interest is in the relative concentrations existing in the critical zone at extinction, uncertainties in the entrained products contribution to the critical zone should have a negligible effect on the relative inhibition parameters calculated from equation (11). Here assuming that the jet stream and entrained combustion products contributions are equal, propane and inhibitor concentrations in the critical zone were calculated. The results are shown in Figure 4 where normalized propane mole fraction in the critical zone at extinction is plotted against inhibitor mole fraction for a main stream velocity of 7.6 m/s. Propane mole fraction is normalized by the fuel mole fraction entering the critical zone at extinction for an all air jet.

Figure 4 shows that HBr and CH_3Br are the most effective inhibitors, with the effectiveness of CH_3Br falling slightly below that of HBr at mole fractions greater than 0.3%. This decrease in effectiveness is attributed to the added hydrocarbon fuel associated with CH_3Br. Next in decreasing order of effectiveness is HCl, followed by the two physical inhibitors N_2 and Ar. The lower specific heat for Ar ($C_{pAr}/C_{pN_2} \simeq 0.7$) would account for the difference between N_2 and Ar effectiveness.

To quantify the results shown in Figure 4, equation (11) was used to calculate relative inhibition parameters using N_2 as a reference case. This was accomplished by computing the slopes of the curves in Figure 4 relative to the slope of the N_2 curve. Only the linear part of the curve (mole fraction less than 0.003 was used for CH_3Br. A standard linear least squares fitting

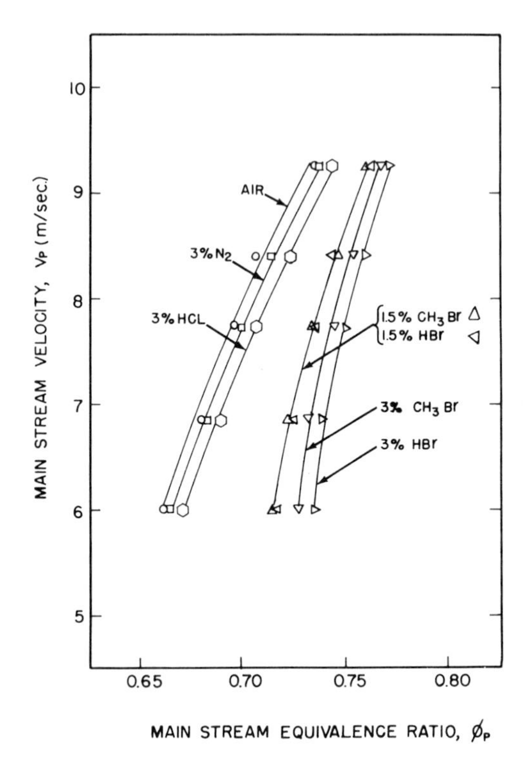

Figure 3. Effect of inhibitor addition on opposed reacting jet stability limits. Inlet temperature = 300°K; jet velocity = 96 m/s.

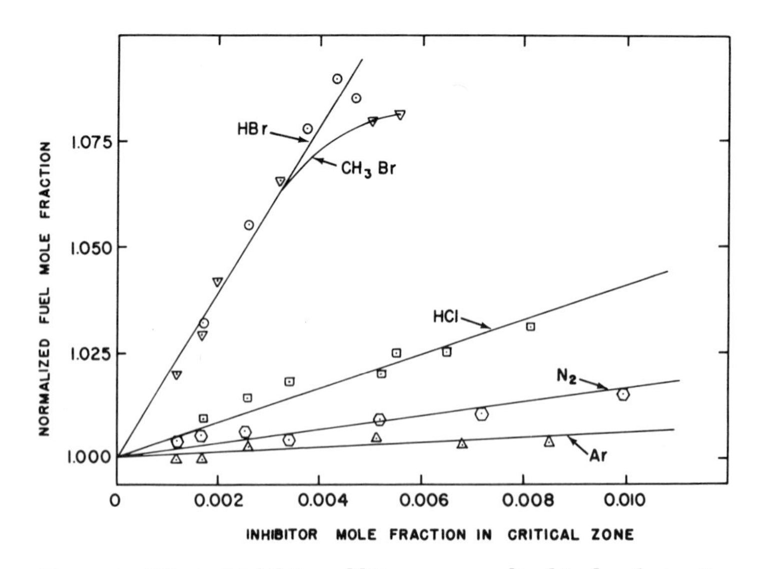

Figure 4. Effect of inhibitor addition on normalized fuel mole fraction in critical zone at extinction. Main stream velocity = 7.6 m/s; inlet temperature = 300°K; jet velocity = 96 m/s.

technique was used to compute the slopes, and the standard deviations were used to place error limits on the relative inhibition parameters. The results are shown in Table IV where $\phi_{v(max)}$ and $\phi_{v(min)}$ indicate the maximum and minimum values, respectively, of the relative inhibition parameter. These values were computed using the average slope and adding or subtracting the standard deviation. Note that the value shown in parenthesis for CH_3Br is the experimentally determined relative inhibition parameter reported in (1).

Table IV

Inhibitor	$\phi_{v(max)}$	$\bar{\phi}_v$	$\phi_{v(min)}$	
N_2	1.00	1.00	1.00	
Ar	0.55	0.37	0.22	
HCl	2.90	2.44	2.08	
HBr	15.0	12.95	11.3	
CH_3Br	15.0	12.95	11.3	(8.83)

Computational Results

A useful comparison between the ORJ and the well stirred reactor can be obtained through curves similar to those of Figure 4. The slopes of these curves were used to determine relative inhibitor effectiveness. The governing differential equations for the well stirred reactor were solved with temperature as the dependent variable while mass flow rate $(\dot{m}/V)$ was independently increased until extinction was reached. A series of curves were obtained for fuel mole fraction at extinction as a function of $(\dot{m}/V)$ for various inhibitor concentrations. The experimental fuel mole fraction at extinction for an all air jet was used as a reference point. An equivalent $(\dot{m}/V)$ was found for the well stirred reactor which would give extinction at this fuel mole fraction. Fuel mole fraction at extinction was then found as a function of mole percentage inhibitor added for the equivalent $(\dot{m}/V)$.

Figure 5 shows the results for the physical inhibitors Ar and N_2. The computed curves fall below the experimental curves. Two curves were computed for N_2, one using mechanism (1) and the other using mechanism (2) for propane oxidation. The curve using the Edelman-Fortune global mechanism (mechanism 2) are somewhat closer to the experimental value; however the differences between

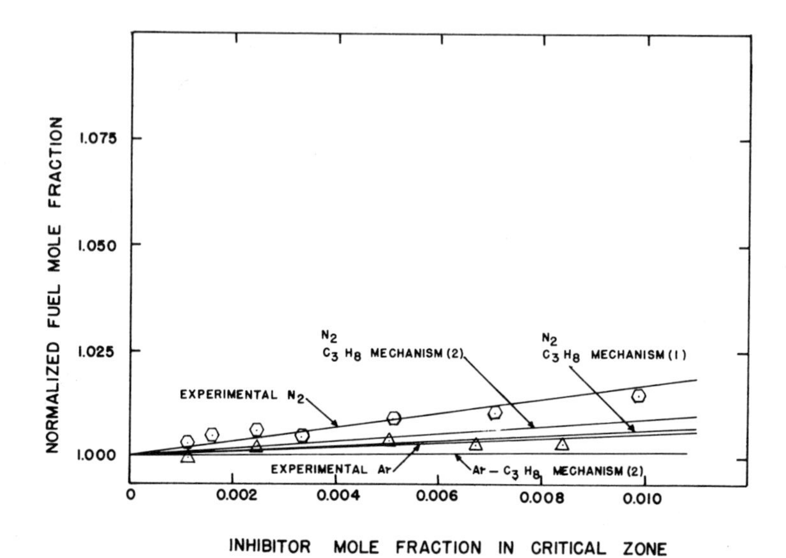

Figure 5. Comparison of experimental values of normalized fuel mole fraction in critical zone at extinction and theoretical values predicted for a well stirred reactor for physical inhibitors N_2 and Ar. Main stream velocity = 7.6 m/s; inlet temperature = 300°K; jet velocity = 96 m/s. Mechanism (1) = infinite propane oxidation rate; mechanism (2) = finite oxidation rate (5).

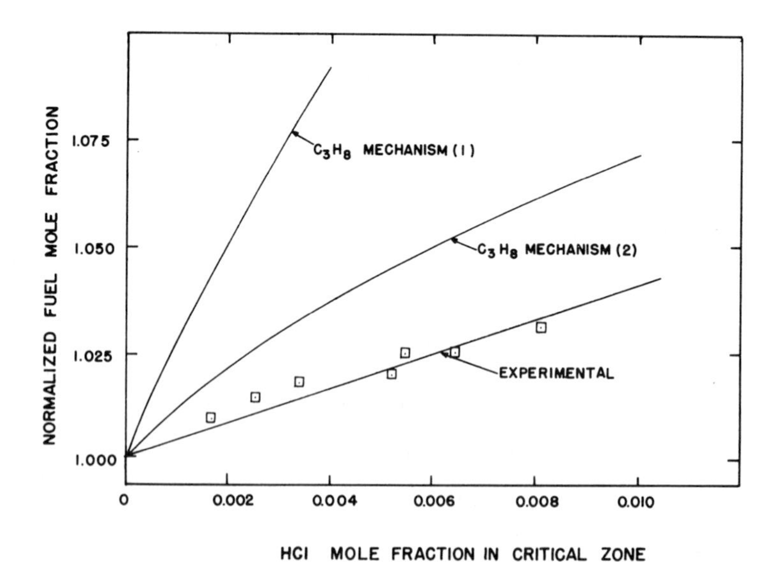

Figure 6. Comparison of experimental values of normalized fuel mole fraction in critical zone at extinction and theoretical values predicted for a well stirred reactor for HCl addition. Main stream velocity = 7.6 m/s; inlet temperature = 300°K; jet velocity = 96 m/s. Mechanism (1) = infinite propane oxidation rate; mechanism (2) = finite propane oxidation rate (5).

the two computed curves isn't significant enough to draw any conclusions about the suitability of one mechanism over the other. The relative inhibition parameter for Ar (using mechanism 2) was computed as 0.21. This falls slightly outside the range of experimental values where the lower limit was $\phi v(min) = 0.22$.

Figure 6 shows the computed curves for HCl inhibition using the rate coefficients of Table II and mechanisms (1) and (2) for propane oxidation. The agreement between the computed and experimental curves is poor, although when relative slopes are compared the agreement is somewhat improved. The lack of agreement between theory and experiment will be discussed in the next section.

In an effort to determine the relative importance of the four chlorine-hydrogen reactions on predicted inhibitor effectiveness, the well stirred reactor calculations were repeated for several variations of the chlorine-hydrogen kinetic data using mechanism (2) for propane-oxidation. The rate coefficients were varied by factors assumed to represent extreme bounds on their values. The results are shown in Figure 7. The lowest curve is the case where the dissociation-recombination reactions were eliminated from the mechanism. Extinction limits (reflected by changes in slopes) were found to be most sensitive to variation in the rate coefficients for the reaction $H + Cl + M \rightarrow HCl + M$.

Well stirred reactor calculations were made for the case of HBr inhibition using the rate coefficients of Table II and mechanisms (1) and (2) for propane oxidation. Figure 8 shows that the computed results are significantly lower than the experimental values. Figure 9 shows the effects of rate coefficient variation on the computed curves. As with HCl, inhibitor effectiveness was found to be most sensitive to the $H + X + M \rightarrow HX + M$ recombination rate coefficient. Increasing the rate coefficients by an order of magnitude resulted in some improvement but the slope was still lower than the experimental value.

Discussion

The large discrepancy between computed and experimental results can most likely be attributed to the kinetic data used in the calculations rather than errors in the interpretation of the experimental results resulting from uncertainties in the model of the fluid mechanics of the experimental system.

The ORJ was assumed to have a homogeneous reaction zone that could be approximated as a well stirred reactor. It was assumed that inhibitor and fuel concentrations in the zone could be computed from jet and mainstream inlet conditions. These approximations have been used previously (20,21) to model the extinction characteristics of an ORJ.

An additional argument against the large discrepancy being attributed to the assumed fluid mechanics is that relative

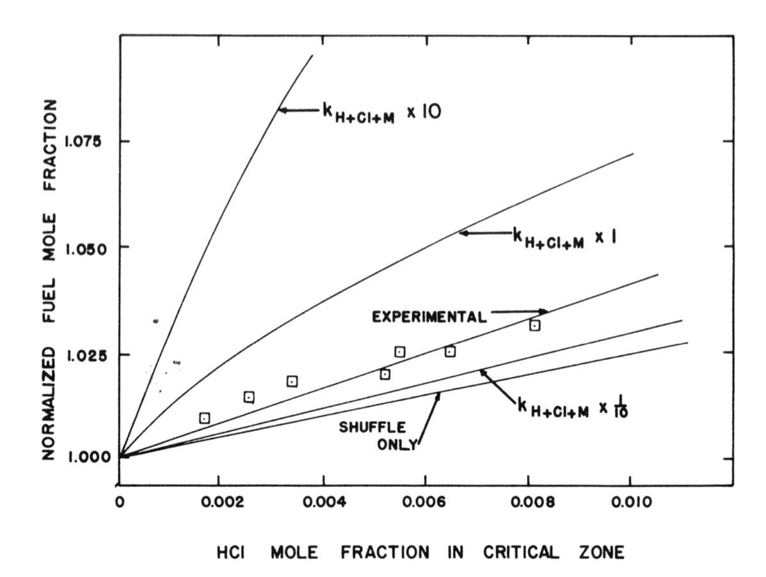

Figure 7. Comparison of experimental values of normalized fuel mole fraction in critical zone at extinction and theoretical values predicted for a well stirred reactor for HCl addition. Main stream velocity = 7.7 m/s; inlet temperature = 300°K; jet velocity = 96 m/s. Mechanism (2) is used for propane oxidation.

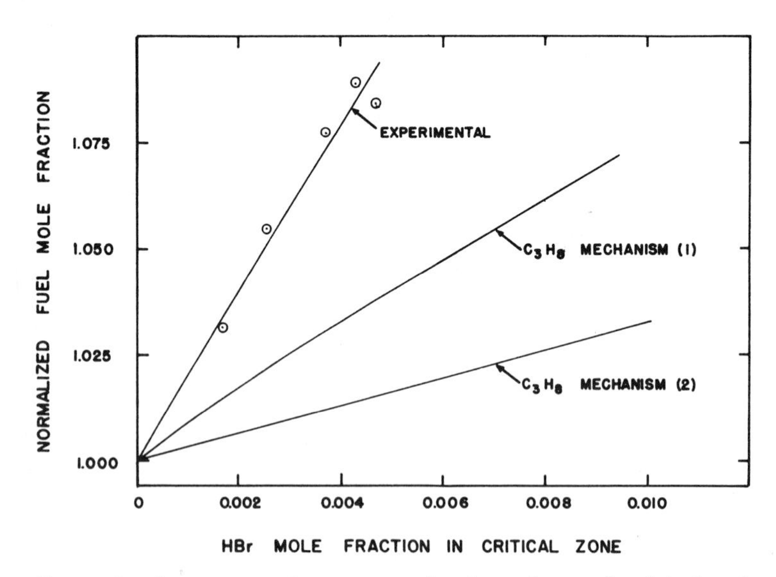

Figure 8. Comparison of experimental values of normalized fuel mole fraction in critical zone at extinction and theoretical values predicted for a well stirred reactor for HBr addition. Main stream velocity = 7.6 m/s; inlet temperature = 300°K; jet velocity = 96 m/s. Mechanism (1) = infinite propane oxidation rate (5).

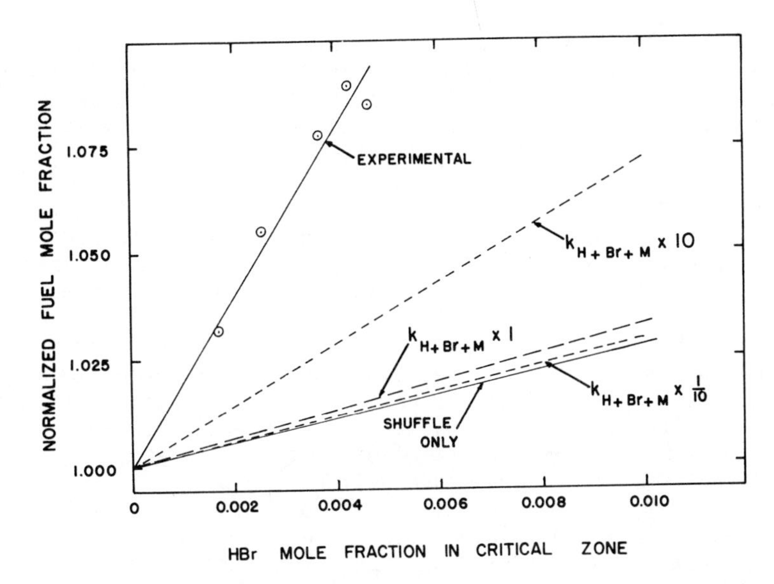

Figure 9. Comparison of experimental values of normalized fuel mole fraction in critical zone at extinction and theoretical values predicted for a well stirred reactor for HBr addition. Main stream velocity = 7.6 m/s; inlet temperature = 300°K; jet velocity = 96 m/s. Mechanism (2) is used for propane oxidation.

inhibition parameters were computed and some of the errors are likely to cancel.

The kinetic mechanism can be separated into three parts 1) propane oxidation, 2) the $CO + H_2$ mechanism and 3) the halogen-hydrogen mechanism. The $CO + H_2$ mechanism is quite well known, and the effects of possible errors in the halogen-hydrogen kinetics can be determined; however, the propane oxidation mechanism is the least satisfactory of the three. Calculations were performed to illustrate the effects of the propane oxidation mechanism by comparing mechanisms (1) and (2). The two mechanisms gave nearly identical results for the physical inhibitors but very different results for the chemical inhibitors. These results are not at all surprising since the chemical inhibitors are believed to react with radical intermediates and neither propane mechanism represents the generation of radical species adequately. Of the two mechanisms (1) and (2), the latter would intuitively appear better. The difficulty in absolutely assessing the relative merits of (1) and (2) can be attributed to lack of understanding of the actual mechanism. Presently it is not possible to say whether or not these mechanisms can be used for bounds on actual behavior. Work is in progress which is investigating the propane oxidation, and the fluid mechanic approximations.

With the assumption that mechanism (2) was more nearly correct than mechanism (1), calculations were performed to assess the relative importance of the various hydrogen-halogen reactions. Inhibitor effectiveness was particularly dependent on the rate of the H + X + M recombination reaction rate coefficient. The rate coefficients for recombination reactions for both HCl and HBr were determined by extrapolating dissociation data and calculating the recombination rate coefficient via the equilibrium constant. This is admittedly a poor way to use high temperature kinetic data. The extrapolation, the quoted uncertainties in the actual high temperaure results, and the quoted lack of knowledge about third body efficiencies are the basis of the large uncertainty factors (10 and 1/10) used here. These rates have not been measured at low temperatures and are not amenable to theoretical calculation because they are prototypes of the difficult case of light and heavy atom recombination.(23)

As a first step toward determining the relative importance of the hydrogen-halogen reactions, both dissociation reactions were removed for the case of 1% inhibitor at an equivalence ratio of 0.625. In the HCl system, this resulted in a significant decrease in predicted inhibitor effectiveness. To determine the relative importance of the two shuffle reactions (17 and 18) forward and backward rates were varied by factors of 2.5 and 1/2.5. Increasing the rate of $H_2 + Cl \rightleftharpoons H + HCl$ resulted in a very small increase in inhibitor effectiveness. This small effect is reasonable, since this reaction is very fast in both directions compared to other reactions and is essentially in a state of partial equilibrium. Similarly, decreasing this rate resulted in a small decrease in inhibitor effectiveness.

Variation of the rate of $H + Cl_2 \rightleftharpoons HCl + Cl$ had no effect on the results. Comparison of the forward and reverse rates for this reaction at 1200°K showed that the reverse reaction rate is approximately 4 orders of magnitude slower than the forward rate. The forward rate is also about 4 orders of magnitude slower than $H_2 + Cl \rightleftharpoons H + HCl$ due to the very low concentrations of Cl_2 present. Thus the important H atom abstraction reaction is $H + HCl \rightleftharpoons H_2 + Cl$.

Similar observations were made in the HBr system with only the two shuffle reactions (21 and 22) present. Reaction 21 ($H_2 + Br \rightleftharpoons H + HBr$) is very fast in both directions in comparison with the other reactions and therefore variations by factors of 2.5 and 1/2.5 in its rate had little effect on inhibitor effectiveness.

Addition of the two dissociation/recombination reactions:

$$HX + M \rightleftharpoons H + X + M$$

$$X_2 + M \rightleftharpoons X + X + M$$

resulted in a substantial increase in inhibitor effectiveness. An effort was made to determine which reaction was responsible for this increase.

In the HCl system, variation in the rate of reaction 20 ($Cl_2 + M \rightleftharpoons Cl + Cl + M$) by factors of 1/10 and 10 had no effect on inhibitor effectiveness. Examination of the rates of this reaction at 1200°K for the case of 1% HCl and $\phi = 0.625$ revealed a rate of approximately 10^{-13}mole/cm^{3}sec for the recombination reaction near reactor blowout. This compares with rates for HCl recombination and the shuffle reaction $HCl + H \rightleftharpoons H_2 + Cl$ of approximately 10^{-3} and 10^{-2} moles/cm^{3}sec, respectively. Typical species concentrations as a function of reactor residence time are shown in Figures 10 and 11. Figure 10 shows the effect of inhibitor addition is indeed a decrease in active radical concentrations. Examination of Figure 11 reveals that it is not until residence times significantly greater than those associated with near extinction conditions are reached that a buildup in Cl_2 occurs and reaction 20 becomes important.

Slightly different results were found in the HBr system. Decreasing the rate coefficient for reaction 24 ($Br_2 + M \rightleftharpoons Br + Br + M$) up to a factor of 1/10 had no effect. Once again, this can be attributed to the relatively slow dissociation and recombination rates at 1200°K (approximately 10^{-10}mole/cm^{3}sec and 10^{-5}mole/cm^{3}sec, respectively, as compared with the HBr recombination reaction 23 ($R \simeq 10^{-4}$mole/cm^{3}sec) and reaction 21 ($R \simeq 10^{-3}$moles/cm sec). However, increasing the rate coefficient for Br atom recombination by a factor of 10 increased inhibitor effectiveness. This is not unexpected since an order of

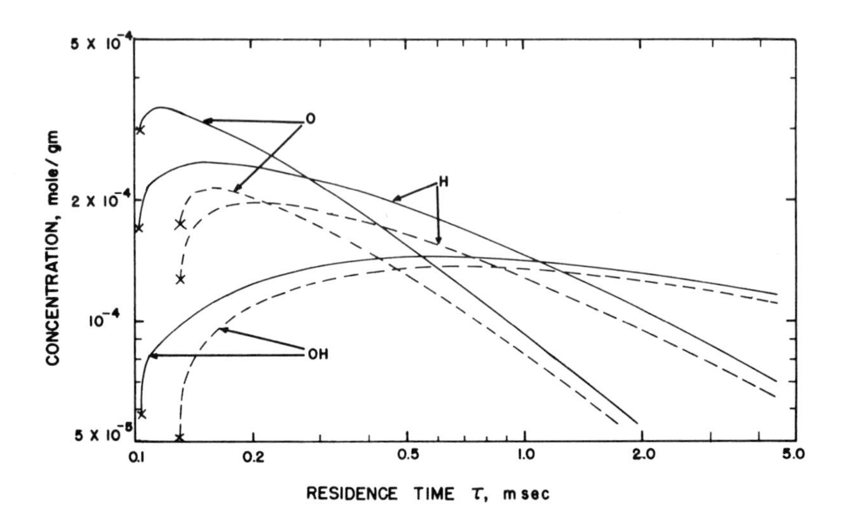

Figure 10. Radical concentration as a function of residence time in an adiabatic well stirred reactor. Equivalence ratio = 0.625; inlet temperature = 300°K; pressure = 1 atm. Solid line for uninhibited case, dashed line for 1% HCl.

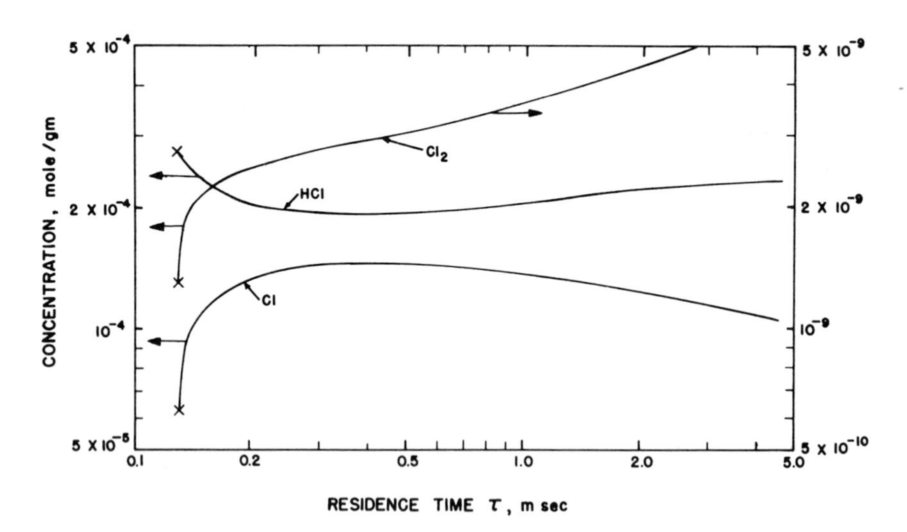

Figure 11. Inhibitor species concentration as a function of residence time in an adiabatic well stirred reactor for 1% HCl addition. Equivalence ratio = 0.625; inlet temperature = 300°K; pressure = 1 atm.

magnitude increase in this rate made it comparable with reaction 23. Two effects then become important. First, the concentration of Br atom is decreased which tends to drive reaction 21 (H_2 + Br = H + HBr) in a direction favoring H atom abstraction, and second, the concentration of Br_2 is increased. Increasing the rate of Br atom recombination by a factor of 8 results in an order of magnitude increase in Br_2 concentration. This makes the rate of reaction 22 (H + $Br_2 \rightleftharpoons$ HBr + Br) comparable with reaction 21 and provides a second route for H atom abstraction.

An examination of Figure 7 for HCl shows the sensitivity of inhibitor effectiveness to variations in the rate of reaction 19. It was felt that the reverse of 19 (e.g. HCl recombination) rather than the dissociation reaction was important. To investigate this the forward rate coefficient was set equal to zero while the reverse reaction rate was varied by factors of 10 and 1/10. No change was found in the dependence shown in Figure 7. Further examination of the rates for the recombination and dissociation reactions revealed that the dissociation reaction was a factor of 10^{11} slower than the recombination rate for the case of 1% HCl and ϕ = 0.625. It was thus concluded that HCl recombination was the important reaction in determining inhibitor effectiveness.

Figure 9 for HBr shows a similar dependence of HBr effectiveness on the rate of the HBr dissociation/recombination reaction. Setting the dissociation rate equal to zero verified that this dependence was indeed due to H + Br + M recombination and not HBr dissociation. Typical temperatures encountered in the present investigation were in the range of 1100°K to 1400°K. If the mechanism used in the present investigation is correct and the H + X + M recombination reaction is as important as it seems, then the uncertainties associated with these rate coefficients at the combustion temperatures of interest will have to be resolved before accurate modeling of flame inhibition is possible.

Conclusions

1. Using Fristrom and Sawyer's inhibition parameter as an index, the inhibitors considered were ranked experimentally as follows

$$HBr > CH_3Br > HCl > N_2 > Ar$$

2. Due to uncertainties in the reaction mechanism considered and rate coefficient data, agreement between experimental and theoretical values of the inhibitor effectiveness was unsatisfactory.

3. In the mechanism considered

$$H_2 + X + HX + H \quad (1)$$

$$H + X_2 = HX + X \quad (2)$$

$$HX + M = H + X + M \quad (3)$$

$$X_2 + M = X + X + M \quad (4)$$

the reverse of reaction (3) was found to be most important in determining inhibitor effectiveness. Reaction (1) was in a state of partial equilibrium, and reactions (2) and (4) were found to have little effect on inhibition.

4. Since the $H + X + M \rightarrow HX + M$ reaction it is important in determining inhibitor effectiveness, uncertainties in this rate coefficient at combustion temperatures must be removed.

Acknowledgements

Helpful discussions with Dr. C.T. Bowman are gratefully acknowledged.

Literature Cited

1. Fristrom, R. and Sawyer,R., AGARD Conference Proceedings No. 84 on Aircraft Fuels, Lubricants, and Fire Safety, AGARD-CP-84-71, Section 12, North Atlantic Treaty Organization (1971).
2. Jones, A. and Prothero, A., Comb. and Flame (1968) 12, 457.
3. Pratt, D.T. and Bowman, B.R., to be published, (1973), Engineering Extension Service, Washington State University.
4. Marteney, P.J., Comb. Sci. Tech. (1970) 1, 461.
5. Edelman, R. and Fortune, O., AIAA Paper 69-86, (1969) Seventh Aerospace Sciences Meeting.
6. Baulch, D.L., Drysdale, D.D., Horne, D.G., and Lloyd, A.C., "Evaluated Kinetic Data for High Temperature Reactions," Butterworths, London (1972).
7. Garvin, D. and Hampson, R.F., Chemical Kinetics Data Survey VII. Tables of Rate and Photochemical Data for Modeling the Stratosphere (Revised), NBSIR 74-430 (1974).
8. Malte, P.C. and Pratt, D.T., Paper No. 73-37, (1973), Fall Meeting, Western State Section, The Combustion Institute, El Segundo.
9. Baulch, D.L., Drysdale, D.D. and Lloyd, A.C., "Critical Evaluation of Rate Data for Homogeneous Gas Phase Reactions of Interest in High Temperature Systems," Nos. 1 and 2,

Dept. of Phys. Chem., The University of Leeds, England (1968).
10. Bowman, C.T., United Aircraft Laboratories, to be published.
11. Wilson, W.E., O'Donovan, J.T. and Fristrom, R.M., "Twelfth Symposium (International) on Combustion," 929, The Combustion Institute, Pittsburgh (1969).
12. Clyne, M.A.A. and Walker, R.F., JCS Faraday Trans. I, (1973) 69, 1547.
13. Dodonov, A.F., Lavrovskaya, G.K., and Morozov, I.I., Kinetika: Kataliz,(1970) 11, 821.
14. Seery, D.J., and Bowman, C.T., J. Chem. Phys.,(1968) 49, 1271.
15. Lloyd, A.C., Int. J. Chem. Kinet.,(1971) 3, 39.
16. Fettis, G.C. and Knox, J.H., "Progress in Reaction Kinetics," (Ed. G. Porter), Pergamon Press, London, 2, 1 (1964).
17. Cohen, N. Giedt, R.R., and Jacobs, T.A., Int. J. Chem. Kinet., (1973) 5, 425.
18. Warshay, M., J. Chem. Phys., (1971), 54, 4060.
19. Brown, N.J., published in this volume.
20. Schaeffer, A.B., Ph.D. Dissertation, (1957), Northwestern University, Evanston, Illinois.
21. Bellamy, L., Ph.D. Dissertation, (1966), Tulane University, New Orleans, Louisiana.
22. Fuhs, A.E., ARS Journal, (1960) 30, (3), 238.
23. Shui, V.H., Appleton, J.P. and Keck, J.C.,"Thirteenth Symposium (International) on Combustion," 21, The Combustion Institute, Pittsburgh, (1971).

Appendix: Inhibition Parameter

A simplified flame model was proposed by Fristrom and Sawyer (1) and employed to aid in understanding the mechanism of inhibition in premixed flames. In this model the flame is divided into two regions. In the first recombination is considered dominant; in the section chain branching is taken to be the most important process. The recombination region precedes the branching region and the two regions are divided by the temperature at which the branching and recombination reactions proceed at equal rates. The reaction rate in the branching region is controlled by the limiting slow reaction, $H + O_2 \rightarrow OH + O$. Inhibition is attribution to bimolecular "deactivation", $H + HX \rightarrow H_2 + X$. Diffusion transport dominates convection transport.

This model predicts that the fractional reduction in flame velocity brought about by the addition of inhibitor should be proportional to concentration of inhibitor normalized by the concentration of consumed oxygen. It then follows that an appropriate inhibition "effectiveness" parameter should be

$$\phi_v = \frac{(O_2)}{(I)} \frac{\delta v}{v}$$

Limited experimental data suggest that this inhibition parameter may be extended to other experiments and configurations (in addition to burning velocity depression in premixed flames for which it was derived). Included are experiments involving extinction, quenching distance, pressure limits, and blow off limits, including non-premixed flames. For the original premixed flame model, the inhibition parameter is also equal to the ratio of the rate constants for the bimolecular inhibition and branching reactions.

DISCUSSION

K. L. WRAY: I would like to compliment you on a very interesting paper. Your attempts to understand your data within the framework of a theoretical model is surely the way to go. I do question, however, the need to put uncertainty limits of two orders of magnitude on the three body recombination rates. If this uncertainty comes about by extrapolating the high temperature dissociation rates obtained by different workers using shock tube techniques and applying detailed balancing, then I would like to point out this is not a reasonable approach. I am sure that a theoretical calculation could narrow the uncertainty bounds considerably.

R. W. SCHEFER: We would certainly agree that extrapolation of high temperature dissociation data and use of microscopic reversibility to compute recombination rate coefficients is undesirable. The rather large factors of uncertainty that we have associated with these rate coefficients reflects this. These factors (10 and 1/10) indicate uncertainties in: 1) the high temperature data, 2) relative efficiencies of third bodies, and 3) the extrapolation procedure. The recombination reactions H + X + M are not amenable to theoretical calculations since they represent the difficult case of light and heavy atom recombination and their rate coefficients are required in a temperature regime ($\tilde{=}$ 1200°K) where the uncertainties in the results of a theoretical calculation may be as large as those used here.

8

The Effect of CF_3Br on a $CO-H_2-O_2-Ar$ Diffusion Flame

L. W. HUNTER, C. GRUNFELDER, and R. M. FRISTROM

Applied Physics Laboratory, The Johns Hopkins University, Silver Spring, Md. 20910

This paper reports the effects of added CF_3Br on the temperature, composition and luminosity of two low pressure carbon monoxide-oxygen diffusion flames. Our motive in studying this system is to provide complementary information on the mechanisms of inhibitors in combustion systems which can both deflagrate and detonate. Companion studies made at the Combustion Laboratory of Louvain-la-Neuve on the effect of this inhibitor on the structure of detonation waves have already been published (1,2). The system was chosen because of its relatively well understood chemistry and to meet the requirement of undergoing both deflagration and detonation. Since the two modes differ drastically in propagation velocity and structure it was felt that interesting comparisons should result. The fundamental physical and chemical processes must be the same in deflagration and detonation, and the initial and final states differ principally in the relative importance of kinetic energy. Clearly the relative importance of reactions and transport processes and their interactions must show substantial differences which should be illuminated to some degree in the action of inhibitors. One of the key discoveries of the detonation studies was that additives affected the propagation velocity only slightly, as might be expected from thermodynamic considerations, but made drastic changes in the structure of the reaction zone. In the present study we will discuss some of the details of the effect of CF_3Br on the diffusion flame structure. Unfortunately, our choice of geometry has made a complete quantitative interpretation of our data impractical but a substantial number of interesting qualitative and semiquantitative observations have been made which we feel add to the understanding of this inhibition system. One virtue of studying diffusion flames is that the inhibitor may be added separately to either fuel or oxidizer stream.

The chemistry of carbon monoxide combustion is described in a number of sources (3,4,5). Its principal peculiarity is that hydrogen free CO-O_2 mixtures react at a very slow rate even at high temperatures (>2000K). Thus, it is customary to add small amounts of hydrogen or hydrogen containing compounds to catalyze the combustion. With premixed flames the burning velocity is proportional to the concentration of added hydrogen and independent of the compound of addition, water and hydrogen being equally effective. Thus we have a combustion system which shows both negative and positive catalysis (6).

Experimental Details

The system was an asymmetric opposed jet diffusion flame (Figure 1) at subatmospheric pressures. The flow conditions required for stability of this burner made the geometry closer to that of the classical axial jet diffusion flame in a stagnant atmosphere than that of the balanced opposed jet flat diffusion flame (7,8). The gases used were from commercial cylinders and nominally of high purity (> 99%). The Ar showed a 5% H_2 impurity. Several temperature runs were marred by a small air leak into the oxidizer flow; this had only a minor effect on the temperatures. Since radiation corrections were already uncertain to about 25°K, the results were accepted. Flows were metered using calibrated critical flow orifice flowmeters (9). Burner pressure was monitored with a manometer and held to 0.1 Torr.

Temperatures were measured with bare 0.001" diameter, Ir-Ir, 10% Rh couples using conventional techniques (9). No evidence of catalytic activity was found in the traverses, so uncorrected values (uncertain to about 25°K) were used since no quantitative interpretation of the data is planned.

Axial composition profiles were measured using microprobe sampling techniques combined with mass spectrometric analysis on a Bendix Time of Flight Mass Spectrometer. A continuum flow system with a small effusive leak into the mass spectrometer was employed so that the spectrometer would respond linearly to the partial pressures of the sampled gases. To reduce absorption problems, teflon lines were used; retention

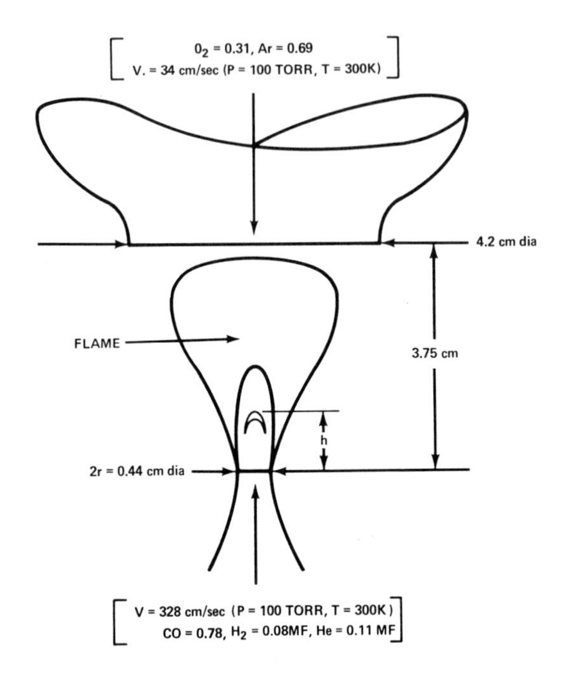

Figure 1. Low pressure diffusion flame apparatus for inhibition studies

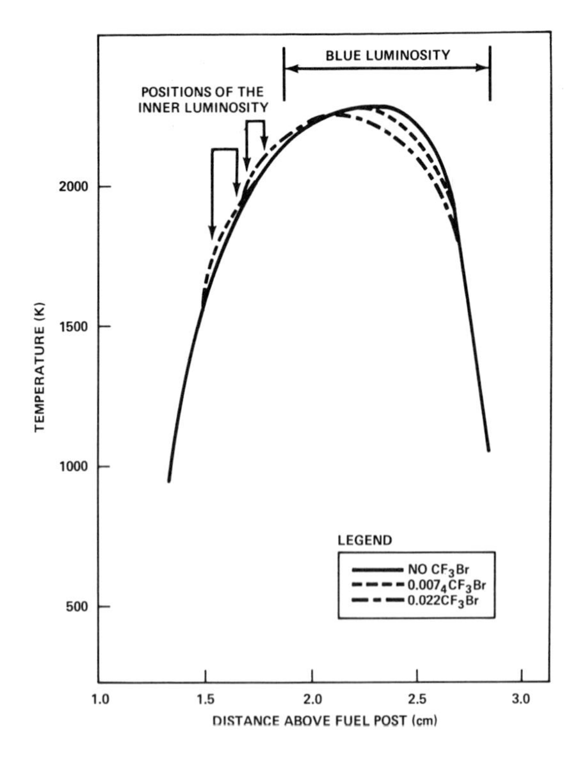

Figure 2. Temperature profiles of several $CO–O_2$ diffusion flames. P = *100 torr.*

times for H_2O, Br_2 and HBr were only a few seconds. The relatively long time interval between sampling and arrival at the spectrometer (ca 0.5 sec) resulted in radical and atomic species being detected as recombination products so that atom concentrations must be inferred from secondary evidence; radical concentrations were low enough so that no substantial bias in stable species measurements was expected. These problems are discussed in Reference 9. The instrument was calibrated directly for major species, but this was not possible for several minor species (COF_2, Br_2 and HF). In these cases, values were estimated from the literature or by analogy with similar compounds. Since this does not affect the relative profiles, from which our qualitative information is derived, the procedure was considered satisfactory.

Absolute positions of the thermocouple bead and the probe were determined by calibrated micrometer drives read to the nearest 0.001 cm. The positions relative to the luminous regions of the flame were determined by a cathetometer which could be read to 0.01 cm.

Data

We present measurements of temperature, composition and positions of luminosity in flames of two pressures (50 and 100 torr) with and without CF_3Br. The CF_3Br was used over a range of concentrations. Most of the measurements were made with CF_3Br injected on the fuel side although a few observations were made adding to the oxidizer flow. Little was done with the latter measurements because large flows were required and the supply of CF_3Br was limited. It should be noted that the large CF_3Br flows required on the oxidizer side were due principally to the flow asymmetry in our system rather than inherent large concentrations required for effects.

The data are presented in Figures 2-5. Figure 2 presents axial profiles comparing the temperatures of 100 torr flames with and without CF_3Br (0.022 MF and 0.0074 MF) added to the fuel jet. Figures 3A and 3B compare the axial composition profiles of the major reactants and products with and without added CF_3Br at 50 torr and 100 torr. Figures 4A and 4B present the profiles of the CF_3Br and product halogen species at 50 torr and 100 torr. Figures 5A and 5B present the luminosity positions associated with CF_3Br destruction both as a function of pressure at constant composition

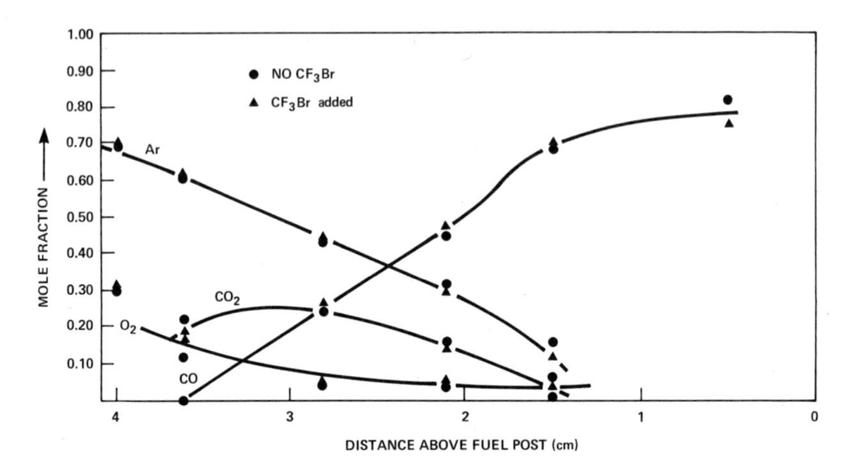

Figure 3A. Major reactant and product species in CO–O_2 diffusion flame with and without added CF_3Br. P = *50 torr.*

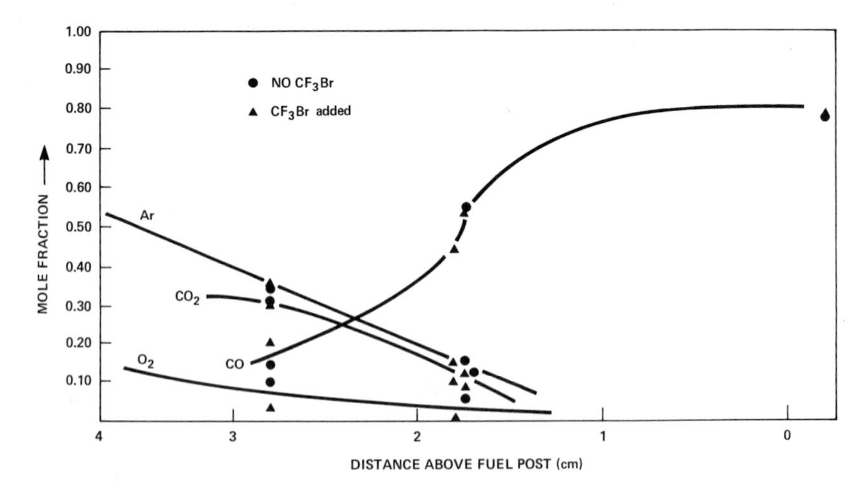

Figure 3B. Major species in CO–O_2 diffusion flames with and without added CF_3Br. P = *100 torr.*

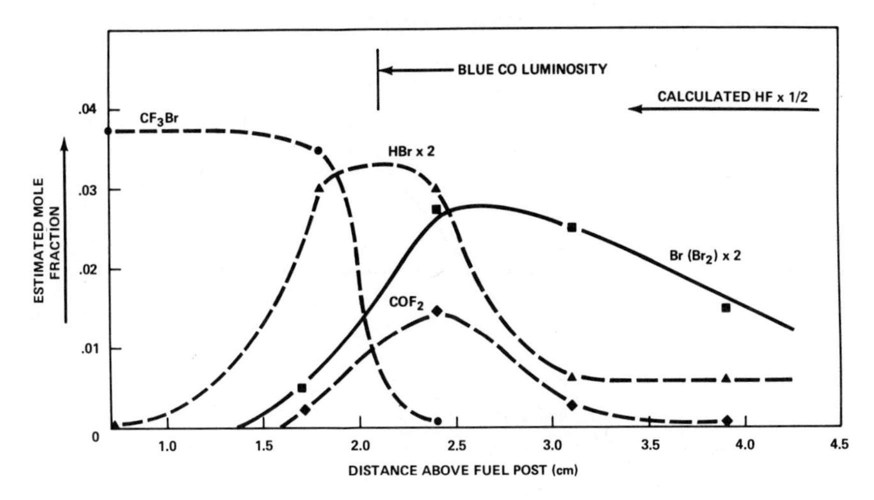

Figure 4A. CF_3Br inhibited CO–O_2 diffusion flame. P = *50 torr.*

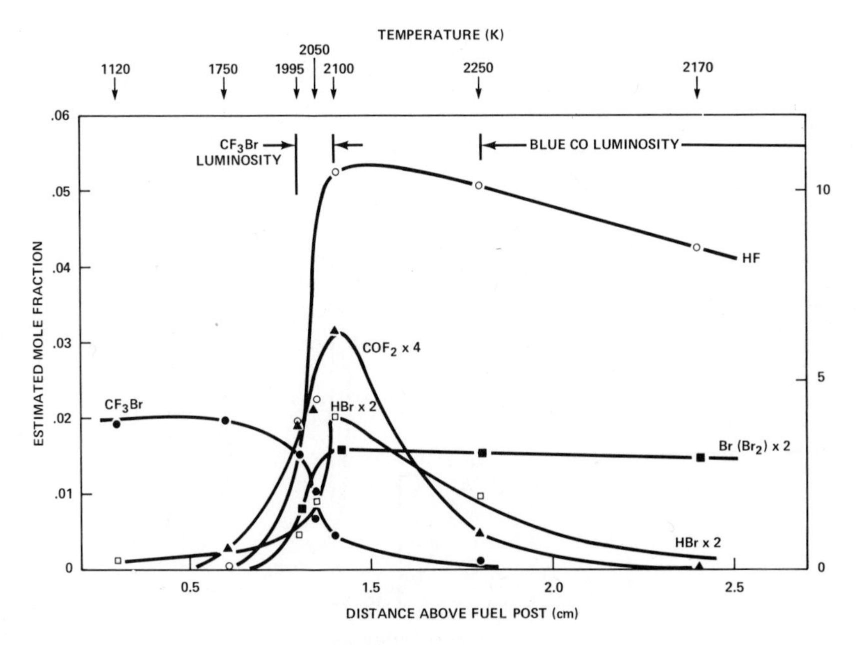

Figure 4B. CF_3Br inhibited CO–O_2 diffusion flame. P = *100 torr.*

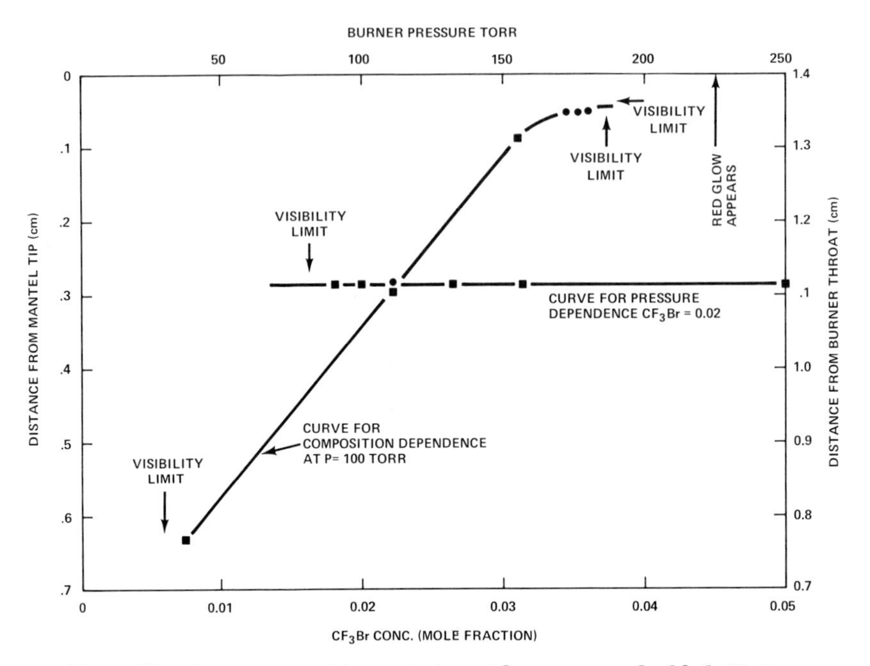

Figure 5A. Luminous position variation with pressure and added CF_3Br

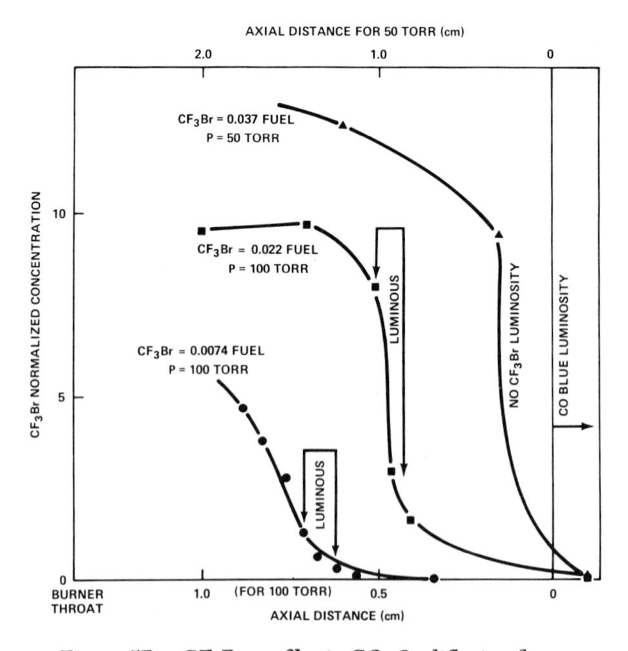

Figure 5B. CF_3Br profiles in $CO–O_2$ diffusion flames

((CF_3Br) = 0.022 of fuel flow) and as a function of added CF_3Br at constant pressure (P = 100 torr).

Discussion

The effect of CF_3Br on the propagation rate of this diffusion flame is not strong. We attempted to measure the effect by comparing the reaction zone volumes both as measured by temperature profiles and by visually observing the size of the CO blue luminosity. The changes were below our detection ability (±20% volume or 3% in linear dimension). This is perhaps not surprising in retrospect since the maximal amount of CF_3Br in the reaction region (assuming stoichiometric reaction between fuel and oxidizer) was less than 1%. A major contributing factor may be the observed decrease in effectiveness of inhibitors as pressure is reduced (10,11) since our studies were made at 100 and 50 torr respectively. The composition profiles of the major reactants and products are relatively unaffected by the CF_3Br addition.

The profiles of halogen species derived from the destruction of the CF_3Br show an order of appearance which strongly suggests a mechanism which is discussed below. The temperature profiles suggest appreciable heat release associated with the destruction of CF_3Br probably by H atoms.

The luminosity has been investigated only visually as to its spatial position, and not spectroscopically; therefore, no positive identification has been made of the emitting species. The behavior of luminosity with pressure and CF_3Br concentration provide clues on the flame geometry and perhaps may allow an estimate of radical concentrations.

Consideration of the geometry of the flame suggests that radial gradients are large compared with axial gradients. This suggests that the system should be analyzed as a fuel jet reacting in an oxidizing atmosphere. The first analysis of such systems was by Burke and Schumann (12) and a simplified analysis, a modification of which will be used here, was suggested by Jost (5).

Temperature Profiles. Axial temperature profiles are shown in Figure 2 for the uninhibited flame and for two levels of addition of CF_3Br to the fuel jet. The peak temperatures are almost independent of CF_3Br addition as might be expected from the small amount of additive, and are a few hundred degrees lower than those expected assuming stoichiometric combustion.

Thermodynamic calculations are consistent with these results (Table I). There is no evidence in these profiles of any major change in reaction zone volume upon addition of CF_3Br, although a small but significant temperature "hump" appears in the region of CF_3Br destruction. This can be attributed to the heat released by the destruction of CF_3Br presumably by reaction with H atoms to produce HBr and HF (Table II). These reactions are rapid compared with the main oxidation of CO to CO_2 so that heat release would be expected to be localized producing the type of "hump" observed. Since the principal reactant is H atoms formed in other parts of the flame, part of this high heat of reaction per molecule compared with that of the fuel is due to transport of energy by diffusion of hydrogen atoms. Phenomena of this type occur in other systems. For example in Göttingen (13) it was found in the case of hydrogen oxygen flames with small amounts of added acetylene, the acetylene acts as a catalytic recombination agent for hydrogen atoms and drastically shifts the temperature profile of the flame toward the cold boundary (6,13). A general discussion of diffusional transfer of heat in reacting systems has been given by Hirschfelder (14).

Composition Profiles. Composition profiles were measured for the accessible species at two pressure levels (50 and 100 torr) with and without the addition of CF_3Br. The results are presented in Figures 3-5 which compare profiles of reactants and products with and without the addition of CF_3Br, the relations between the halogen species resulting from the reactions of CF_3Br and the relation between the position of CF_3Br destruction and the flame luminosity.

Figures 3A and 3B show axial profiles of the major reactants (CO and H_2) and products (CO_2 and H_2O). There appears to be no significant difference between these profiles with and without additive. Combined with the similar results of the temperature profiles, this reinforces the conclusion that at these pressures the addition of small amounts of CF_3Br does not affect the major reaction pathways or reaction rates. It should be noted that carbon monoxide and oxygen change almost linearly along the axis; this can be explained qualitatively by the simple diffusional model discussed in the conclusions. The hydrogen impurity in the Ar, being a catalyst for CO reactions, perhaps simplified the system in making reactive radicals available throughout the entire reaction zone rather than requiring that they originate and diffuse from the fuel

TABLE I.

ADIABATIC FLAME CONDITIONS ASSUMING STOICHIOMETRIC REACTION
[0.35 CO + 0.175 O_2 + 0.38 Ar + 0.05 He + 0.03 H_2]

Pressure (torr)	Added CF_3Br (mole percent)	Flame Temperature (°K)	Products (Mole Percent)																	
			Ar	Br	Br_2	C_MF_N	CF_3Br	CF_2O	CO	CO_2	F	F_2	H	H_2	HBr	HF	HO	H_2O	O	O_2
100	0.77*	2537	48.0	0.81	0	0	0	0	14.5	25.3	0.01	0	0.34	0.17	0.04	2.56	0.78	1.88	1.01	4.5
100	0	2510	46.5	0	0	0	0	0	16.3	25.3	0	0	0.58	0.32	0	0	1.14	3.04	1.37	5.38
50	0.77	2485	47.9	0.82	0	0	0	0	14.9	24.7	0.01	0	0.39	0.18	0.04	2.55	0.79	1.84	1.13	4.72
50	0	2564	46.7	0	0	0	0	0	15.9	25.9	0	0	0.50	0.30	0	0	1.13	3.11	1.24	5.21

*Corresponds to 2.2 mole percent in incoming fuel jet

TABLE II. THERMOCHEMISTRY AND MECHANISMS OF H_2, CO and CF_3Br at 2000°K

	Reactions	Δ H at 2000°K
Elementary Reactions	$H + CF_3Br \rightarrow HBr + CF_3$	-22.27
	$H + CF_3 \rightarrow HF + CF_2$	-52.88
	$CF_2 + O_2 \rightarrow O + OCF_2$	-47.26
	$H + OCF_2 \rightarrow HF + OCF$	- 9.81
	$H + OCF \rightarrow HF + CO$	-106.08
	$CO + OH \rightarrow CO_2 + H$	-21.07
	$H + O_2 \rightarrow OH + O$	+15.67
	$O + H_2 \rightarrow H + OH$	+ 2.09
	$OH + H_2 \rightarrow H_2O + H$	-14.8
Net Reactions	$CO + \frac{1}{2} O_2 \rightarrow CO_2$	-66.42
	$H_2 + \frac{1}{2} O_2 \rightarrow H_2O$	-60.15

diffuse blue luminosity, associated with CO oxidation, found in CO flames and in hydrocarbon flames which involve CO as an intermediate. Due to the low pressure this luminosity extended over almost a two centimeter region (Figures 1 and 2). The addition of CF_3Br to the fuel jet introduced an unexpected second luminosity associated with the disappearance of CF_3Br (Figures 5A and 5B). It was a bright thin cap which showed a most interesting behavior. Its position moved up the fuel jet almost linearly with increasing CF_3Br concentration until reaching a position within a millimeter of the CO blue luminosity, Figure 5A, where it stayed fixed for a short concentration range disappearing quite sharply at an inlet fraction of 0.035. When the additive was increased to 0.045, a red halo appeared outside the blue CO mantle. The dark cone inside the blue mantle appeared unaffected by CF_3Br addition. In another experiment, the flow rate was held constant but the pressure was varied by changing the pumping speed; then the position of the cap remained fixed. At decreasing pressure the CF_3Br luminosity became fainter and more diffuse, and disappeared at a pressure of 75 torr. Adding small amounts of CF_3Br to the oxidizer flow gave no visible effect until a flux fraction of 0.007 was reached. At this point a red halo appeared around the blue CO mantle similar to that observed when excess CF_3Br was added on the fuel side. It should be noted that this concentration is similar to the overall concentration when injected from the fuel side if it is assumed that the fuel reacts stoichiometrically with the oxidizer.

A possible explanation of this behavior can be made by assuming that the flame is an axially symmetric fuel jet entering an oxidizing atmosphere containing the CO-O_2-H_2 reaction zone which we tentatively identified with the blue luminous mantle. This mantle is at high temperature ($\simeq$ 2000K) and acts as a source of H atoms which attack the CF_3Br. Because of the excess of H_2 and CO, OH and O cannot penetrate the jet, but heat can be conducted from the sides and oxygen, water, and H atoms can diffuse inward. Conversely, CO and hydrogen diffuse laterally outward. If one assumes the tip represents the point where the flux of H atoms is sufficient to destroy all of the CF_3Br, the height of this point should vary linearly with inlet flux of additive. This should continue until the CF_3Br cap reaches the tip of the dark region inside the CO mantle. With further CF_3Br, the luminosity can be no longer seen presumably because it diffuses into the mantle.

The basic assumption is that CF_3Br reacts principally with H atoms in a region where radicals are not generated. They therefore must diffuse laterally from the CO-O_2-H_2 reaction region where H atoms are assumed to be generated uniformly and independently of CF_3Br. We identify this zone with the outer blue luminous mantle. It is observed that the geometry is approximately cylindrical. We assume that the tip of the luminous region represents the point where the inlet flux of CF_3Br is balanced by the total flux of H atoms into the cylindrical fuel jet (four being required for completion (Table II)):

$$(1) \quad 4\pi r^2 V_o (CF_3Br)_o = 2\pi rhV_H (H)_f (298/T_f) = 2\pi hD_H (H)_f (298/T_f)$$

Since detailed axial concentration and temperature profiles are not available we approximate the axial flux of H atoms with the term $D(H)_f/r$. This is a common order of magnitude approximation. Solving (1) for h we get:

$$(2) \quad h = 2r^2 V_o D_H^{-1} (H)_f^{-1} (F_f/298)(CF_3Br)_o$$

This predicts that h should vary linearly with inlet concentration of CF_3Br. It should be noted that if the pressure is changed while keeping the inlet mass flow constant, the only two pressure dependent factors in the coefficient of $(CF_3Br)_o$ are V_o and D_H. They both are proportional to inverse pressure, and since they appear as a ratio, the coefficient is pressure independent (neglecting the small change in $(H)_f$). Hence, the position of the luminous cap should be pressure independent. This is in agreement with the observations (Figures 5A and 5B).

The model allows one to estimate the concentration of H atoms from the observed height of the CF_3Br cap at a known inlet mole fraction. By rearranging Eq. (2), we have

$$(3) \quad (H)_f = 2r^2 V_o D_H^{-1} h^{-1} (CF_3Br)_o (T_f/298).$$

Using an approximate diffusion coefficient for H atoms under flame conditions, $D = 282\ cm^2\ sec^{-1}$, and the fact that $h = 1.1$ cm when $(CF_3Br)_o = 0.022$ MF, the value of $(H)_f = 0.01_6$ is obtained. This is four times the calculated adiabatic equilibrium value of 0.0034 MF. The true value is expected to be somewhat larger than the equilibrium value but of the same order of magnitude.

The somewhat simpler Jost model ($r^2 = 2Dt$ where t

= H/V_o) gives an estimate of the height H of the dark zone inside the CO mantle. Here we assume that the CO must diffuse out of the jet to react. This differs from our model for CF_3Br, which is presumed to be attacked in place by the more rapidly diffusing H atoms. The observed axial concentration profiles of CO and CF_3Br are consistent with these assumptions. Using an approximate diffusion coefficient for CO, D = 7.6 cm^2 sec^{-1} at flame conditions, and the values r = 0.22 cm, V_o = 250 cm sec^{-1}, and T_f = 2450°K (calculated), the height of the dark zone is estimated to be 0.8 cm. This value is comparable with the observed value of 1.8 cm.

The notation used in these equations may be summarized as follows:

- V - flow velocity (cm sec^{-1})
- D - diffusion coefficient
- r - radius of the fuel jet (cm)
- h - height of the luminous cap (cm)
- T - flame temperature (°K)
- () - mole fraction
- o - a subscript indicating inlet conditions
- f - a subscript indicating flame conditions
- t - time (sec)
- H - height of dark zone (cm).

In the calculations, the diffusion coefficients at flame conditions were estimated from values at room temperature and pressure by the formula (9)

$$D = D_{298} P_f^{-1} (T_f/298)^{1.67} \quad (4)$$

where P (atm) is the flame pressure.

V. Summary

We have reported the appearance and behavior of a low pressure carbon monoxide-oxygen diffusion flame with and without added CF_3Br. Measurements were made of axial temperature and composition and the spatial behavior of the various luminosities. Based on the observed composition profiles a mechanism has been proposed for the destruction of CF_3Br in this system. The Burke-Schumann-Jost model for diffusion flames has been applied in a modified form to explain the behavior of the luminous cap associated with CF_3Br destruction. To further justify this simple interpretation of the system a calculation was made of the height of the dark cone of the flame and its pressure dependence. The reaction region. An interesting experiment would be the rigorous exclusion of hydrogen containing species from each stream separately. Since this requires control below parts per million it appears to be too difficult to be worthwhile with the available less than perfect apparatus.

The profiles of CF_3Br and its products at the two pressures (50 and 100 torr) were qualitatively similar, but the reaction thicknesses scaled with inverse pressure. This is common behavior in flames controlled by bimolecular reactions (9). Since the initial concentrations of CF_3Br differed, the relative positions of the reaction zones also differed.

The species which were found from the reaction of CF_3Br in this flame included HBr, HF, Br_2 and COF_2. Our sampling techniques do not allow direct detection of atoms and radicals, but we are fairly confident that the major part of the analyzed molecular bromine was sampled in the form of atomic bromine. This species is almost surely formed as atoms by the reaction of hydrogen atoms with HBr (Table II) and thermodynamic calculations indicate that in the temperature regime where this species is detected the maximum value attained at equilibrium would be less than 1% of the observed bromine (Table I). Therefore, the Br_2 curve is interpreted as representing atomic bromine. Both COF_2 and HBr were intermediates in this flame as might be expected from thermodynamic considerations (Table I). There was no evidence of perfluoro compounds such as CF_4 and C_2F_6, nor did C_2F_4 appear. The first product to appear was HBr followed by bromine atoms. The CF_3 must be rapidly attacked since no evidence of its recombination products was found. Attack was probably by hydrogen atoms forming HF and CF_2. The latter radical reacts rapidly with molecular oxygen under flame conditions (15) to form COF_2. Under our conditions COF_2 is rapidly attacked presumably by H atoms with the ultimate formation of HF and CO (Table II). A mechanism consistent with these observed facts is presented in Table II. No evidence of BrF or F_2 was found mass spectrometrically. This probably reflects a relatively low F atom concentration and even lower molecular fluorine concentration in the sample. This would be expected from the stability of HF (Table I) and the fact that there appears to be no obvious kinetic mechanism for maintaining a high nonequilibrium concentration of F atoms.

Luminosity. Two luminosities are associated with this flame; the first was the expected typical results were in reasonable agreement with the experiment. A final piece of corroborating evidence was obtained by estimating the maximum concentration of H atoms in the flame from the positional behavior of the CF_3Br luminous cap. A reasonable value was obtained.[3]

Acknowledgements

This work was supported by the National Science Foundation RANN (Research Applied to National Needs) Program, under grant GI-440088.

The authors thank Luigi Perini for performing the computer calculations.

One of use (RMF) would like to thank the NATO Scientific Community for supporting exchange of information between the Applied Physics Laboratory and Laboratorie de Physico-Chemie de la Combustion, Université de Louvain-la-Neuve, Belgium.

Literature Cited

1. Libouton, C., Dormal, M., and Van Tiggelen, P.J., "15th Symposium on Combustion," The Combustion Institute, Pittsburgh, Pa. (in press 1975).
2. Van Tiggelen, P.J. and Fristrom, R.M., (Comparative Inhibition Studies in Detonation and Deflagration in the CO-O_2 System) submitted to the Second European Symposium on Combustion (scheduled September 1975).
3. Minkoff, G.J. and Tipper, C.F.H., "The Chemistry of Combustion Reactions," Butterworths, London (1963).
4. Fristrom, R., "Survey of Progress in Chemistry," Vol. 3, ed. A.F. Scott, Academic Press (1966).
5. Jost, W., "Explosion and Combustion Processes in Gases," (in German 1938), translated by H.O. Croft, McGraw-Hill Book Co., New York (1946).
6. Fristrom, R., "Katalyse" (in German), ed. K. Haufe, W. de Gruyter and Co., Berlin (in press 1975).
7. Lewis, B. and von Elbe, G., "Combustion, Flames and Explosions," Academic Press, New York, 2nd ed. (1961).
8. Potter, A.E., Heimel, A., and Butler, J., in "8th Symposium on Combustion," The Combustion Institute, Pittsburgh, Pa., pg. 1027 (1962).
9. Fristrom, R. and Westenberg, A.A., "Flame Structure," McGraw-Hill, New York (1965).
10. Lask, G. and Wagner, H.G., in "8th Symposium on Combustion," Williams and Wilkins, Baltimore, pg. 432 (1962).
11. Bonne, U., Jost, W. and Wagner, H.G., Fire Research Abstracts and Reviews, 4, (6) (1962).
12. Burke, S. and Schumann, T.E.W., in "1st Symposium on Combustion," reprinted Combustion Institute, Pittsburgh, Pa., pg. 1 (1964).
13. Hoyermann, K.H., Priv. Comm., Thesis, Gottingen, W. Germany, paper in preparation for submission to Combustion and Flame (1975).
14. Hirschfelder, J.O., J. Chem. Phys. 26, 274 (1957).
15. Matula, R., "Combustion Kinetics of Reentry Vehicle Ablation Materials," Air Force Office of Scientific Research Report No. DUCK 731, September 1973.

DISCUSSION

R. G. GANN: In your sampling system with COF_2 and water co-flowing to the mass spectrometer, wouldn't you expect some serious hydrolysis of the COF_2?

L. W. HUNTER: We do not expect the hydrolysis to be important since our pressures were low, the residence time between flame and spectrometer was short and the H_2O was only slightly absorbed.

C. HUGGETT: What was the concentration of H_2 in your flame relative to the CF_3Br concentration?

L. W. HUNTER: The fuel contained eight mole percent hydrogen and in addition there was some hydrogen present in the oxidizer stream. The total hydrogen available was thus in stoichiometric excess of CF_3Br.

C. HUGGETT: This flame will have a relative low low H_2O content. This may account for the observance of COF_2 in this case. The proposed mechanism suggests that CF_3Br provides a very effective sink for hydrogen atoms, consuming about 4 H atoms per CF_3Br molecule. It would be interesting to observe the behavior of the flame in the region where the hydrogen concentration is reduced to the region of equivalence with the CF_3Br concentration.

This mechanism also suggests that CF_3Br should be as effective or somewhat more effective than CF_2Br_2, on a molar basis. Why is this not the case? Perhaps it would be found to be true in flames containing low concentrations of hydrogen.

L. W. HUNTER: In comparing the effectiveness of CF_3Br and CF_2Br_2 as inhibitors it is usually assumed that chemical inhibition is a result of reduction of radical concentration in the primary reaction zone. One can visualize two general mechanisms for this: (1) scavenging of radicals, i.e., the reaction of radicals to produce stable molecules or unreactive radicals; (2) catalytic recombination of radicals. The first is a bimolecular process and there is a stoichiometric relation between radical concentration and radical completion. In the second process the limiting slow step is a termolecular recombination, but the processes can recur with a long chain length. For example,

Scavenging: $H + CF_3Br \rightarrow HBr + CF_3$

Recombination: $H + Br + M \rightarrow HBr + M^*$ (slow)

$HBr + OH(H,O) \rightarrow H_2O + Br$ (fast, with regeneration of Br)

In comparing the two molecules, if one assumes our mechanism is correct, both molecules have the same scavenging potential, i.e., four H atoms per molecule. Insofar as recombination is concerned, CF_2Br_2 produces two Br atoms and should show twice the recombination capacity of CF_3Br.

By comparing the effectiveness of these two or similar compounds it might be possible to establish the relative contributions of the two mechanisms. Parenthetically, since scavenging is bimolecular and while the slow step in catalyzed recombination is termolecular, the two mechanisms might be distinguished by studying the pressure dependence of inhibition.

J. BIORDI: I would simply like to state that your proposed mechanism for the disappearance of CF_3Br and subsequent fate of the CF_3 and CF_2 species is entirely consistent with our observations in stoichiometric premixed methane flames containing CF_3Br, and, in fact, we used a similar mechanism to account for these observations. Recently, we have detected the CF_2 radical in these flames, although the CF_3 radical remains undetectable.

It seems to me that the reaction $CF_3 + H \rightarrow HF + CF_2$, being a true sink for H atoms, must provide some inhibition in this system. What sort of behavior would you have expected of your flame system if it were "inhibited" in the usual sense?

L. W. HUNTER: We agree that the reaction you quote should provide some inhibition. Our thoughts on this matter are presented in the response to Dr. Huggett's question.

We would expect that in our system an inhibited flame would show an increase in the size of the mantle and a drop in the peak flame temperature. The concentration profiles of the major stable species should also show some change. We didn't observe these

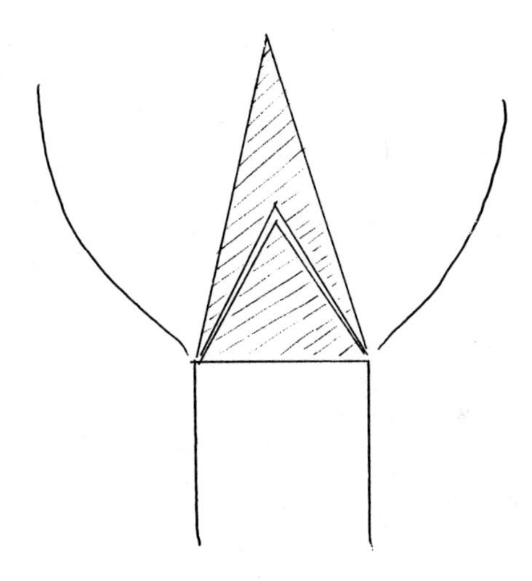

Ideal geometry

changes because of our low pressure and because H_2 (a catalyst of CO reactions) was present in excess over the CF_3Br.

F. A. WILLIAMS: Your model for the inner flame seems to involve a more or less continuous diffusion process which might lead to a gradual increase in luminosity instead of the apparently fairly sharp cap that you observed. Would you like to comment on this?

L. W. HUNTER: This is an interesting question, to which we do not have a complete answer. Our model may be too naive to allow detailed structural calculations since our original aim was to explain the peculiar behavior of the tip. Ideally if the jet velocity profile is uniform and the reaction of H atoms with CF_3Br and its fragments is rapid we should expect a sharp conical luminous zone representing the interface between H and CF_3Br within a larger conical dark zone representing the onset of CO reaction (see figure). We observe instead a crescent shaped inner luminosity inside a slightly tapered cylindrical dark region with hemispherical cap. (Refer to our first figure). We feel that the deviations from the expected ideal conical geometry in the two cases are due to a nonuniform velocity profile across the fuel jet and that our failure to observe a continuous inner luminous surface extending to the lip of the fuel nozzle is that when the CF_3Br and CO luminosities lie close to one another, the CF_3Br luminosity is obscured. Note that when the inner luminous cap approaches the tip of the dark zone it stays stationary about two millimeters below the tip of the dark zone over a short composition range and then disappears at a ralatively reproducable concentration rather than traveling smoothly into the blue CO luminosity as concentration of CF_3Br is increased.

YEN-HSIUNG KIANG: In this paper, only one low concentration of H_2 in the fuel is used. Because of this low H_2 concentration, all of the bromine appears as Br_2 in the product gas. We have done some test work on the burning of brominated hydrocarbons in a commercial burner, the data show that when the $[H_2]/[Br_2]$ ratio increases, $[Br_2]$ in the product gas decreases and [HBr] increases. I think the possible reactions are

$$Br_2 + H \rightarrow HBr + Br$$

$$1/2Br_2 + H_2O \rightarrow HBr + OH$$

due to higher $[H_2]$. Further works on varying the H_2 concentration in the feed and determining corresponding $[Br_2]/[HBr]$ ratios in the product gas will be quite interesting.

L. W. HUNTER: In this system we believe bromine occurs in the atomic form. The equations you present correspond to overall stoichiometry and in this sense we agree with your comments. The elementary reactions in flames are not between whole molecules.

9

The Effect CF_3Br on Radical Concentration Profiles in Methane Flames

JOAN C. BIORDI, CHARLES P. LAZZARA, and JOHN F. PAPP

Pittsburgh Mining and Safety Research Center, Bureau of Mines, U. S. Department of the Interior, Pittsburgh, Pa. 15213

That complicated reaction systems involving chain reactions may be extremely sensitive to small amounts of substances that interfere with the chain propagation is well known in chemistry (1). Flames are examples of such reaction systems and the recognition that certain molecules retard flame propagation out of proportion to their thermal or mechanical influence leads to the supposition that their effect is a matter of chemical reactivity. The idea of "chemical extinguishment" becomes very attractive when it is appreciated, as pointed out by Friedman and Levy (2), that chemical reactivity varies by orders of magnitude and so offers the promise of extinguishing agents significantly more effective than those now in use. Aside from screening tests, the development and efficient use of such agents requires an understanding of the mechanisms by which uninhibited flames propagate and of the mechanisms by which chemical agents impede that propagation.

The survey of extinguishment mechanisms by Friedman and Levy (2) has been followed by periodic reviews (3-8) which, in their entirety, cover physical as well as chemical extinguishment, heterogeneous and homogeneous systems of fuel, oxidant, and extinguishing agent, and a variety of measurements of agent effectiveness. We wish to narrow this discussion primarily to volatile halogenated inhibitors in premixed flames studied principally by the technique of flame microstructure, and, further, to scrutinize the available data for its implications on the effects of these molecules on the primary chain propagating radical species in the flame. The papers that, in our view, are most pertinent to this discussion are summarized in Table I, and they have been divided into those providing indirect and direct evidence regarding atom and radical concentrations in the presence of inhibitor. It may be seen that a wide variety of conditions have been employed in these studies. This fact precludes the formulation of a generalized mechanism whose consequences can be predicted and directly tested.

Rosser, Wise and Miller (9) measured the effect of various halogenated inhibitors on CH_4-air flame speeds and observed Br containing inhibitors to be more effective in rich than in lean mixtures. Their effectiveness (in a 10% CH_4-air flame) was proportional to the number of Br atoms in the inhibitor molecule. They also verified that small quantities of inhibitors could significantly reduce flame speed without affecting the final flame temperature. More precise measurements of methane-air-CH_3Br burning velocities gave similar results (10) and were considered to support the mechanism suggested in (9). In that mechanism the critical species were the halogen atom and the halogen acid, the former considered to arise from the thermal decomposition of the inhibitor and the latter from hydrogen atom abstraction (from the fuel or any other hydrogen containing species) by the halogen atom. The acid in turn may react with any available radical species, H, O, OH, CH_3, to regenerate the halogen atom. In effect, it is a matter of substituting a halogen atom for the usual chain propagating radicals with the result that reactions usually associated in a critical way with flame propagation, $H+O_2 \rightarrow OH+O$, $O+CH_3 \rightarrow H_2CO \ldots \rightarrow CO$, and $CO+OH \rightarrow CO_2+H$ are reduced in rate, and that reduction is observed macroscopically as a reduction of burning velocity. This interference was considered to occur when the temperature was high. The halogen atoms may initiate a competing chain, but they cannot serve the function of H, O, or OH in the reactions cited. Although recognizing the possibility, the authors did not think a reduction in the rate of diffusion of H, O and OH was a critical feature of combustion inhibition.

One of the first structure studies of inhibited flames (11) supported this mechanism to the extent that the concentration of CO was found to be reduced in a lean, atmospheric CH_4-O_2-Ar flame containing 0.36% HBr, relative to the clean flame. It was suggested that formaldehyde should also be markedly affected by HBr. Such observations were later made in structure studies of lean (12) and stoichiometric (13) low pressure methane flames, the latter with CF_3Br inhibitor.

The most concentrated effort in determining the structure of inhibited flames was that of Wilson and co-workers (12, 14), who reported the structures of two clean and four inhibited (CH_3Br, HBr, HCl, Cl_2) flames, all lean CH_4-O_2 maintained at 0.05 atm. Common to all the inhibited flames was an early appearance of new products related to the inhibitor, e.g. CH_3X when the inhibitor was HX. The maximum H_2 concentration was greater in flames containing HX than in clean flames. The data could be analyzed for individual net chemical reaction rates, K_i, and in each inhibited flame the profile of K_{CH_4} was sharper, narrower, and shifted to a higher-temperature part of the flame. Profiles for OH and H were calculated from the

TABLE I

Summary of Studies Providing Indirect, (A), or Direct, (B), Information on the Effect of (Mostly) Halogenated Hydrocarbons on Radical Concentrations in Flames

	Flame	Stoichiometry	Inhibitor (Mole %)	Press., atm.	Technique	Ref.
A.	CH_4-air	range	Variety of Br, I, Cl contg. organics (≤ 0.5)	1.0	Burning vel. red'n.	9
	CH_4-air	range	CH_3Br (≤ 0.4)	1.0	Burning vel. red'n.	10
	CH_4-O_2-Ar	lean	HBr(0.36)	1.0	Structure	11
	CH_4-O_2	lean	CH_3Br(1.71), HBr(1.88), HCl(2.69), Cl_2(1.71)	0.05	Structure	12, 14
	CH_4-O_2-Ar	stoich.	CF_3Br (0.3)	0.042	Structure	13
	H_2-O_2-N_2	range	CH_3Br(1-3), CH_4(1-3)	0.092	Structure	16
	H_2-N_2O-Ar	rich	CH_3Br(5)	0.132	Structure	16
	H_2-O_2-N_2	rich	HBr (0.027,0.1)	1.0	Burning vel. red'n. and computat.	17
	H_2-N_2O-N_2	rich	HBr, CH_3Br, C_2H_4	1.0	Limit	17
	H_2-O_2-N_2	range	HBr, HI, HCl	1.0	Limit	18

TABLE I (Cont'd)

Flame	Stoichiometry	Inhibitor (Mole %)	Press., atm.	Technique	Ref.
A. Cont'd.					
C_3H_8-O_2-Ar	lean	HBr(0.2-0.6)	1.0	Structure	20
(C_3H_8-C_3H_6)-air	rich	$C_2F_4Br_2$(1.2)	1.0	Structure	24
B. CH_4-O_2-Ar	stoich.	CF_3Br (0.3, 1.1)	0.042	Molecular beam-MS sampling	29, this work
CH_4-air	stoich.	$Fe(CO)_5$ (0.001-0.01)	0.079	Optical absorption	25
CH_4-air	rich	$NaHCO_3$	1.0	Optical absorption	26

appropriate K_i assuming a rate constant for $CO+OH \rightarrow CO_2+H$ and $H+O_2 \rightarrow OH+O$, respectively. The maximum [OH] was found to be greater in the HBr than in the clean flame, and $[H]_{max.}$ the same, though shifted to a higher temperature.

Wilson accounted for these observations by proposing that the principal inhibiting reactions take place early, in the low temperature region of the flame. After Burdon, Burgoyne and Weinberg (15), they suggest that the major effect results from the efficient competition of reactions such as $H+RX \rightarrow HX+R$ and $H+HX \rightarrow H_2+X$ with the chain branching reaction, $H+O_2 \rightarrow OH+O$. Such competition will prevail until the inhibitor is completely consumed or the temperature has risen high enough to render the competition inefficient. Thus, the concentration of radicals would be expected to be reduced in the early part of the flame, but not necessarily in the later part.

Biordi et al (13) reported structure studies of low pressure CF_3Br-inhibited stoichiometric flames. Observations similar to Wilson's were made regarding the early decay of inhibitor, but the nature of these inhibitor-related products suggested that O atom and CF_3 radical reactions might be more important than generally supposed. High concentrations of Br atoms were observed directly. In detail, CF_3Br was not considered to conform to either of the suggested inhibition mechanisms.

The great practical interest in methane as a fuel accounts for the emphasis on it in these studies, but the simpler and better known chemistry of H_2/O_2 combustion might be expected to offer clearer insights into the way in which that chemistry is disrupted by inhibitors. Only three papers have been addressed specifically to this problem and two of them have also examined some aspects of inhibition in flames supported by N_2O, where the propagation is via a simple rather than branching chain mechanism.

Fenimore and Jones (16) examined the structure of H_2-O_2-Ar and H_2-N_2O-Ar flames containing CH_3Br. In each case H atom profiles were calculated from the observed net reaction rate for a species considered to react solely with H atom in the flame, assuming a value for the rate coefficient. These profiles showed [H] at high T to be the expected equilibrium concentration and to be unaffected by CH_3Br in the H_2-N_2O-Ar flames. The lean H_2-O_2-Ar flames, on the other hand, did show a reduction of [H] with CH_3Br; by inference from K_{O_2}, so did the rich H_2-O_2-Ar flames. While observations using CH_4 in these flames suggested that the CH_3 radicals formed from CH_3Br acted as inhibitors, the presence of bromine was also important. The authors postulated that part of the effect might be due to "a more complete recombination of the active radicals in the presence of the halogen."

The feasibility of this idea was demonstrated by two complementary studies based upon computational methods. Dixon-Lewis and co-workers (17) solved the combined heat conduction, diffusion and chemical reaction equations to determine flame velocity and species concentration profiles for a near limit rich H_2-O_2-N_2 flame at atmospheric pressure (T_f=1050°K) with and without 0.1% HBr. Lovachev and colleagues (18) carried out calculations of overall reaction times using a set of kinetic isothermal equations that describe H_2/O_2 oxidation. The conditions examined were 1 atm and 1000°K, lean and rich H_2/O_2 mixtures, without and with 1% HBr. Although there is a somewhat different interpretation of the effects observed, both groups come to the conclusion that consideration and evaluation of all of the following reactions are necessary to simulate observed macroscopic effects of HBr on this system:

$$H(O,\ OH)+HBr \rightleftarrows H_2(OH,H_2O)+Br \quad (1)$$

$$Br+HBr \rightleftarrows Br_2+H \quad (2)$$

$$Br+HO_2 \rightleftarrows HBr+O_2 \quad (3)$$

$$Br+Br+M \rightleftarrows Br_2+M \quad (4)$$

$$Br+H+M \rightleftarrows HBr+M \quad (5)$$

The effect of reaction (1) alone was found to be insufficient, by an order of magnitude. Termolecular reactions are not usually considered to be important in the chemistry of the primary reaction zone of the flame. However, observations that several halogenated inhibitors show reduced effectiveness at pressures below 1 atm. imply that termolecular reactions have a role in the inhibition mechanism (21, 22, 23). The computed profiles of Dixon-Lewis show a reduction in $[H]_{max}$ by 25% with 0.1% HBr as well as a shift downstream. Other features of these profiles are interpreted to imply that there is no real competition between reaction (1) and the chain branching reaction, $H+O_2 \rightarrow OH+O$.

That such a competition does keep chain branching in check below some critical temperature is, however, the basis on which Butlin and Simmons (19) were able to correlate their measured rich limit O_2-HBr concentrations for H_2-O_2-N_2 flames, having shown the [HBr] to be independent of $[H_2]$ at the rich limit. Assuming the critical temperature to be the final flame temperature and the ratio $[O_2]/[HX]$ to be the same as in the cold gas, such a competition would predict a linear relation between log $[O_2]/[HX]$ and $^1/T_f$. This was observed for HBr and HI and the slopes were in agreement with the difference in activation energies of reaction (1) (and the analogous HI reaction) and the branching reaction. The lean limit did not conform to this model.

There are two pertinent studies dealing with fuels other than methane or hydrogen. Pownall and Simmons (20) examined the structure of lean 1 atm C_3H_8-O_2-Ar flames containing HBr. In contrast to Wilson, they observe that the maximum net reaction rates of fuel and oxygen, though reduced by the additive, are shifted to lower temperatures and that the calculated (from $CO+OH \rightarrow CO_2+H$ and K_{CO_2}) $[OH]_{max.}$ is reduced with the inhibitor. The authors interpret the data to mean that HBr promotes reaction at low temperature (~600°K) by providing a facile initiation of the normal low temperature hydrocarbon oxidation mechanism, namely $HBr+O_2 \rightarrow HO_2+Br$, and $Br+C_3H_8 \rightarrow HBr+C_3H_7$ followed by $C_3H_7+O_2 \rightarrow HO_2+C_3H_6$ and $HO_2+C_3H_8 \rightarrow C_3H_7+H_2O_2$. As the temperature increases, back diffusion of H atoms becomes important. These react preferentially with the additive, prevent chain branching, and thereby reduce OH, as well as provide a net inhibiting effect. The reaction of Br atoms with CH_4 is so slow that the low temperature promotion does not occur with methane as a fuel.

Ksnadopulo et al (24) reported calculated H atom profiles in the low temperature region (<450°K) of 1 atm rich ($C_3H_8+C_3H_6$)-air flames containing $C_2F_4Br_2$. Calculations were based upon observed net reaction rates for C_3H_8 and $C_2F_4Br_2$ and an assumed mechanism. The H atom profile is reduced sharply with the inhibitor at the lowest temperature, but shows a plateau before achieving its maximum concentration. The authors suggest that the Br and HBr in this system "significantly attenuate the flow of H atom into the preflame region" and attribute this to the cumulative effect of a reaction scheme similar to (1)-(5) and including the attack of Br atom on the fuel. The plateau suggested to them an additional, though unspecified, source of H atoms in the flame when the inhibitor was present.

Direct measurement of radical concentrations in flames containing inhibitors are few and frequently confined to the post-reaction, or recombination zone where spectroscopic techniques having relatively low spatial resolution may be usefully employed. The available techniques account also for the fact that it is usually the hydroxyl radical that has been monitored. Bonne et al (25) observed that the $[OH]_{max.}$ was essentially unaffected by addition of $Fe(CO)_5$ to a low pressure CH_4-air flame, but that the decay of OH in the recombination zone was catalyzed by the presence of the additive. Iya et al (26) made similar observations at atmospheric pressure with $NaHCO_3$ and, having been able to associate the degree of inhibition with the concentration of Na in the gas phase, concluded that the idea that inhibition may arise from competing homogeneous inhibitor catalyzed reactions is a tenable one. Actually, a considerable body of literature is available regarding the catalysis by small ($<\sim 10^{-5}$ mole fraction) amounts of metal additives on H and OH recombination in $H_2/O_2/N_2$ flames (27, 28). A mechanism for these observations that appears to be emerging as general involves catalysis by the metal oxide with the hydroxide as an intermediate, but there are some important subtleties in individual behavior of certain metals that need to be understood, and the role of heterogeneous catalysis is still not well established.

To our knowledge, the only direct measurements of flame radical concentrations for halogenated inhibitors are the results reported by us from flame microstructure studies of low pressure 0.3% CF_3Br inhibited methane flames (29). Briefly, that data showed that at low temperature H atom and methyl radical concentrations were reduced in the inhibited relative to a clean flame, but that maximum H, O, and OH concentrations were unaffected. No measurements were made in the recombination zone for these inhibited flames.

Introduction To The Present Work

The research summarized above shows that halogenated hydrocarbons do indeed modify the chemical processes in flames and that the modifications are consistent with a delayed or reduced chemical reactivity. The data has been interpreted to mean that a diversion of or substitution for H atoms is responsible (8, 30) although Creitz (31) has suggested that O atom reactions may be controlling. Certain reactions or reaction schemes as well as broader hypotheses have been advanced to show how this might be accomplished.

The evidence available for how the flame propagating radicals are affected in inhibited flames is mostly indirect, though quite compelling. Certain obvious questions present themselves in the event that one can measure directly radical concentrations through the flame. Is there an observable reduction in the concentration of chain carriers in the case of a flame "inhibited" by other criteria, e.g. a reduction of burning velocity? If so, in what region of the flame is this observed? Which radicals are affected, or are they all affected? Is this a matter of degree, or of kind (or both) as the concentration of inhibitor changes? How do observed radical profiles compare with those calculated indirectly? Do the observations conform to the prevailing ideas regarding mechanism?

Using molecular beam sampling and mass spectrometric detection, we have identified and measured the concentration profiles of a variety of active species in methane flames, including H, OH, O and CH_3. In this report we address ourselves to the questions raised for the specific case of a low pressure, stoichiometric CH_4-O_2-Ar flame inhibited by 0.3% and 1.1% CF_3Br. The data to be discussed is part of a larger body of data accumulated for the flames, namely, the temperature, area expansion ratio, and concentration profiles for all of the major and most of the minor stable and radical species. There is some danger in extracting these portions for more or less isolated examination (32), but if one is careful to keep in mind the larger view such an approach should be valuable.

Experimental System and Procedures

The details of construction of the apparatus, its performance and the procedures used in these experiments have been published elsewhere (13, 29, 33-35). However, for the sake of completeness, the essentials of that infomation will be repeated.

Four flames have been examined, two inhibited flames and their uninhibited analogs. They are characterized by the information given in Table II. These flames are stabilized on a cooled, porous plug flat flame burner having a diameter of 10 cm and enclosed in a low pressure housing, Figure 1. The gases fed to the burner are metered using calibrated critical flow orifices and, by providing critical flow conditions also at the exits of the housing, the pressure is maintained constant to better than 0.1% (29).

The flame appears as a luminous disc; its position relative to the burner surface depends upon stoichiometry and the initial mass flow velocity. Radial temperature profiles at various heights above the burner (29) as well as radial velocity profiles at both atmospheric and reduced pressure showed the flame to be a good approximation to a pseudo-one dimensional system (36). The burner is moved in a vertical direction and the distance of travel relative to some fixed point can be measured to ±0.03mm.

The burner housing is equipped with opposing quartz windows (not shown in Figure 1). This permits optical absorption and emission measurements, but the spatial resolution for these measurements is relatively low. A low power microwave discharge through water vapor in an argon carrier provides a convenient line source for OH (37). A 1.0 meter Czerney-Turner grating monochrometer equipped with a 1P28 phototube is the detector.

In order to preserve the identity of atomic and radical species (and also stable species that are strongly adsorbed on glass or metal surfaces), it is necessary to rapidly reduce gas phase and wall collisions in the flame gas sample. Several groups of investigators have successfully applied molecular beam sampling techniques to this problem (38-42), but the development of these techniques to permit sufficiently precise and accurate measurements of profiles of both stable and radical species for quantitative kinetic analyses is relatively recent (29, 43). It is a matter of devising a sampling probe that provides good sensitivity for radicals while offering acceptably small perturbation to the flame.

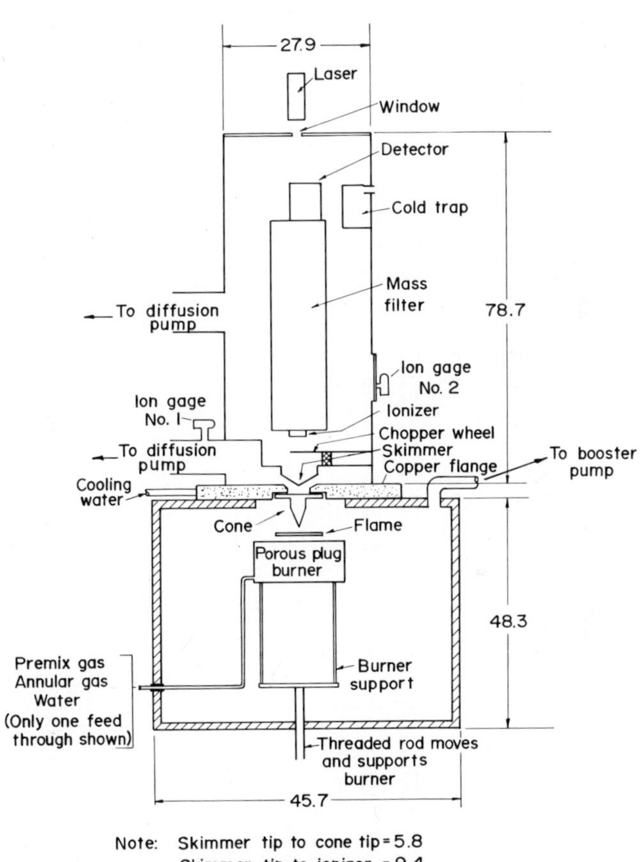

Figure 1. Mass spectrometer—low pressure flame system, schematic

TABLE II

Characteristics of Flames Examined At 0.042 atm.

	Flame I		Flame II		Flame III		Flame IV	
	Mole pct	Flow, gm sec^{-1}	Mole pct	Flow, gm sec^{-1}	Mole pct	Flow, gm sec^{-1}	Mole pct	Flow, gm sec^{-1}
CH_4	10.3	0.0182	10.3	0.0182	10.1	0.0107	10.1	0.0108
O_2	21.6	0.0763	21.6	0.0765	21.5	0.0456	21.2	0.0455
Ar	68.1	0.3005	67.8	0.3002	68.4	0.1811	67.6	0.1808
CF_3Br	0	0	0.3	0.0047	0	0	1.1	0.0110
v_o[a], cm sec^{-1}	79.3		79.5		47.6		48.0	
T_{max}[b] °K	1868		1911		1781		1966	
T_{Ad} °K	2379		2374		2375		2358	

a. Calculated using $T_{initial}$=298°K.

b. As determined in the absence of the sampling probe (34).

The probe finally adopted for these studies was of quartz with an outside angle of ~38° and having an orifice at the tip of 87μ. Profiles of composition and temperature determined with this type of probe compared well with those determined using conventional microprobes (44). In addition, reduction of these profiles according to Hirschfelder's model (45) of flame propagation gave results comparable to microprobes and permitted determination of the rate constant for the reaction $H+O_2 \rightarrow OH+O$, in good agreement with the accepted values at high temperature (44, 46).

As shown in Figure 1, the beam forming apparatus consists of two differentially pumped stages leading to a quadrupole mass spectrometer. Figure 2 is a block diagram of the system (Extranuclear Laboratories EMBA II)[1] employed for the detection of the beam. A small portion of the molecular beam formed in the 1st chamber is permitted to pass into the 2nd chamber and is directed along the axis of the ionizer and mass filter. A toothed chopper wheel located upstream of the ionizer modulates the beam, and phase sensitive detection permits distinction between ions originating from molecules within the beam and those randomly scattered off surfaces in the mass spectrometer. This permits detection of neutral particles from the source even if their density in the ionizer is several orders of magnitude lower than the background gas density. The mass filter may be tuned to observe a single mass or to sweep a mass range (maximum width 0 - 168 amu under current operating conditions). This may be accomplished manually or automatically by programming the quadrupole power system. Sweep speeds are limited, when detecting modulated beams, by signal-to-noise conditions and by the frequency of modulation. An electron multiplier is used to detect the mass analyzed ions. The signal may be displayed on an oscilloscope and/or dual channel strip chart recorder.

The sampling system-mass spectrometer is calibrated directly for the major stable species from mixtures of known composition passed through the burner, without ignition, and sampled otherwise in exactly the same manner as with the flame. The procedures used to identify and to calibrate for or estimate the concentrations of minor stable species are specified elsewhere (33, 46).

The identification and monitoring of radicals through the flame and the determination of their concentrations require different procedures. Identification is made by a combination of measured ionization efficiency curves and appearance potentials, behavior of profiles through the flame, and reactions that have been previously identified in methane flames (34).

---

[1]Reference to specific trade names is made to facilitate understanding and does not imply endorsement by the Bureau of Mines.

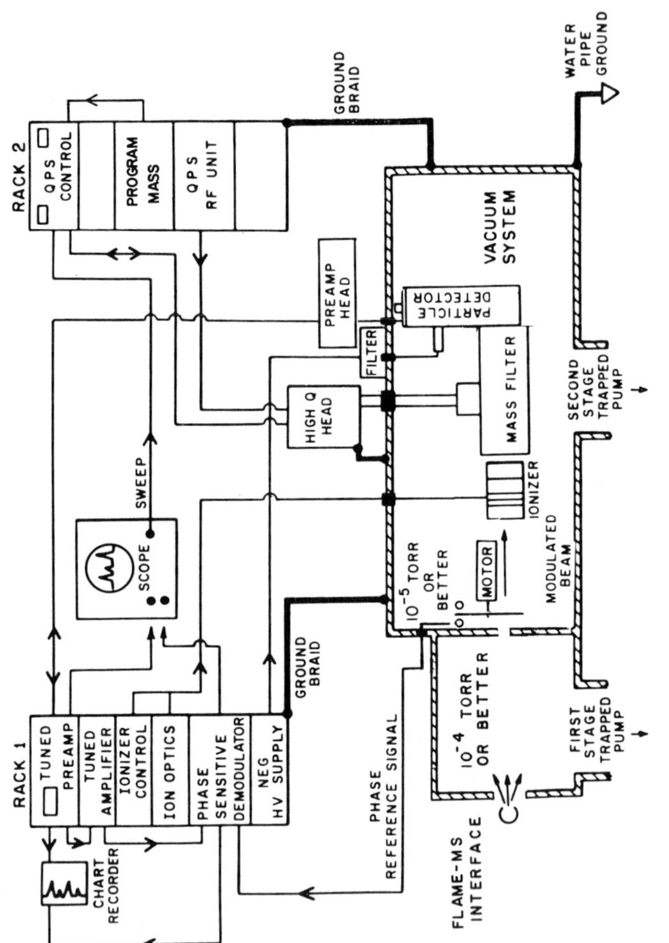

Figure 2. Modulated molecular beam—mass spectrometric detection system

Some of the appearance potentials that have been measured (relative to the argon constituent of the flame) are shown in Table III. At normal operating electron energies (50-70eV), the base peak of a free radical is usually also a fragment ion peak of one or more stable components of the sampled flame gas. Low electron energies are therefore required for detection of the radical. For profile measurements this electron energy should be low enough that contributions from fragment ions are either negligible, or can be corrected quantitatively. Alternatively, if the resolution of the mass spectrometer is high enough to separate the peaks of interest, higher electron energies may be used. We use both procedures.

For H, OH, and CH_3, profiles are measured at low electron energies, 1-2eV above their appearance potentials. For H and CH_3, the nominal operating electron energies for which there is no detectable contribution from expected interfering fragment ions is determined experimentally and correspond to actual electron energies of ~15.6eV and ~11.1eV, respectively. Hydroxyl radical profiles are measured also at 15.6eV, but a small correction for $C^{13}H_4^+$ is required in those parts of the flame where the methane concentration is non-zero.

Except at distances close to the burner, O atom profiles are also observed using 15.6eV. However, when methane is present it is necessary to resolve the CH_4^+ and O^+ ions in order to define the O atom behavior. The required resolution is 400 and is achievable with the mass spectrometer used.

Absolute radical concentrations are determined by measurement of their intensity relative to a structually similar stable molecule (e.g., OH and H_2O). The intensities for each species of the pair are measured at electron energies above their respective ionization potentials by the same amount. Assuming the ionization cross section as a function of energy has the same shape for the two species, their relative ionization cross sections at 70eV may be used to relate the measured relative intensities to relative concentrations. Absolute ionization cross sections are available for O, H, H_2, O_2 (48), H_2O (49), and CH_4(50); for OH and CH_3, they were estimated by additivity (51, 52).

Because of the uncertainties in the individual values of the measured or estimated cross sections, this procedure would not be expected to give concentrations for atom and radical species accurate to better than a factor of two. For H, O, and OH, however, it is possible to achieve better accuracy. It can be demonstrated that the reactions characteristic of the H_2/O_2 system are very nearly balanced[2] in the high temperature region of this flame (34, 43). This fact permits determination of the

[2]A reaction $a+b \rightarrow c+d$ is said to be balanced if the ratio $\frac{[c][d]}{[a][b]}$ is equal to the equilibrium constant for the reaction.

TABLE III

Appearance Potentials Measured for Some Species Observed in Clean and CF_3Br Inhibited Methane Flames[a]

Mass No.	Appearance Potential, eV	Assigned to-	Literature[b] Values, eV
1	13.7 ± 0.1	H	13.6
2	15.3 ± 0.1	H_2	15.4
15	10.2 ± 0.3	CH_3	9.8
16	13.7 ± 0.4	O	13.6
17	13.1 ± 0.1	OH	13.2
18	12.4 ± 0.1	H_2O	12.6
29	9.5 ± 0.1	HCO	<9.8
50	11.6 ± 0.2	CF_2	11.8
79	11.7	Br	11.8

a. An appearance potential of 15.76 eV for argon was used in the calibration of the electron energy scale.

b. Values quoted are from optical spectroscopic determinations where available (47).

concentration of each of the radicals in terms of known stable species concentrations and equilibrium constants. Concentration of O, H, and OH determined by the procedure outlined were within 20% of their expected concentrations, assuming balanced reactions. We estimate the absolute radical concentration for H, O and OH to be accurate to ±50%.

Temperature profiles were determined separately from species concentration profiles using Pt-Pt 10% Rh thermocouples constructed from 1 mil wire. These were coated with silica to minimize catalytic recombination on the wire surface. The emissivity of the silica coated bead, which was usually 1-2 mil in diameter, was measured (ϵ=0.22 ±.02) and a radiation correction after the formulation by Kaskan (53) was applied. The final flame temperature determined by thermocouple measurements for flame I was in good agreement with that determined by OH absorption measurements (44).

For each of the four flames listed in Table II concentration profiles of H, OH, O, and CH_3 were measured. The measurements extended from the preheat zone, through the primary reaction zone and well into the secondary reaction zone, to distances about 13 cm from the burner surface. Temperature measurements were determined to Z = 10-11cm. In each case the distance limitation is imposed by apparatus dimensions. Because the gradients in concentration and temperature are steepest in the primary reaction zone (roughly between 0.2 and 0.8cm from the burner surface), more experimental points are taken in this than other regions of the flame.

Results and Discussion

On account of the heat loss to the burner, the flames studied are non-adiabatic. The calculated adiabatic temperatures are given in Table II. The burning velocity for these "quenched" flames, once they are stabilized, is simply the volumetric flow rate divided by the burner surface area, and the inhibiting effect of additives is therefore not observable as a reduction in burning velocity. Kaskan (53) has shown, for quenched flat flames at 1 atm, that the log of the mass burning velocity is a linear function of the reciprocal of the maximum flame temperature, and has proposed (54) and demonstrated (55) that a rise in flame temperature at constant burning velocity is a measure of inhibition for quenched flames. We have done similar experiments with 0.042 atm. flames and find analogous behavior (56). Thus the rise in temperature of flame II relative to flame I and flame IV relative to flame III is indicative of inhibition of these flames by CF_3Br according to the commonly used criterion for judging effective flame inhibition. The temperature rise can be related to a relative reduction in burning velocity for the adiabatic flame

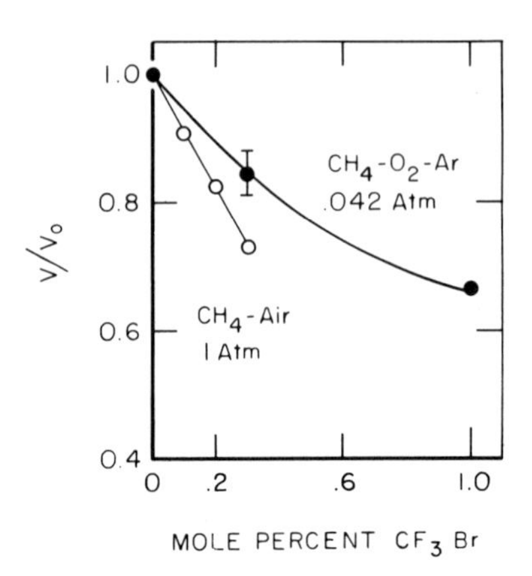

Figure 3. Relative reduction of burning velocity of methane flames as a function of initial CF_3Br concentration at low (23) and atmospheric (9) pressure

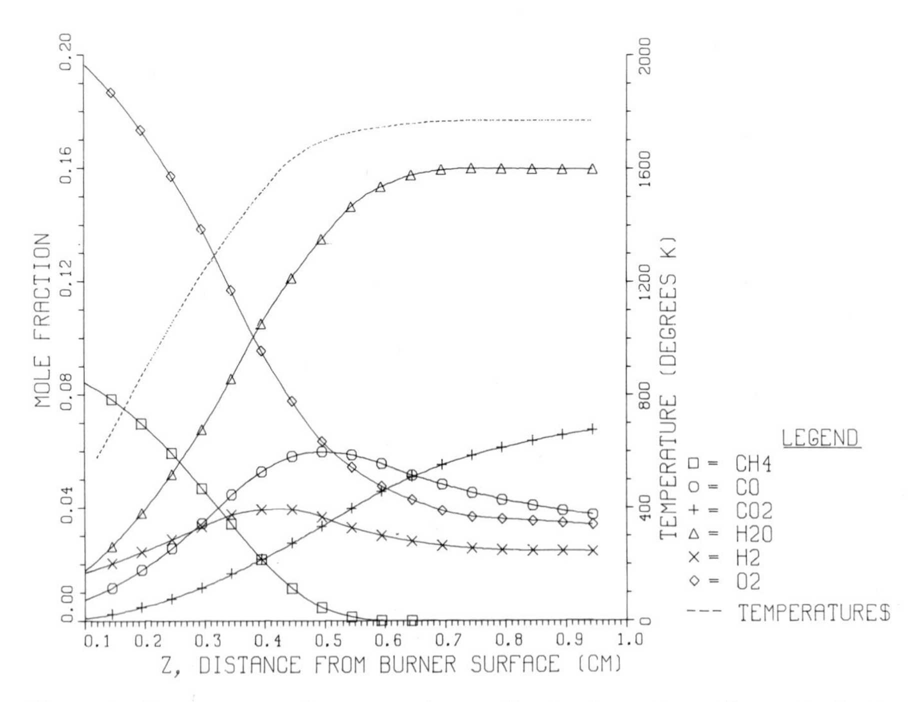

Figure 4. Temperature and concentration profiles for the major stable species in the uninhibited flame, I (see *Table I*)

(56) for comparison with observations at atmospheric pressure. Figure 3 shows the results from flames II and IV, and those calculated from data taken at 1 atm for a flame of similar composition (9). As has been observed for some other halogenated hydrocarbon inhibitors (57), the effectiveness of CF_3Br is less at reduced pressure. It is almost half as effective in reducing the burning velocity at 1/20 atm. as it is at 1 atm.[3]

For convenience in presenting the data, the flame is divided into two regions, $Z<1cm$ and $Z>1cm$. At this pressure, the maximum flame temperature is achieved and all the rapid reactions consuming CH_4 and O_2 and producing H_2O, CO and, to a large extent, CO_2 occur at distances <1cm from the burner surface. This is illustrated in Figure 4 which shows the profiles of temperature and the major stable species for flame I. Figures 5-8 show the profiles for the major atom and radical species of all four flames over this region of the flame, together with either the CH_4 or the CF_3Br profile for orientation. In these and subsequent illustrations of radical profiles, individual data points have been left out in several instances for clarity. However, at least one representative set of data points have been included for each radical in each range of Z in the flame.

In each case, the profiles for the inhibited flames are shifted downstream relative to the clean flames. This is true of all the species profiles, not simply the radicals, and is a reflection of the inhibiting action of CF_3Br, just as is the rise in temperature. This point is so important to clear thinking about inhibited flames studied on cooled burners that it is worthwhile to repeat Botha and Spalding's (59) description of flame stabilization on a porous plug burner. The linear flow velocity, v_0, of the gas issuing from burner is set at some value below the adiabatic burning velocity, S_a, of the mixture. Upon ignition, since $S_a>v_0$, the initial, adiabatic flame travels upstream toward the porous plug. Heat is transferred to the plug, which, however, rises only a few degrees in temperature on account of the efficient cooling. The loss of heat from the gas reduces the burning velocity of the flame, and hence "the flame rapidly and automatically takes up a position of equilibrium a short distance away from the disk where it loses just enough heat to reduce the flame speed to the stream velocity (59)." The effect of heat removal by the plug may be viewed as identical to that of pre-cooling the reactants. Addition of a chemical inhibiting agent has the effect of reducing the adiabatic burning velocity and so a smaller reduction in burning velocity of

---

[3]A diminution of the effectiveness of bromine, for $0.002 \leq X_{Br_2} \leq 0.012$, in reducing the burning velocity of 8.5% CH_4-air has been observed also as the pressure was _increased_ from 1 to 70 atm. (58).

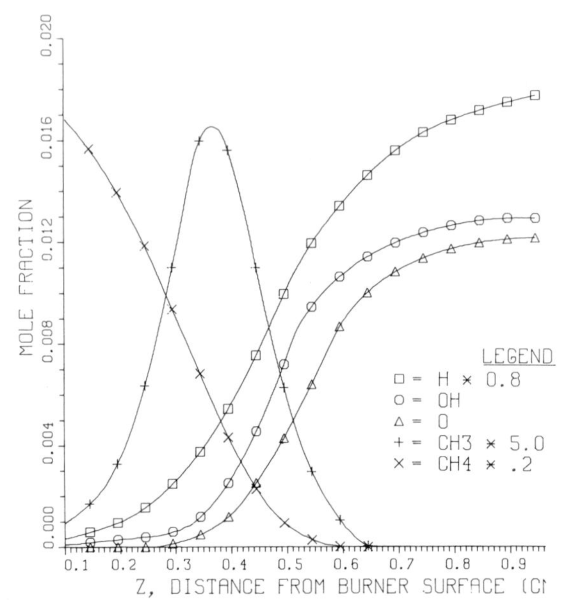

Figure 5. Concentration profiles for the major radical species of the uninhibited flame, I (see *Table I*), *for* Z < *1 cm*

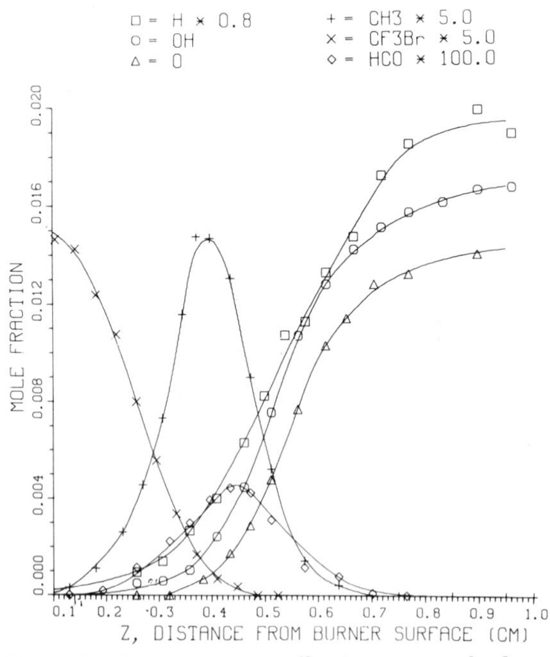

Figure 6. Concentration profiles for major radical species, CF_3Br, and HCO in the 0.3% CF_3Br containing flame, II (see *Table I*), *for* Z < *1 cm*

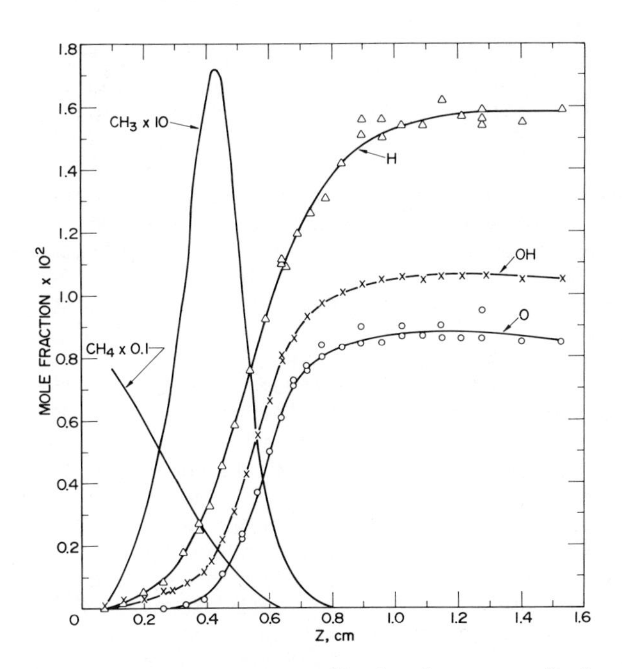

Figure 7. Concentration profiles for the major radical species and methane in the uninhibited flame, III (see Table I), for Z < 1 cm

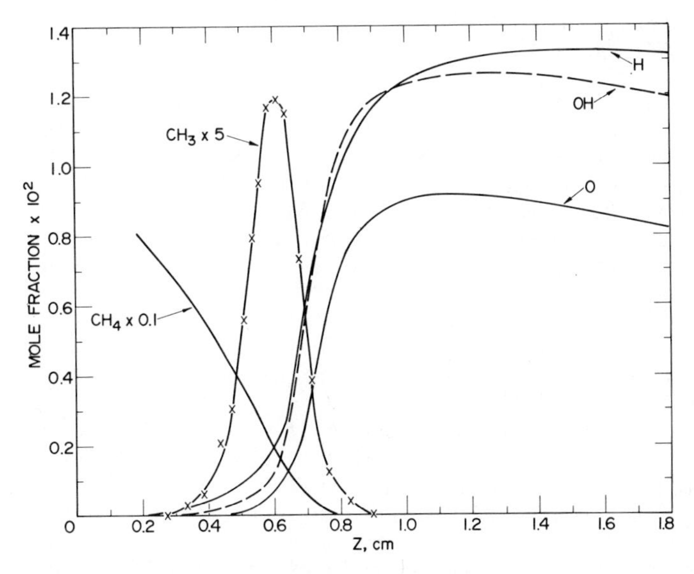

Figure 8. Concentration profiles for the major radical species and methane in the 1.1% CF_3Br containing flame, IV (see Table I), for Z < 1 cm

the flame formed on ignition is required to equal the stream velocity. Therefore, the position of equilibrium is further away from the porous plug, and this is just what is observed in the profiles of flames II and IV relative to I and III, respectively.

In each flame the maximum H, O, and OH concentrations are found just downstream of the primary reaction zone and these maxima are always greater than the concentrations calculated assuming thermodynamic equilibrium at the maximum flame temperature. Table IV shows the ratio of observed maximum to calculated equilibrium concentrations for H, O, OH and also Br where appropriate. In the equilibrium calculations the temperature used was the maximum flame temperature determined in the absence of the sampling probe. The sampling probe is expected to offer a "cooling" effect (34) and, indeed, a correction is made for this effect in analyzing the profiles for reaction rate data in the primary reaction zone as reflected by the temperature profile in figure 4. The ratios in Table IV should be considered to be conservative estimates of the radical overshoot since if, in the presence of the probe used to sample these radicals, a temperature 100°K lower than used were appropriate, the ratios would be larger. The effect is greatest for the H atom overshoot which would be 3-4 times larger if each flame temperature were 100°K lower than supposed. This uncertainty in the appropriate temperature for use in the equilibrium calculations is not critical, however, as it does not obscure the fact of radical overshoot, its order of magnitude, or the dependence of these on final flame temperature. The overshoot is greatest in the flame having the lowest final flame temperature, flame III, and smallest in that having the highest final flame temperature, IV.

Large supraequilibrium concentrations for radicals have been observed and have been well characterized with respect to temperature and stoichiometry for H_2-O_2 flames (60). Less work has been done on hydrocarbon flame radicals. Fenimore and Jones (61) found [H] in the burnt gas to be that expected from equilibrium calculations for 1 atm. rich hydrocarbon flames; Reid and Wheeler (62) found that an initial excess H in 1 atm. propane-air flames decayed to a steady-state value, presumably $[H]_{eq}$. They found $[H]/[H]_{eq.}$ increased as the flames were made leaner. Others have observed modest excesses of OH and O for 1 atm. C_2H_2-air (63), very large excesses at reduced pressure for all three radicals in lean C_2H_2-O_2(64), and $[H]/[H]_{eq}$ decreasing from 10 to 1 as the CH_4-air composition at 1 atm. was changed from rich to lean (65), in contrast to what might have been expected from the propane study (62).

The phenomenon of radical overshoot is an interesting topic for study in itself, but it may be peculiarly important for understanding the broader aspects of chemical flame inhibition, particularly its limitations. Babkin and V'Yun (58) used the

TABLE IV

Observed Excess Radical Concentrations in Stoichiometric Clean and CF_3Br- Containing CH_4-O_2-Ar Flames at 0.042 atm.

Flame[a/]	CF_3Br, Mole pct	T_{max}, °K	$\frac{[H]_{max}}{[H]_{eq.}}$	$\frac{[O]_{max}}{[O]_{eq.}}$	$\frac{[OH]_{max}}{[OH]_{eq.}}$	$\frac{[Br]_{max}}{[Br]_{eq.}}$
I	0	1868	250	100	10	-
II	0.3	1911	150	100	10	1
III	0	1781	600	150	10	-
IV	1.1	1966	50	30	5	1

a. See Table II.

effect of bromine on the burning velocity of CH_4-air flames at pressure >1 atm. as an indicator of the difference between the actual and equilibrium concentration of active centers as the pressure is increased. According to these authors, if the inhibitor effects a general reduction of radical concentration in the flame, it presumably cannot reduce their concentration below the thermodynamic equilibrium concentration at the final flame temperature. If the temperature (or the pressure) is high enough that the difference between actual and equilibrium radical concentration is small, the chemical inhibitor may be relatively ineffective. This could explain the facts that chemical inhibitors are less effective in oxygen than air flames, and that the observed decrease in burning velocity is not directly proportional to the amount of inhibitor added.

There is not yet sufficient data on radical concentration in normal or inhibited hydrocarbon flames to test this hypothesis directly. The data of Table IV does not really provide a test, except inasmuch as it shows that an overshoot does exist in these flames. For the four flames shown, the log of $[H]/[H]_{eq.}$ varies linearly with the reciprocal of the maximum flame temperature, just as has been observed in studies in uninhibited H_2-O_2 flames at both atmospheric and reduced pressure (61, 66).

The positions of the H, O, and OH profiles relative to each other are different between the clean and inhibited flames. In each of the normal flames these three profiles are well separated, H atom being observed earliest in the flame followed by OH and finally O. In the inhibited flames, most clearly illustrated in figure 8 (Flame IV) compared to figure 7 (Flame III), the H atom profile moves closer, as it were, to that for OH, and so the total spatial separation of the three profiles is smaller in the inhibited flames.

This direct observation of a delay in the appearance of H atom in measurable concentration supports the idea that the H atom and its removal is of special significance in chemical inhibition. It is also true, however, that the concentrations of the other chain carriers are affected when sufficient CF_3Br is present. If one plots the radical concentration as a function of temperature rather than distance it is seen that with 0.3% CF_3Br, figure 9, only [H] is reduced at equivalent temperatures below 1600°K with respect to the clean flame. With 1% CF_3Br, figure 10, all three, H, O, and OH, are found in lower concentration in the low temperature region of the flame. These observations conform to the model for inhibition suggested by Wilson (12), even though his experiments were conducted with very lean flames in which O and OH are the most abundant chain carriers in the low temperature region of the flame.

In the high temperature part of the flame where the three characteristic H_2/O_2 system reactions are balanced, it is clear that removal of any one of the radicals will be reflected in

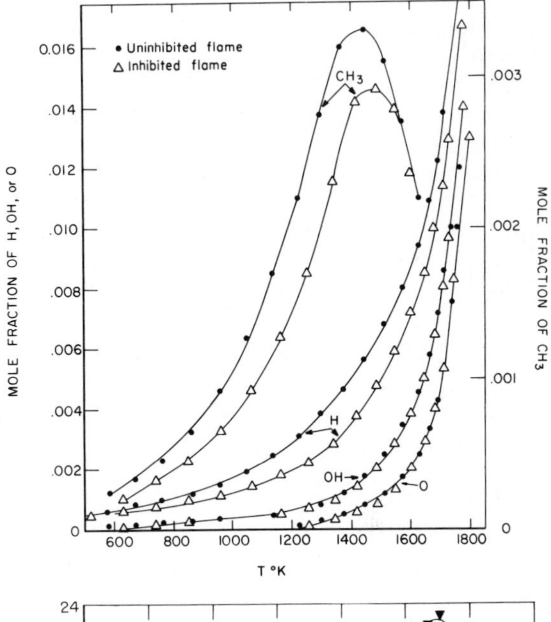

Figure 9. The mole fraction of H, O, OH, and CH_3 as a function of temperature in the uninhibited flame, III, and the 1.1% CF_3Br-containing flame, IV (see Table I).

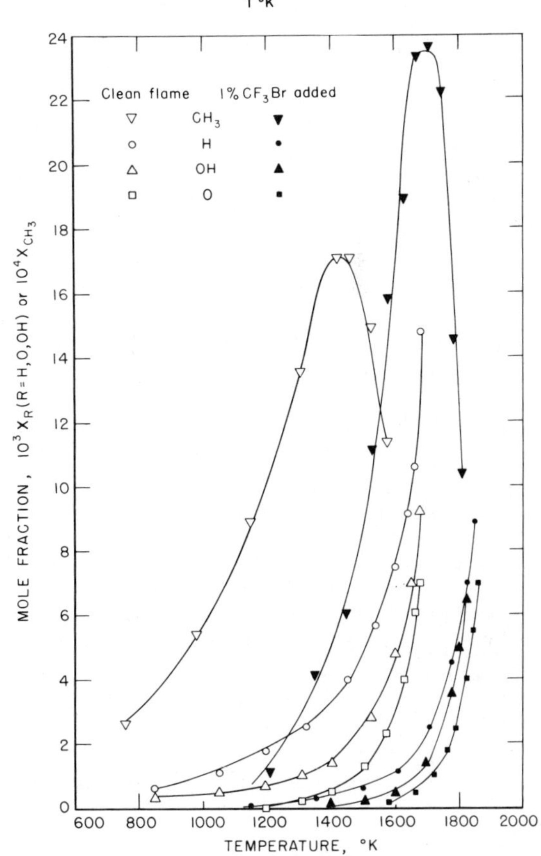

Figure 10. The mole fraction of H, OH, and CH_3 as a function of temperature in the uninhibited flame, I and the 0.3% CF_3Br-containing flame, II (see Table I)

a reduction of the other two. We observe a similar effect at low temperature although the reactions in question are not equilibrated at these temperatures.

For the flame containing 1% CF_3Br, the maximum [H] has decreased, the maximum [OH] has increased, and the maximum [O] is unchanged relative to the companion clean flame, III. This observation is independent of any assumptions made in obtaining absolute concentrations for these radicals. It may be derived from observation of the changes in the relative mass spectral intensitites for the species in question at Z=1.3cm, as the inhibitor flow is turned on and off. These are precisely the sorts of changes, both in direction and in magnitude, that one would expect if the bimolecular reactions of the H_2/O_2 systems are balanced at the final flame temperatures in flames IV and III. Similar observations were reported for flame I and II (29). To within about 20%, the observed absolute H, OH and O concentrations for all four flames just downstream of the primary reaction zone are those one would calculate using the balanced reactions hypothesis and with a knowledge of the equilibrium constants, temperatures, and the appropriate stable species concentrations.

For quenched flames and particularly at low pressure, the observation of a decrease or increase in the maximum radical concentration with inhibitor need not imply an analogous change in their concentrations in the primary reaction zone of the flame. Some apparent anomalies in the literature of inhibition may be resolvable by this consideration, or by comparison of calculated with observed radical profiles. For example, Wilson (12) found a calculated $[OH]_{max}$ greater in HBr (1.88%) inhibited than in clean 0.05 atm. methane flames; Pownall and Simmons (20) found the opposite for HBr (.07%) in 1 atm propane flames. In each case, the differences were about a factor of two, and this situation is sometimes cited as indicative of the need for a more comprehensive theory of chemical flame inhibition. In Wilson's case, although the maximum measured flame temperature was assumed to be the calculated adiabatic flame temperature (for the purpose of establishing the thermocouple radiation correction), the flame was, in fact, a quenched flame. There was heat loss to the burner, although it is not clear how much. Simmons, on the other hand, states that measured final flame temperatures were within 10°K of the "expected" values, but does not give these values for the two flames for which X_{OH} was calculated and compared. Thus, comparisons are limited at the outset. In addition, our experience suggests that the character of calculated OH profiles is dominated by the value of the rate constant chosen and by the neglect of the reverse reaction in the scheme $CO+OH = CO_2+H$, which is used to calculated X_{OH}, and not necessarily by the presence or absence of the inhibitor. We performed the same calculations on our data and compare the calculated hydroxyl profiles with the observed hydroxyl profiles, figure 11. Calculations were made neglecting the reverse reaction, parts A and B of figure 11, and including the reverse, C and D. Rate coefficients for the forward reaction were taken from a recent critical evaluation (67), from Wilson (12), and from Simmons (20). Neglecting the reverse reaction underestimates by 25-100% the calculated maximum OH mole fraction. There is also a misrepresentation of the shape of the OH profile inasmuch as exclusion of the reverse forces X_{OH} prematurely toward zero (fig. 11, A & B) as the reaction $CO+OH=CO_2+H$ becomes balanced. The absolute magnitude of the maximum concentration depends upon the value of the rate constant chosen and can be as much as 2 times greater than the observed. Note that for these two flames the calculated OH is greater in the inhibited flame, as observed, but the magnitude of the change is exaggerated in the calculated profile. One must, at the very least, exercise caution in interpreting these calculated radical profiles. The directly observed effect of CF_3Br on the radical concentration profiles appear to be more subtle than our measurements of specific reaction rate constants at flame temperatures.

The H, O, and OH profiles in the inhibited flames, again most obviously for the 1% CF_3Br addition, are steeper than in the uninhibited. The value of $|dX_{rad}/dZ|_{max}$ is 20-30% greater with inhibitor. No such change in slope is seen for the major reactants and products, their profiles being superimposable by shifting the Z axes. The greater slope for radicals in the inhibited flames results in a higher diffusion velocity for these species so that, whatever the specific chemical mechanism responsible for the radical trapping in the early part of the flame, back diffusion naturally tends to oppose the effect.

The methyl radical profile is narrower in the presence of CF_3Br and shifted downstream. The shift is about the same as for the major stable species when 0.3% CF_3Br is added, but it is nearly twice that expected from similar considerations with the addition of 1% CF_3Br. The maximum methyl radical concentration is actually greater in flame IV than in flame III, but the addition of 0.3% CF_3Br depresses this maximum relative to the uninhibited flame.

This behavior is similar to that observed by Wilson (12) for the methane reaction rate profiles. He interpreted this to be the result of a delayed attack on the fuel which then finds itself at higher temperature before it reacts and, since the temperature is higher, the maximum reaction rate is actually faster in the inhibited flame. The explanation would serve also for our observations in the 1% CF_3Br flame as it further implies a higher maximum concentration for methyl radicals (because the rate constant for H attack on methane, producing methyl radicals, goes up faster with temperature than the rate constant for the reaction destroying methyl radicals). In the 0.3% flame, however, there is al-

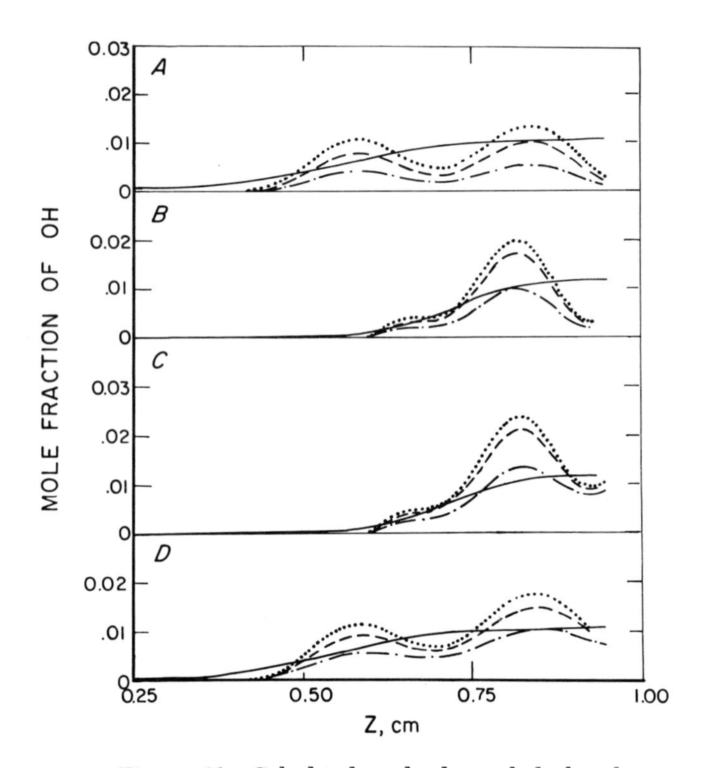

Figure 11. Calculated and observed hydroxyl radical profiles. Solid line, observed X_{OH}; dotted line, calculated using k_f *for* $CO + OH \xrightarrow{k_f} CO_2 + H$ *from (67); dashed line,* k_f *from (20); dash-dot line,* k_f *from (12). A. Flame III, ignoring the reverse reaction; B. Flame IV, ignoring the reverse reaction; C. Flame IV, including the reverse reaction; D. Flame III, including the reverse reaction.*

most no difference in the CH_4 net rate of reaction or the temperature region where this occurs, yet methyl radical is obviously affected by the inhibitor. Our initial postulate was that methyl and trifluoromethyl radicals recombine to give, eventually, the observed difluoroethylene, and this results in a depression of the methyl radical concentration. This reaction probably does occur, but the data from the 1% flame suggests that it is not the most significant reaction of the trifluoromethyl radicals in the inhibition scheme and that peak methyl radical concentrations can be higher in inhibited relative to normal flames, despite the occurrence of the CH_3+CF_3 recombination reaction. For methyl radicals, there is not in this flame, any point of reference analogous to the balanced reaction of the H_2-O_2 system radicals. This means methyl radical absolute concentrations are less accurate than H, O and OH concentrations. The character of the observations made here with regard to methyl radicals do not, however, depend upon any approximations made in estimating absolute methyl radical concentration.

Figures 12 - 15 show the extension of the H, O, and OH, and the temperature profiles into the secondary reaction zone for flames I-IV. The contrast between the appearance of the radical profiles in this region of the flame and in the primary reaction zone is striking. The very slow decay of these species from their (relatively rapidly attained) maxima is a reflection of the by now familiar understanding that the bimolecular reactions occurring in this region of the flame cannot by themselves lead to the final products and that slow, three body recombination reactions must occur in order to reach equilibrium eventually.

Hirschfelder (45) has shown that in the flame distances should scale as $P^{-\alpha/2}$, where α is the order of the dominant chemical reaction. For the primary reaction zone α=2 has given satisfactory experimental comparisons and is reasonable relative to what we know about the sorts of reactions occurring in the primary reaction zone, i.e., they are generally bimolecular.

In the secondary reaction zone, since termolecular reactions determine the course of the radical decay, the decay distance should scale as $P^{-3/2}$. Thus, what is observed here at 13 cm should correspond roughly to observations ~1mm beyond the primary reaction zone at 1 atm, and about 5cm beyond at 60 torr. These are the two pressures for which direct observations have been made of the effect of inhibitors on radical recombination in inhibited flames (see Table I). In both cases the inhibitors were other than halogenated hydrocarbons

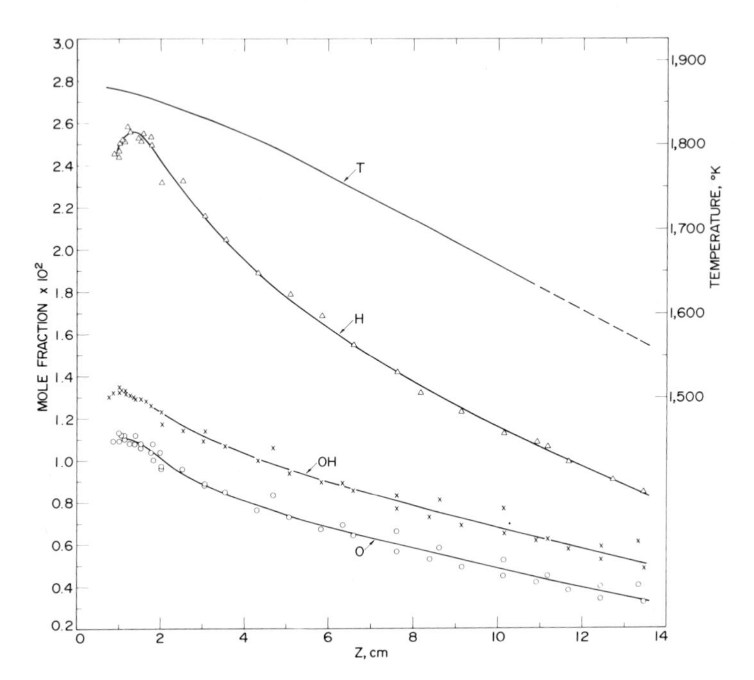

Figure 12. Temperature profile and concentration profiles for H, O, and OH in the uninhibited flame, I (see Table I), *for Z > 1 cm*

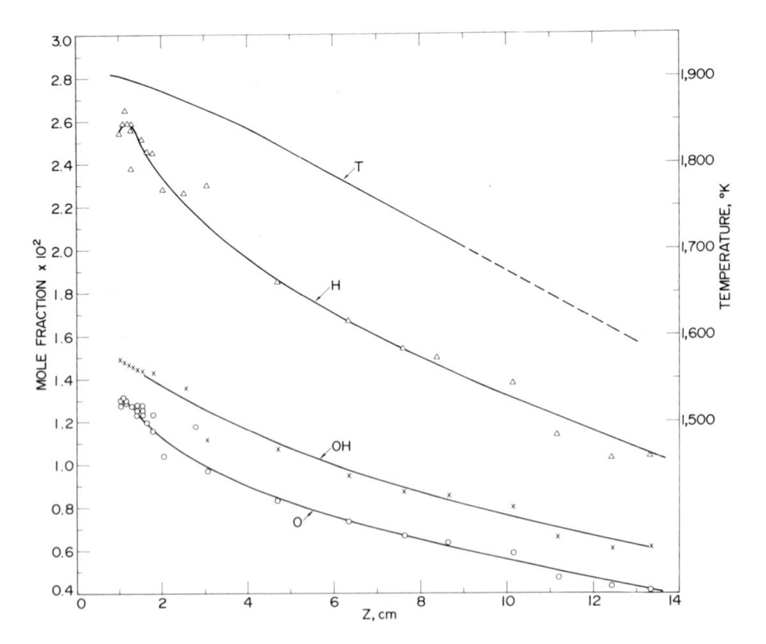

Figure 13. Temperature profile and concentration profiles for H, O, and OH in the 0.3% CF_3Br-containing flame, II (see Table I), *for Z > 1 cm*

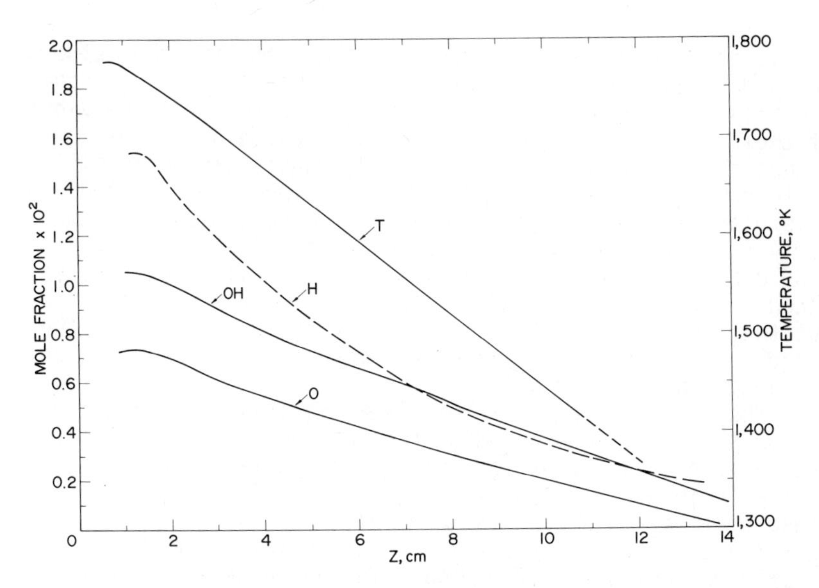

Figure 14. Temperature profile and concentration profiles for H ,O, and OH in the uninhibited flame, III (see *Table I*), *for* Z > *1 cm*

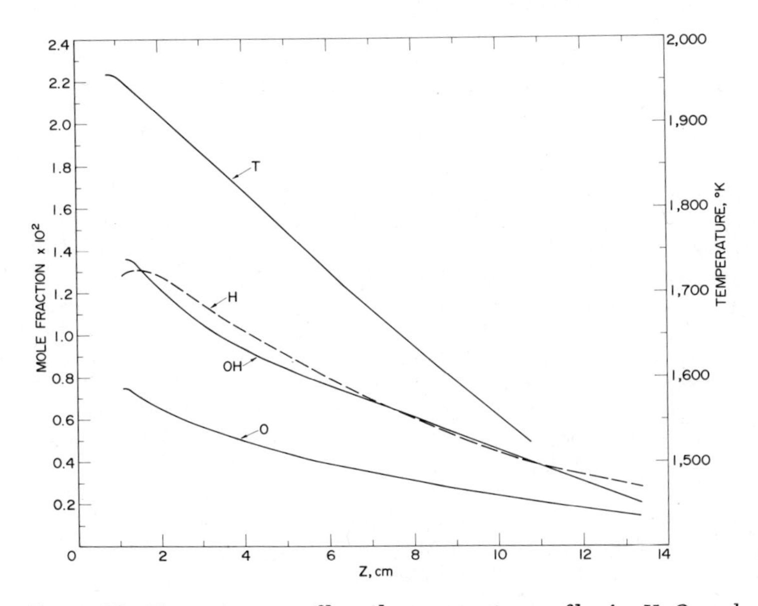

Figure 15. Temperature profile and concentration profiles for H, O, and OH in the 1.1% CF_3Br*-containing flame, IV* (see *Table I*), *for* Z > *1 cm*

and so comparison is limited.[4] In those studies the authors found that in the presence of the inhibitor OH radical recombination was enhanced and this can be observed from the slopes of their radical decay curves over the scaled distances. We do not observe a similar increased OH radical recombination, (although we have postulated additional inhibitor related bimolecular reactions that effectively remove radicals in the early (Z<2.5cm) part of the secondary reaction zone where F_2CO decays to give HF). We find, incidentally, that the balanced reactions remain so over the whole course of this decay, but that is a subject for another discussion. Likewise, the H and O do not suffer enhanced recombination in the presence of the inhibitor at 0.042 atm.

The non-observation of catalyzed radical recombination at low pressure, together with the observed reduction in effectiveness of CF_3Br at low pressure supports the view that radical recombination reactions do play an important role in these systems. One can imagine that, in an atmospheric pressure flame, if the only effect of the inhibitor were to effect a rapid recombination of radicals in the secondary reaction zone, there could result a lower maximum radical concentration on account of the now enhanced diffusion in the +Z direction, away from the primary reaction zone. Since the radicals in the low temperature region of the flame are there by virtue of diffusion, their concentration would be lowered also by this reduction in the maxima. The effect described could, presumably, occur and reduce radical concentration early in the flame even without the aid of competing bimolecular reactions.

To get an idea of the magnitude of the diffusion effect, consider the two uninhibited flames, I and III. Their initial compositions are the same, but the experimental conditions are such that the maximum [H] differs by nearly a factor of two, that in flame I being larger. A comparison of the H atom concentrations as a function of temperature for these two flames (compare figures 9 and 10) shows [H] to be reduced in flame II relative to flame I. The same order of magnitude decrease in [H] appears at low temperature in the 0.3% CF_3Br with respect to the clean flame, even though $[H]_{max}$ is the same in both flames. In this case, the decrease is a result of chemical reaction, i.e. "chemical" flame inhibition.

What this all adds up to, of course, is an increasingly complicated picture of "the mechanism" of chemical flame inhibition. It seems unlikely that a single reaction, such as $H+HBr \rightarrow H_2+Br$, will suffice to describe the essence of flame

[4]A mechanism that may account for the observations of Bonne et al (25) is that of Jensen and Jones (28) who studied $Fe(CO)_5$ in atmospheric H_2-O_2-N_2 flames.

inhibitors effectiveness, as is sometimes the case in, e.g., photochemical systems. Indeed, the reaction mentioned is found to be balanced in the low pressure flames studied here (29). In the absence of any competitive reactions that would force it to proceed in the forward direction (such as additional rapid reaction removing Br atoms, or forming HBr or Br_2), this reaction, once balanced, cannot lead to any net removal of H atoms. In low pressure flames, the termolecular reaction that might perform these functions should be too slow to be effective, but there may be competitive bimolecular reactions, e.g. $Br + HO_2$ $HBr+O_2$ (17).

We have already mentioned some of the reactions that need to be dealt with in any generalized scheme of inhibition, (1) - (5). There are others identifiable as requiring consideration in this particular system,

$$CF_3Br+H \rightarrow HBr+CF_3 \qquad (6)$$

$$CF_3+CH_3 \rightarrow (CF_3CH_3)^* \rightarrow CF_2CH_2+HF \qquad (7)$$

$$\overset{M}{\rightarrow} CF_3CH_3 \qquad (8)$$

$$(CF_3+H)? \rightarrow HF+CF_2 \qquad (9)$$

Reaction (9), and any other reaction producing HF in fluorinated inhibitor systems, is the closest to a true "scavenging" reaction since, once formed, the HF does not react to give back the H atom.

To summarize, the data presented here provides answers to some of the relatively straight forward questions that have been raised about halogenated inhibitors, but leaves others still unanswered. For example, it seems quite clear from the profiles that O atom is not preferentially removed anywhere in these inhibited flames, as was suggested in one theory of inhibition (31). If any radical is preferentially removed in these stoichiometric flames, it is the H atom, but if sufficient inhibitor is present, all the radical concentrations are reduced in the low temperature region of the flame, even if the H_2/O_2 system reactions are not strictly balanced in this region of the flame. This, and the observed delay in methyl radical production followed by a greater $[CH_3]_{max.}$ in the 1% CF_3Br, supports the idea that the halogenated chemical inhibitors act, at least in part, by competing with the fuel and oxidant for radicals in the relatively low temperature region of the flame.

On looking only at radical maxima, one may observe apparently anomolous behavior in the effect of CF_3Br on those maximum concentrations, but this may be simply a reflection of the change in final flame temperature and the subsequent adjustment of radical concentrations to their new values required by equilibration of the four H_2/O_2 reactions at the new temperature. The effect is most pronounced for quenched flames, and needs to be considered when comparisons are made among different studies. Calculated radical profiles can only be expected to resemble the actual profiles when both the mechanism and the rate constants are well defined. In particular, OH profiles calculated from CO_2 net reaction rates are unrealistic if the reverse, CO_2+H, is neglected.

Radical recombination in the hot gas region of the flame is not noticeably enhanced with CF_3Br at low pressure and CF_3Br is about half as effective in reducing burning velocity at 1/20 than at 1 atm. These two observations are consistent with the contention (17, 18) that termolecular reactions are an important part of the inhibition mechanism.

Dedication

This paper is respectfully dedicated to the memory of Walter E. Kaskan who contributed eloquently to the understanding of radical behavior in flames, and whose kindness and encouragement will be missed by us.

Abstract

Low pressure stoichiometric methane-oxygen-argon flames have been stabilized on a cooled, porous plug flat flame burner and have been probed for atom and radical species using molecular beam mass spectrometry. Concentration profiles for OH, H, O, and CH_3 were determined from the preheat zone, through the primary reaction zone, and into the secondary reaction zone for clean and CF_3Br-containing flames. In each flame supraequilibrium concentrations of H, O, and OH are observed. [H] atom is reduced at low temperature ($<\sim 1600°K$) in the 0.3% CF_3Br-containing flames, but with 1.1% CF_3Br, H, O, and OH are all less, at equivalent temperature, than in the companion clean flame. No catalyzed radical recombination occurs in the secondary reaction zone with the inhibitor. These and other observations are discussed with reference to current ideas regarding the mechanisms through which chemical inhibitors effect flame radicals.

Literature Cited

1. Calvert, Jack G., and Pitts, James N., Jr., "Photochemistry," pp. 596-603, John Wiley & Sons, Inc., New York, 1966.
2. Friedman, R., and Levy, J. B., Wright Air Development Center Technical Report 56-568, AD No. 110685, January, 1957, 99 pp.

3. Friedman, R., FRAR (1961) 3, pp. 128-132.
4. Fristrom, R. M., FRAR (1967) 9, pp. 125-160.
5. McHale, E. T., FRAR (1969) 11, pp. 90-104.
6. Geyer, G. B., Fire Tech. (May 1969), pp. 151-159.
7. Creitz, E. C., J. Res. Nat. Bur. Stand. (1970) 74A, pp. 521-530.
8. Hastie, J. W., J. Res. Nat. Bur. Stand. (1973) 77A, pp. 733-754.
9. Rosser, W. A., Wise, H., and Miller, J., "Seventh Internat. Symp. on Combustion," pp. 175-182, Butterworth Scientific Publications, London, 1959.
10. Edmondson, H. and Heap, M. P., Comb. Flame (1969) 13, pp. 472-477.
11. Levy, A., Droege, J. W., Tighe, J. J., and Foster, J. J., "Eighth Internat. Symp. on Combustion," pp. 524-533, Williams and Wilkens, Baltimore, 1962.
12. Wilson, W. E., O'Donovan, J. T., and Fristrom, R. F., "Twelfth Internat. Symp. on Combustion," pp. 929-942, The Combustion Institute, Pittsburgh, 1969.
13. Biordi, J. C., Lazzara, C. P., and Papp, J. F. "Fourteenth Internat. Symp. on Combustion," pp. 367-381, The Combustion Institute, Pittsburgh, 1973.
14. Wilson, W. E., "Tenth Internat. Symp. on Combustion," pp. 47-54, The Combustion Institute, Pittsburgh, 1965.
15. Burdon, M. C., Burgoyne, J. H., and Weinberg, F. J., "Fifth Internat. Symp. on Combustion," pp. 647-651, Reinhold Publishing Corp., New York, 1955.
16. Fenimore, C. P., and Jones, G. W., Comb. Flame (1963) 7, pp. 323-329.
17. Day, J. J., Stamp, D. V., Thompson, K., and Dixon-Lewis, G., "Thirteenth Internat. Symp. on Combustion," pp. 705-712, The Combustion Institute, Pittsburgh, 1971.
18. Lovachev, L. A., Babkin, V. S., Bunev, V. A., V'Yun, A.V., Krivulin, V. N., and Baratov, A. N., Comb. Flame (1973) 20, pp. 259-289.
19. Butlin, R. N., and Simmons, R. F., Comb. Flame (1968) 12, pp. 447-456.
20. Pownall, C. and Simmons, R. F., "Thirteenth Internat. Symp. on Combustion," The Combustion Institute, pp. 585-592, Pittsburgh, 1971.
21. Miller, W. J., Comb. Flame (1969) 13, pp. 210-211.
22. Homann, K. H., and Poss, R., Comb. Flame (1972) 18, pp. 300-302.
23. Biordi, J. C., Lazzara, C. P., and Papp, J. F., Comb. Flame (1975), 24, pp.
24. Ksnadopulo, G. I., Kolesnikov, B. Ya., and Odnorog, D. S., Doklady Akademii Nauk SSSR (1974) 216, pp. 1098-1101.
25. Bonne, U., Jost, W., and Wagner, H. Gg., FRAR (1962) 4, pp. 6-18.
26. Iya, K. S., Wollowitz, S., and Kaskan, W. E., "Fif-

teenth Internat. Symp. on Combustion," pp. 329-336, The Combustion Institute, Pittsburgh, 1975.
27. Bulewicz, E. M., and Padley, P. J., "Thirteenth Internat. Symp. on Combustion," pp. 73-79, The Combustion Institute, Pittsburgh, 1971.
28. Jensen, D. E., and Jones, G. A., J. Chem. Phys. (1974) 60, pp. 3421-3425.
29. Biordi, J. C., Lazzara, C. P., and Papp, J. F. "Fifteenth Internat. Symp. on Combustion," pp. 917-931, The Combustion Institute, Pittsburgh, 1975.
30. Fristrom, R. M., and Sawyer, R. F., paper presented at The Thirty-seventh AGARD Symp., Aircraft, Fuels, Lubricants, and Fire Safety, The Hague, May, 1971.
31. Creitz, E. C., Fire Technology (1972) 8, pp. 131-141.
32. Anon., "The Blind Men and the Elephant," Retold by Lillian Quigley, Charles Scribner's Sons, New York, 1959.
33. Biordi, J. C., Lazzara, C. P., and Papp, J. F., U. S. Bur. Mines RI 7723, 1973, 39 pp.
34. Lazzara, C. P., Biordi, J. C., and Papp, J. F., Comb. Flame (1973) 21, pp. 371-382.
35. Papp, J. F., Lazzara, C. P., and Biordi, J. C., U. S. Bur. Mines RI 8019, 1975, 90 pp.
36. Fristrom, R. F., and Westenberg, A. A. "Flame Structure," 424 pp., McGraw-Hill, New York, 1965.
37. Davis, M. G., McGregor, W. K., Jr., and Mason, A. A., Arnold Engineering Development Ctr. TR-69-95, October, 1969, 77 pp.
38. Foner, S. N., and Hudson, R. L., J. Chem. Phys. (1953) 21, pp. 1374-1382.
39. Milne, T. A., and Greene, F. T., "Tenth Internat. Symp. on Combustion," pp. 153-159, The Combustion Institute, Pittsburgh, 1965.
40. Homann, K. N., Mochizuki, M., and Wagner, H. Gg., Z. Phy. Chem. N. F. (1963) 37, pp. 299-313.
41. Hastie, J. W., Comb. Flame (1973) 21, pp. 187-194.
42. Williams, G. J., and Wilkins, R. G., Comb. Flame (1973) 21, pp. 325-337.
43. Peeters, J., and Mahnen, G., "Fourteenth Internat. Symp. on Combustion," pp. 133-141, The Combustion Institute, Pittsburgh, 1973.
44. Biordi, J. C., Lazzara, C. P., and Papp, J. F., Comb. Flame (1974) 23, pp. 73-82.
45. Hirschfelder, J. J., Curtiss, C. F., and Bird, R. B., "Molecular Theory of Gases and Liquids," pp. 756-783, John Wiley and Sons, Inc., New York, 1954.
46. Biordi, J. C., Lazzara, C. P. and Papp, J. F., U. S. Bur. Mines RI 8029, 1975, 42 pp.
47. Field, F. H. and Franklin, J. L., "Electron Impact and Ionization Phenomena," (Revised Edition), pp. 239-439,

Academic Press, New York, 1970.
48. Kieffer, L. J., and Dunn, G. H., Rev. Modern Phys. (1966) 38, pp. 1-35.
49. Melton, C. E., J. Phys. Chem., (1970) 74, pp. 582-587.
50. Rapp, D., and Englander-Golden, P., J. Chem. Phys. (1965) 43, pp. 1464-1479.
51. Otvos, J. W., and Stevenson, D. P., J. Am. Chem. Soc. (1956) 78, pp. 546-551.
52. Lampe, F., Franklin, J. L., and Field, J. H., J. Am. Chem. Soc. (1957) 79, pp. 6129-6132.
53. Kaskan, W. E.,"Sixth International Symp on Comb.," pp. 134-143, Reinhold Publishing Corp., New York, 1951.
54. Iya, K. S., Wollowitz, S., and Kaskan, W. E., Comb. Flame (1974) 22, pp. 415-417.
55. Hayes, K., and Kaskan, W. E., Comb. Flame, in press.
56. Biordi, J. C., Lazzara, C. P., and Papp, J. F., Comb. Flame, in press.
57. Homann, K. H., and Poss, R., Comb. Flame (1972) 18, pp. 300-302.
58. Babkin, V. S., and V'Yun, A. V. Combustion, Explosion and Shock Waves (1971) 7, pp. 203-206.
59. Botha, J. P., and Spalding, D. B., Proc. Roy. Soc. (1954) A225, pp. 71-96.
60. Fenimore, C. P., "Chemistry in Premixed Flames," pp. 11-42 The Macmillan Co., New York, 1964.
61. Fenimore, C. P. and Jones, G. W., J. Phys. Chem. (1958) 62, pp. 693-697.
62. Reid, R., and Wheeler, R., J. Phys. Chem. (1961) 65, pp. 527-530.
63. Zeegers, P. J. Th., and Alkemande, C. Th. J., Comb. Flame (1965) 9, pp. 247-257.
64. Eberius, K. H., Hoyermann, K., and Wagner, H. Gg., "Fourteenth Internat. Symp. on Combustion," pp. 147-155, The Combustion Institute, Pittsburgh, 1973.
65. Egorov, V. I., Ermolenko, V. I., and Ryabikov, O. B., Doklady Akademii Nauk SSSR (1974), 215, pp. 370-372.
66. Bascombe, K. N., "Tenth Internat. Symp. on Combustion," pp. 55-63, The Combustion Institute, Pittsburgh, 1965.
67. Baulch, D. L. and Drysdale, D. D., Comb. Flame (1974) 23, pp. 215-225.

DISCUSSION

R. G. GANN: Some of your peak flame temperatures are at or even slightly above the softening point of the quartz sampling cone. I therefore assume that the cone is cooler than the flame gas. As you scan through the flame, and thus through a wide temperature range, do you expect the cone - flame temperature difference to affect your relative concentration profiles, especially of radicals?

J. C. BIORDI: The cone is, of course, cooler than the flame gas and we have expended considerable effort in attempting to characterize this temperature perturbation. The empirical characterization is detailed in reference 44 of the paper. It appears that the effect, and this is true of microprobes also, may be treated as a position error, i.e., the sampled gas is more nearly characteristic of the conditions a small distance upstream from the actual probe tip position. The difficulty, then, lies in properly aligning temperature and composition profiles when they are determined, as is done here, in separate experiments. It seemed to us that the only way to assess objectively the errors to be expected by this procedure is to examine the behavior of rate constants of known reactions determined from flame data under conditions of extreme misalignment. The results are now under review. (Biordi, J. C., Lazzara, C. P., and Papp, J. F. "Molecular Beam Mass Spectrometry Applied to Determining the Kinetics of Reactions in Flames. Part II. A Critique of Rate Coefficient Determination." Submitted to Combustion and Flame.)

In regard to the profiles, X_i _vs._ Z, the radical profiles, looking as they do like "products" in these low pressure flames, probably fare better than, say, reactant profiles, since the perturbations by the probe seem to be greatest when the tip of the cone is below the luminous zone.

J. W. HASTIE: We have made several tests for the absence of a significant probe effect on the flame chemistry -- as discussed in the reference, J. W. Hastie, Int. J. Mass Spec. Ion Phys. (1975), _16_, 89.

Insofar as the mass spectrometric relative ionization cross sections can be estimated, we find from the addition to the flame of a halogen source, such as $SbBr_3$ which dissociates completely to yield Br-atoms, that a quantitative extraction of these atoms can be made using conical "lava" probes. We also observe satisfactory agreement between mass spectrometrically obtained H-atom flame profiles and those obtained with a non-perturbing optical spectroscopic technique -- at least in the reaction zone-burnt gas region of atmospheric pressure H_2-O_2-N_2 flames. Further, our observed HO_2 flame profiles are in accord with literature rate data for the pertinent HO_2-formation and loss-reactions. We should emphasize, however, that such observations are only semi-quantitative owing to the absence of more quantitative independent kinetic data.

N. J. BROWN: Errors in radical concentrations were quoted to be ±50%. Do these errors also obtain for relative radical concentrations? If so, do the trends observed for radical profiles in the inhibited flame related to the uninhibited flame still remain after errors in concentration are accounted for?

J. C. BIORDI: The ±50% refers to the limits of accuracy for absolute radical concentrations. The precision of the measurements, mass spectrometrically, is more like ±10%, so ratios of radical concentrations could be off by as much as 20%. However, in every case we verified, qualitatively and quantitatively, the observed trends by simply looking at the relative mass spectral intensities for the radical species in question while turning the inhibitor flow off and on. For H, O, and OH it is sufficient to go out to some value of Z $\sim$ 1-1.5 cm where in both flames the profile is more or less flat. For CH_3 the comparison was made at the position corresponding to the maximum in each flame.

A. S. GORDON: Did you measure the HBr profiles through the flame?

J. C. BIORDI: Yes, we did measure HBr profiles. These, as well as the profiles of other species related to the inhibitor for flame II were reported at the 15th Combustion Symposium and, in more detail, in ref. 36.

A. S. GORDON: Since your work is at low pressure

you suppose that three body recombination reactions are not very important, yet you do observe some inhibiting effect of CF_3Br. How does the inhibition you observe come about?

J. C. BIORDI: In the early part of the flame the reaction producing HBr, i.e., $H + CF_3Br \rightarrow HBr + CF_3$, as well as reaction (1), $H + HBr \rightarrow H_2 Br$, before it becomes balanced, will act to reduce the H atom concentration. If, as we suggest, HF is produced by a reaction such as (9), $CF_3 + H \rightarrow HF + CF_2$, its very early appearance in the inhibited flame also implies H atom scavenging. In addition, removal of methyl radicals is occurring to produce CH_2CF_2, presumably via a recombination of methyl and trifluoromethyl radicals. All of these reactions will cause inhibition of the flame's usual propagation reactions. Now if HBr can be regenerated, or, in fact, if any reaction occurs which causes reaction (1) to proceed in the forward direction, then the inhibiting action becomes catalytic. At present, the view seems to be that $H + Br + M \rightarrow HBr + M$ is the most likely reaction that can regenerate HBr. We suppose (without the benefit of a measured rate constant for this reaction, though) that this reaction is relatively unimportant in our low pressure flame. However, there is still the possibility that reaction (3), $Br + HO_2 \rightleftarrows HBr + O_2$, could be significant even at these pressures.

10

Inhibition of the Hydrogen–Oxygen Reaction by CF_3Br and CF_2BrCF_2Br

GORDON B. SKINNER

Department of Chemistry, Wright State University, Dayton, Ohio 45431

About ten years ago we published results of shock-tube experiments in which ignition delays of H_2 - O_2 - Ar mixtures containing either CF_3Br or CF_2BrCF_2Br, were measured (1, 2). Substantial increases in the ignition delay times were found for the gas mixtures containing halocarbons. These studies also included some product analyses of partially reacted H_2-O_2-Ar-halocarbon mixtures, and product analyses of shock heated Ar-halocarbon mixtures (with no H_2 or O_2 present).

An attempt was made to interpret the ignition delay data using a set of eleven elementary reactions for hydrogen-oxygen itself, five additional reactions for the CF_3Br experiments, and four additional reactions for the CF_2BrCF_2Br experiments. We were able to adjust the rate constants to give quite good agreement of the calculations with experiment, but it was recognized that rate constants for many of the reactions in the hydrogen-oxygen scheme were uncertain, so that the rate constants deduced for the inhibition reactions might well be substantially in error. This suspicion was strengthened by the fact that some of the deduced rate constants were substantially larger than would be expected from general chemical considerations.

Now that much more accurate date are available for the hydrogen-oxygen reaction, it appeared to be worthwhile to carry out a new kinetic analysis of our data.

The Hydrogen-Oxygen-Argon System

The 13-reaction scheme we have used for hydrogen oxidation is shown in Table I. This was developed during a study of methane oxidation (3,4) and while it does not contain all reactions of possible importance for hydrogen oxidation, it includes those most important for stoichiometric and richer mixtures between 900 and 1600 K. While the evidence concerning the initiation step is not entirely conclusive yet, the date of Jachimowski and Houghton (5) strongly suggest that our reaction 1, rather than

Table I Reactions and Rate Constants Assumed for the $H_2 - O_2$ Reaction with Argon Diluent

Reaction	A(forward)[a] A(reverse)	E(forward)[b] E(reverse)	Ref.
1. $H_2+O_2 = 2OH$	1.36E + 13 4.48E + 11	48.15 29.85	(5)
2. $OH+H_2 = OH+H$	2.20E + 13 8.40E + 13	5.15 20.10	(6)
3. $H+O_2 = OH+O$	2.20E + 14 1.30E + 13	16.80 0.00	(6)
4. $O+H_2 = OH+H$	5.90E + 13 2.50E + 13	11.20 9.05	(6)
5. $O+H_2O = 2OH$	6.80E + 13 6.80E + 12	18.35 1.13	(6)
6. $H+OH+Ar = H_2O+Ar$	2.60E + 15 4.90E + 16	-2.30 117.20	(6),c
7. $H+OH+H_2O = H_2O+H_2O$	4.35E + 16 8.00E + 17	-2.30 117.20	(6),c
8. $H+H+Ar = H_2+Ar$	2.40E + 14 1.00E + 15	-2.30 102.20	(6),c
9. $H+H+H_2O = H_2+H_2O$	4.00E + 15 1.60E + 16	-2.30 102.20	(6),c
10. $O+O+Ar = O_2+Ar$	1.28E + 13 1.28E + 15	-2.30 121.90	(7),c
11. $H+O_2+Ar = HO_2+Ar$	1.68E + 15 1.20E + 15	45.90 -1.00	(6),d
12. $HO_2+H_2 = H_2O_2+H$	1.10E + 12 2.55E + 12	18.70 3.75	(6),e
13. $H_2O_2+Ar = 2OH+Ar$	1.20E + 17 9.00E + 14	45.50 -5.07	(6)

(a) Units: $mole^{-1}$ cc sec^{-1} for bimolecular reactions and $mole^{-2}$ cc^2 sec^{-1} for termolecular reactions. All reactions are assumed to be elementary and to maintain the same order under all conditions considered. "E" means "times 10 to the".

(b) Units: Kcal.

(c) Equation in table matches references rate constant at 1200K. E of -2.30 Kcal. corresponds to a 1/T dependence of the rate constant in this temperature range.

(d) Equation in table gives rate constants 0.8 times those of reference.

(e) Equation in table gives rate constants 1.5 times those of reference.

Table II Comparison of Observed and Calculated Ignition Delay Times for H_2 - O_2 - Ar Mixtures

Mole %		Temp.	P	Ignition Time Microseconds		Reference
H_2	O_2	K	atm	Obs	Calc.	
8	2	1000	5	2630	3400	(1)
8	2	1050	5	410	325	(1)
8	2	1100	5	135	144	(1)
4	2	1200	1	230	230	(8)
4	2	1600	1	52	43	(8)

Table III Reactions Considered to be Involved in Inhibition

	Reaction	A(forward)[a] A(reverse)	E(forward)[b] E(reverse)	Ref.
14.	$CF_3Br+H = CF_3+HBr$	2.50E + 13 4.55E + 11	3.40 24.00	
15.	$CF_3+H = CF_3H$	5.00E + 13 2.80E + 15	4.00 110.30	
16.	$CF_3+H_2 = CF_3H+H$	3.50E + 11 5.20E + 12	9.20 11.20	(9)
17.	$HBr+H = Br+H_2$	1.00E + 14 2.70E + 15	2.90 19.70	(9)
18.	$Br_2+Ar = 2Br+Ar$	1.18E + 13 8.70E + 12	32.40 -11.40	(10)[c]
19.	$H+Br_2 = HBr+Br$	1.40E + 14 8.00E + 13	0.90 41.70	(9)
20.	$CF_3Br = CF_3+Br$	1.68E + 13 3.83E + 11	66.30 0.50	
21.	$CF_3Br+Br = CF_3+Br_2$	6.00E + 13 1.87E + 12	23.00 1.00	
22.	$C_2F_6 = 2CF_3$	2.10E + 17 1.60E + 13	92.00 0.00	(15)
23.	$HBr+Ar = H+Br+Ar$	1.00E + 15 7.20E + 14	83.80 3.70	(16,17)[c]
24.	$C_2F_4Br_2+H = C_2F_4Br+HBr$	5.00E + 13 4.15E + 10	3.00 20.80	
25.	$C_2F_4Br_2 = C_2F_4Br+Br$	2.00E + 13 1.46E + 09	68.60 1.00	
26.	$C_2F_4Br_2+Br = C_2F_4Br+Br_2$	1.00E + 14 4.00E + 11	20.00 4.00	
27.	$C_2F_4Br+H = C_2F_4+HBr$	2.50E + 13 1.80E + 11	2.00 74.40	
28.	$C_2F_4Br = C_2F_4+Br$	1.00E + 13 7.10E + 09	50.00 36.70	

Table III (cont'd.)

	Reaction	A(forward)[a] A(reverse)	E(forward)[b] E(reverse)	Ref.
29.	$C_2F_4Br+Br = C_2F_4+Br_2$	5.00E + 13 1.72E + 12	8.00 46.10	

(a) Units: Sec^{-1} for unimolecular reactions, $mole^{-1}$ cc sec^{-1} for bimolecular reactions and $mole^{-2}$ cc^2 sec^{-1} for termolecular reactions. All reactions are assumed to be elementary and to maintain the same order under all conditions considered. "E" means "times 10 to the".

(b) Units: Kcal.

(c) In the calculations these rate constants were multiplied by 1.2 to allow for the fact that most of the reactants present would be more efficient energy transfer agents than Ar.

$$H_2 + O_2 = HO_2 + H$$

may be the most important one. Most of the other rate constants, as indicated in the table, have been taken from the review by Baulch, Drysdale, Horne and Lloyd (6), but a few modifications within their recommended error limits have been made to give better agreement of calculated ignition delays with experimental values. Table II shows the extent of agreement of ignition delay times calculated by this scheme with literature data.

In our experimental work the end of the ignition period was defined as the moment when the intensity of OH emission at 309-310 nm reached its maximum. It has been shown for both hydrogen and methane oxidation that this maximum OH emission occurs at nearly the same point in the reaction as the maximum OH concentration (which can be detected by absorption in the same region) and the maximum rate of temperature rise. In our calculations we have defined the end of the ignition period as the point where the OH concentration reaches a maximum value, except for the last two values reported in Table II, where both experimental and calculated ignition times correspond to the point where the concentration of OH reaches 10^{-9} mole/cc.

Experimental temperatures were based on measurements of incident shock speeds. Pressures behind the reflected shock waves were also measured. These were in good agreement (within a few percent) of the calculated values right behind the reflected shock waves, but usually the pressure rose several percent during the ignition period, apparently due to a slight pressure gradient behind the incident shock wave due to a small amount of attenuation. (Since the pressure rise occurred behind shocks in pure argon, it could not be due to chemical reactions.) Temperature changes due to these pressure changes were calculated assuming isentropic conditions, and the reported temperatures are arithmetic averages of those calculated during the ignition periods. No allowance for heats of reaction was made in these calculations.

Inhibition by CF_3Br

The reactions considered potentially important for inhibition of the H_2-O_2 reaction are shown in Table III. Of the ten reactions considered for CF_3Br (14-23), five (16-19 and 22) could be specified within factors of less than two from literature data. These five sets of rate constants were not varied during the analysis. Forward and reverse rate constants of these and all other reactions involved in CF_3Br inhibition could be checked for thermodynamic consistency by means of the JANAF tables (11). These tables also were used to calculate reverse rate constants for those reactions where forward rate constants had to be estimated.

For Reaction 15 a forward rate constant identical to one we developed for the reaction

$$CH_3 + H = CH_4$$

was assigned. The methane rate parameters were arrived at from an RRKM calculation that reconciles as well as possible the data of Hartig, Troe and Wagner (12) and Skinner and Ruehrwein (13). The Arrhenius parameters are applicable in a dilute argon mixture of 5 atm. pressure in our temperature range, a region in which the rate constants as written for second- and first-order reactions would show a dependence on the argon concentration. Actually, this reaction was not very important until the end of the ignition period, and hence did not have a strong influence on the results of the calculations.

Rate constants for Reaction 20 have been reported by Sehon and Szwarc (14) and discussed by Benson and O'Neal (15). The latter authors recommend an Arrhenuis A of $5x10^{13}$ sec^{-1} and Arrhenius E of 66.3 Kcal. We have used the somewhat lower A value of Table III since it leads to better agreement with our experimental pyrolysis results. This reaction was not included in our original scheme since it seemed likely from the observed result that CF_3Br disappeared, at a given temperature, much faster in the presence than the absence of the H_2 and O_2, that during the inhibition Reaction 14 removes CF_3Br much faster than Reaction 20 does. While this conclusion still appears correct, we have realized that Reaction 20, coupled with Reactions 16 and 17 (reverse), can be an effective initiating reaction even if its rate is not very large, since the rate of Reaction 1 is not very large either.

Rate constants for Reaction 21 are essentially those reported for

$$CH_3Br + Br = CH_3 + Br_2$$

by Trotman-Dickenson and Milne (9). The similarity in rate constant would be expected from both the general similarity of the reactions and the close similarity of the thermochemistry.

Reaction 23, and especially its reverse, the combination of H and Br to form HBr, was included mainly because of its possible effect on CF_2BrCF_2Br inhibition, in which it was expected that relatively large concentrations of Br atoms would be produced, and it had only a small effect on CF_3Br inhibition. Rate constants were obtained from data of Westberg and Greene (16) and Cohen, Giedt and Jacobs (17) who studied HBr dissociation in the 1500-2500K range. The two studies differed by about a factor of two in rate constants, and an extrapolation from 1500K down to our 900-1300K temperature range was needed, so an uncertainty of a factor of 3 in these rate constants could be present. As mentioned above, this uncertainty is not very significant for CF_3Br inhibition because the Br atom concentration remains low.

In reaction 14, which we consider to be the chief inhibiting one, a low Arrhenius activation energy (no more than 5 Kcal) and a high Arrhenius A (above $2x10^{13} mole^{-1} cc\ sec^{-1}$) would be expected by comparison with similar reactions such as 19. Within these limitations, rate constants for this reaction were considered variable parameters, while rate constants for the other reactions (15, 20, 21 and 23) were considered variable within limited ranges of experimental uncertainty.

Calculations on the pyrolysis of CF_3Br were made first. From the product analyses of our original study, it seemed that the principal course of the reaction is

$$2CF_3Br = C_2F_6 + Br_2$$

where, at equilibrium, a considerable amount of the bromine would be present as Br atoms. The equilibrium constant for the overall reaction changes slowly, from 0.76 at 1100K to 0.62 at 1300K, so that (allowing for formation of Br atoms) 50 to 60% of the CF_3Br would be decomposed at equilibrium. This calculation indicates that equilibrium is being approached in our higher-temperature runs, but that at lower temperatures just above 1100K, equilibrium was not being approached and the observed conversions were determined by the kinetics.

In our experimental study, final concentrations of both CF_3Br and C_2F_6 were measured for 1% and 4% mixtures of CF_3Br in argon, at 5 atmospheres total pressure, heated for times of about 14 milliseconds. At low conversions, it is appropriate to give greater weight to the C_2F_6 analyses, since at 5% conversion a 2% error in CF_3Br would correspond to a 40% error in C_2F_6. From smooth curves of the data, weighted as indicated, it appears that 6.0% of the CF_3Br in the 1% mixture would be converted to C_2F_6 in 14 milliseconds and 1120K, while 4.3% of the 4% mixture would be converted. These, then, are the "experimental" date to which calculated conversions should be compared.

Pyrolysis of CF_3Br involves only a small subset of the reactions we have assumed for inhibition. These are:

$$CF_3Br = CF_3 + Br \quad \text{(Reaction 20)}$$

$$CF_3Br + Br = CF_3 + Br_2 \quad \text{(Reaction 21)}$$

$$C_2F_6 = 2CF_3 \quad \text{(Reaction 22)}$$

$$Br_2 + Ar = 2Br + Ar \quad \text{(Reaction 18)}$$

Of course, Reaction 22 will proceed mainly in the reverse direction during the pyrolysis. With the rate constants of Table III, it was calculated that 5.6% of the CF_3Br in the 1% mixture, and 5.8% in the 4% mixture, would be converted to C_2F_6. This seems to be good agreement with the experimental results.

Some features of the pyrolysis process brought out by the calculations may be noted. Most of the Br atoms produced by Reaction 20 reached further by Reaction 21 rather than by the reverse of 18. As a result, Reaction 18 proceeded mainly in the forward direction, producing Br atoms, and in the end several times as much CF_3Br disappeared via Reaction 21 than via 20. The reverse rate of Reaction 22 was always at least 100 times as great as the forward rate, but for all the other reactions the forward and reverse rates had come within a factor of 2 of one another after 14 milliseconds of reaction. At 14 milliseconds, the Br_2 concentrations in the 1% and 4% mixtures were 2.1 and 4.8 times the Br concentrations, respectively.

For inhibition of the H_2-O_2 reaction by CF_3Br, experimental data to be matched by the calculations were the delay times shown in Figure 1 and the analyses of partially reacted mixtures. Specifically, we found that heating an inhibited gas mixture at 1000K for 14 milliseconds resulted in conversion of 17% of the CF_3Br to CF_3H. With the rate constants of Table III, good agreements of calculated ignition times and CF_3H yield was obtained, as shown in Figure 2. The rate expression for Reaction 14 appears to be in a reasonable range, with both A and E of the right magnitude. It should be pointed out, however, that the calculations are insensitive to small changes in A and E that result in unchanged rate constants in the middle of our temperature range. For example, an increase of E by 1.6 Kcal.(to 5.0 Kcal) and a corresponding increase in A of a factor of 2 result in essentially unchanged ignition delay times and a very small drop in the calculated conversion of CF_3Br to CF_3H. Changes in the absolute magnitude of the rate constant are, of course, more important. Doubling of the rate constant for Reaction 14 increased the calculated ignition delay from 3.0 to 5.0 milliseconds at 1200K, while decreasing the calculated conversion of CF_3Br to CF_3H at 1000K from 11% to 6%.

Biordi, Lazzara and Papp (18) have reported an Arrhenius A of 1.22×10^{14} $mole^{-1} cc\ sec^{-1}$ and an E of 8,430 cal. for Reaction 14. At 1150K their rate constant is 0.55 times that of Table III, which might be considered reasonably good agreement. Considering the uncertainties in the many other reactions involved in the combustion process, as well as in the experimental data, an uncertainty of about a factor of 3 in k_{14} may be estimated, in the temperature range we have been considering.

It is of interest to note the rates, and consequently the roles and relative importance, of the various reactions in the inhibition process. Table IV shows rates (in mole cc^{-1} sec^{-1}) for a calculation for a mixture of 8% H_2, 2% O_2, and 0.8% CF_3Br at 5 atm., initial temperature 1200K, at 1.5 milliseconds, which is half-way through the ignition period. At this time, about 96% of the original H_2, 97% of the O_2, and 69% of the CF_3Br remain, while the temperature has risen by 20 degrees. Those reactions

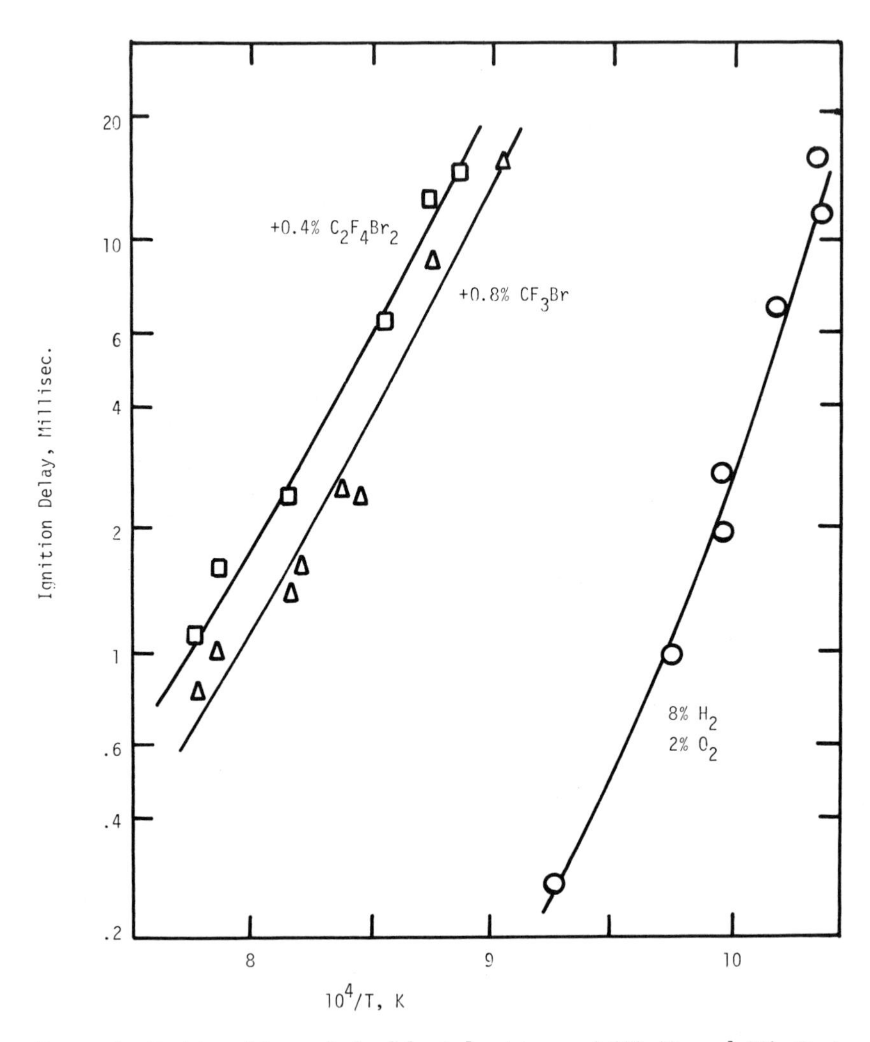

Figure 1. Ignition delays of shock-heated mixtures of 8% H_2 and 2% O_2 in argon alone and with added CF_3Br or $C_2F_4Br_2$, at a total pressure of 5 atm

not listed contribute to only a minor extent at this stage of the process. It is clear that Reaction 20 is a much more important chain initiator than Reaction 1. Even so, over 10 times as much CF_3Br disappears by Reaction 14 as by Reaction 20. Much more H atom is produced and consumed by Reactions 14, 16 and 17 than by the hydrogen-oxygen combustion reactions, with the result that the CF_3Br disappears much more rapidly than the H_2 and O_2. Once the CF_3Br is gone, the slight inhibiting effect of Br_2 and HBr are incapable of controlling the growth of H (and other active species) so their concentrations grow rapidly and ignition occurs.

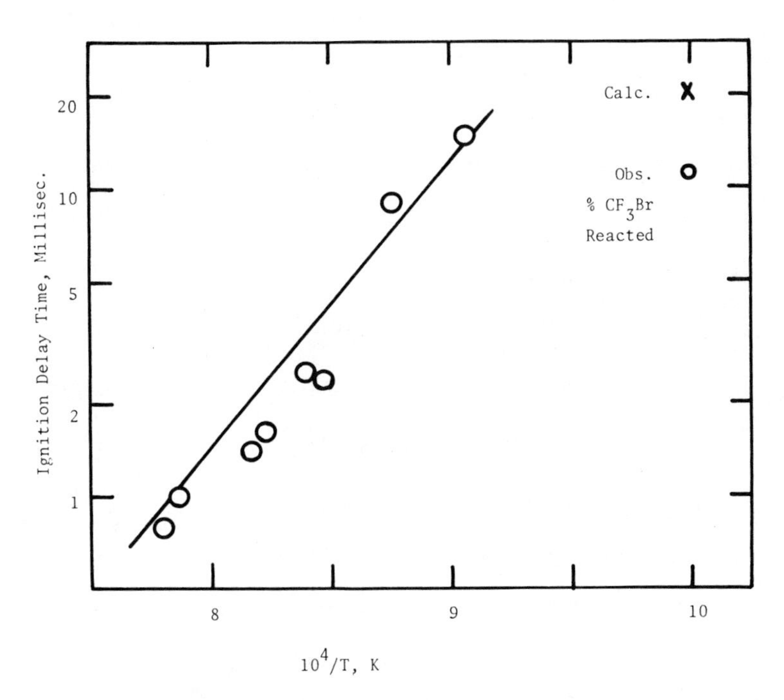

Figure 2. Observed (O) and calculated (———, X) ignition delays and product yields for mixture of 8% H_2, 2% O_2, and 0.8% CF_3Br at 5 atm

Table IV Rates of Elementary Reactions at 1.4 Milliseconds for Heating of Mixture of 8% H_2, 2% O_2, 0.8% CF_3Br at 5 Atm. and an Initial Temperature of 1200K

	Reaction	Rate, Mole^{-1} cc sec^{-1} Forward	Reverse
1.	$H_2+O_2 = 2OH$	1.22E-07	1.77E-17
2.	$OH+H_2 = H_2O+H$	3.13E-05	3.44E-06
3.	$H+O_2 = OH+O$	1.25E-05	2.18E-10
4.	$O+H_2 = OH+H$	1.25E-05	1.08E-10
11.	$H+O_2+Ar = HO_2+Ar$	7.23E-07	4.81E-06
12.	$HO_2+H_2 = H_2O_2+H$	3.10E-06	2.60E-09
13.	$H_2O_2+Ar = 2OH+Ar$	3.05E-06	3.08E-12
14.	$CF_3Br+H = CF_3+HBr$	1.02E-04	3.09E-09
15.	$CF_3+H = CF_3H$	6.32E-07	3.82E-12
16.	$CF_3+H_2 = CF_3H+H$	6.85E-05	5.00E-07
17.	$HBr+H = Br+H_2$	2.21E-04	2.26E-04
19.	$H+Br_2 = HBr+Br$	6.08E-07	2.40E-10
20.	$CF_3Br = CF_3+Br$	6.05E-06	2.53E-07
21.	$CF_3Br+Br = CF_3+Br_2$	9.19E-07	1.47E-07
22.	$C_2F_6 = 2CF_3$	1.36E-07	1.99E-05

Inhibition by CF_2BrCF_2Br

Our kinetic analysis for this inhibition process has been less satisfactory than that for CF_3Br, largely because of the absence of thermodynamic or kinetic data for C_2F_4Br, an expected intermediate in the reaction scheme. Even for $C_2F_4Br_2$, thermodynamic data are incomplete. The National Bureau of Standards tables (19) include an estimate of the heat of formation of $C_2F_4Br_2$ at 298 K ($\Delta Hf°$=-186.5 Kcal). We lowered this value to -189 Kcal to make it more compatible with JANAF data for related compounds, which tend to be a little lower than the NBS ones where comparisons can be made. The heat of formation of C_2F_4Br was estimated by assuming that the difference between its heat of formation and that of $C_2F_4Br_2$ is the same as the difference between that of CF_3 and CF_3Br (43 Kcal) since in both cases a Br atom is being removed. This leads to -146 Kcal for $\Delta Hf°$ for C_2F_4Br at 298 K. Entropies of $C_2F_4Br_2$ and C_2F_4Br at 298 K were estimated, respectively, at 88 and 82 cal deg^{-1} $mole^{-1}$, by comparison with data on a variety of related compounds in the NBS tables.

Once these estimates were made, equilibrium constants for each of the inhibiting reactions was obtained by calculating $\Delta S°$ and $\Delta H°$ for each reaction at 298K, and assuming these quantities are constant with temperature. This approximation was tested on some of the reactions involving CF_3Br for which complete data are available, and appears to be reasonably accurate for compounds of this type.

Reactions 24-29 of Table III, along with 17, 18, 19, and 23, were considered to be of potential importance for inhibition by $C_2F_4Br_2$. This mechanism is a close parallel to the CF_3Br scheme, with Reaction 24 corresponding to 14, 25 to 20, and 26 to 21, while 27-29 are C_2F_4Br analogues of 24-26. As with CF_3Br, it seemed desirable to use the pyrolysis data if possible. Steps involved in pyrolysis would be expected to be

$$C_2F_4Br_2 = C_2F_4Br + Br \quad \text{(Reaction 25)}$$

$$C_2F_4Br_2 + Br = C_2F_4Br + Br_2 \quad \text{(Reaction 26)}$$

$$C_2F_4Br = C_2F_4 + Br \quad \text{(Reaction 28)}$$

$$2Br + Ar = Br_2 + Ar \quad \text{(Reaction 18, reverse)}$$

with the overall reaction being

$$C_2F_4Br_2 = C_2F_4 + Br_2$$

under our experimental conditions, although a few percent of the bromine would be expected to end up as Br atoms.

Equilibrium calculations indicate that approximately 17% of the $C_2F_4Br_2$ should be decomposed at 1000K, and 50% at 1200K, which are temperatures near the ends of the experimental range. Since experimental conversions at these temperatures were 15% and 50% (for pyrolysis of 2% $C_2F_4Br_2$ in argon at 5 atm total pressure and a reaction time of 15 milliseconds) it seems clear that equilibrium was being approached in all of our experiments, so that quantitative kinetic data cannot be deduced from them.

From Figure 3 it may be seen that $C_2F_4Br_2$ pyrolysis is a good deal more rapid than that of CF_3Br. It does not seem likely that Reaction 25 will be faster than Reaction 20, since our estimation of its activation energy is about 2 Kcal. higher than that of Reaction 20, and the Arrhenius A should not be much different. The probable cause is a linear reaction chain involving Reactions 26 and 28, which parallel the reactions

$$C_2H_5 = C_2H_4 + H$$

$$H + C_2H_6 = C_2H_5 + H_2$$

in ethane decomposition. This sequence leads to several molecules of $C_2F_4Br_2$ being decomposed for each one that decomposes via Reaction 25.

Because of the lack of kinetic data, rate constants for Reaction 25 were set by comparison with 20. The slightly higher A value reflects the dual reaction path available for 25, partially compensated by the expected smaller moment of inertia change in the formation of an activated complex for $C_2F_4Br_2$ decomposition, as compared to CF_3Br. Similarly, rate constants for Reaction 26 were set by comparison with those of 21. For Reaction 29, which proved to be unimportant since concentrations of both C_2F_4Br and Br were low in inhibition experiments, A was set at half that for Reaction 26 because of the single reaction path, while E was lowered since Reaction 29 is exothermic.

The rate constant for Reaction 22 cannot be derived, even approximately, from any of our data, and it is difficult to estimate from general considerations. One appealing possibility is that the reaction may be so fast that all of the C_2F_4Br produced by reactions 24-26 will immediately decompose to C_2F_4 and Br. This will eliminate consideration of Reactions 27-29, a great simplification. Under these conditions, Reactions 24 to 26 would be written as

$$C_2F_4Br_2 + H = C_2F_4 + HBr + Br \quad \text{(Reaction 24A)}$$

$$C_2F_4Br_2 = C_2F_4 + 2Br \quad \text{(Reaction 25A)}$$

$$C_2F_4Br_2 + Br = C_2F_4 + Br_2 + Br \quad \text{(Reaction 26A)}$$

Since these reactions are no longer elementary, it was assumed that reverse rate constants would be 0 (a reasonable assumption, since in calculations with the full set of reactions the reverse rates for 24-26 were much smaller than the forward rates). It was found that with forward rate constants for Reactions 25A and 26A as given in Table III for Reactions 25 and 26, calculated ignition delays were low for all reasonable rate constants for Reaction 24A. For an Arrhenius A of 4.0 X 10^{14} and an activation energy of 3.0 Kcal (which seems an upper limit for the rate constant) calculated ignition delays, as indicated in Figure 4, are low by a factor of 2 to 3. For what seem the "most reasonable" kinetic parameters for Reaction 24A, the A and E shown in Table III, calculated delay times are low by a factor of 4 to 5. Comparison with the 11% of $C_2F_4Br_2$ lost in 15 milliseconds at 935K showed that with the upper limit calculation 4% of the $C_2F_4Br_2$ was lost, while with the "most reasonable" calculation 5% of the $C_2F_4Br_2$ was lost.

An important reason for the low ignition delay times in the above calculations is the rapid production of Br atoms by decomposition of C_2F_4Br, and the subsequent reaction of these Br atoms with H_2 via Reaction 17 (reverse) to produce H atoms. A finite rate of dissociation of C_2F_4Br will slow this process, thereby lengthening the ignition delay time. The approach used was to assign an Arrhenius A of 10^{13} sec^{-1} to Reaction 28 (this being just half the A value for Reaction 25) and to try increasing values of the Arrhenius E until a noticeable effect on the ignition delay times was observed. It was found that E had to be increased to 45 or 50 Kcal to produce an appreciable difference in the calculated delay times. The value of 50 Kcal. leads to good agreement between calculated and observed ignition delays when combined with the "most reasonable" A value of 5 X 10^{13} for Reaction 24, and these figures have been listed in Table III. Good agreement with observed ignition delays can also be obtained however, by using larger A values for Reaction 24, up to 4 X 10^{14}, with somewhat smaller E values for Reaction 28, so our calculations do not define a unique set of rate constants for the inhibition reactions.

The slower rates of decomposition of C_2F_4Br result in substantially smaller conversions of $C_2F_4Br_2$ to C_2F_4 than were found for "instantaneous" decomposition - less than 0.1% decomposition at 935K for the rate equations that gave ignition delays in agreement with experiment. This poor agreement at 935K suggests that slow decomposition of C_2F_4Br may not be the right approach to interpreting our results. An activation energy of 50 Kcal. does seem very high for Reaction 28. The reaction is endothermic by only about 15 Kcal, so that an activation energy of about 20 Kcal would seem more reasonable.

One alternative approach would be to consider possible changes in certain of the other reaction rate constants. For example, a reduction for Reaction 17 would reduce the rate of conversion of Br atoms to H, while increases for Reactions 18 and 23 would lead to reductions in Br and H atom concentrations.

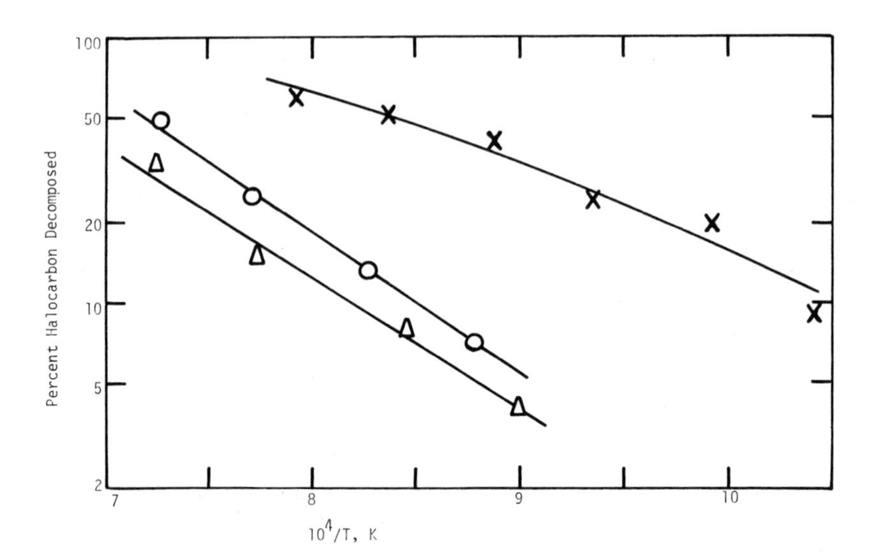

Figure 3. Percent of halocarbons disappearing in pyrolysis experiments with total pressure 5 atm, heating time averaging 15 msec, with argon as diluent. Mixtures are: 4% CF_3Br, △; 1% CF_3Br, O; 2% $C_2F_4Br_2$, X.

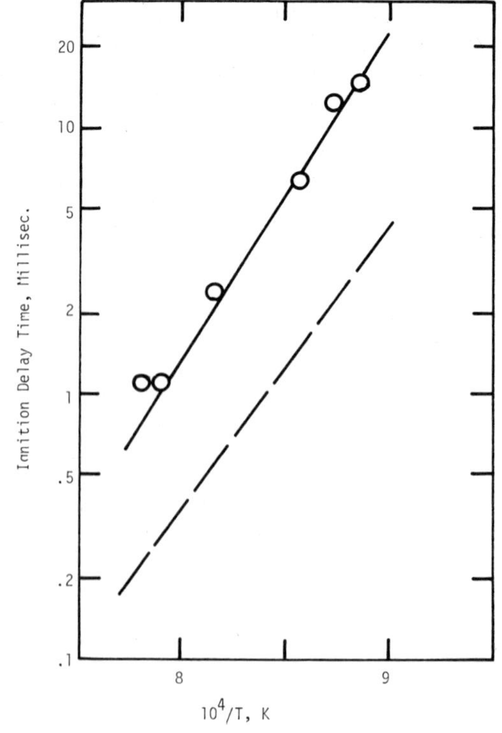

Figure 4. Ignition delays for mixture of 8% H_2, 2% O_2, and 0.4% $C_2F_4Br_2$ at 5 atm: observed, O; calculated from rate constants of Table IV, ———; Calculated using rate constants of Table IV for Reactions 24, 25, and 26, and instantaneous decomposition of C_2F_4Br, - - -

These changes would increase ignition delay times, but at the same time reduce the already low amount of $C_2F_4Br_2$ decomposed in the calculation at 935K. Another approach would be to consider additional reactions leading to formation of COF_2 (which was observed in somewhat larger relative concentrations in $C_2F_4Br_2$ inhibition than in CF_3Br inhibition) and of HF and CO, which were not measured but are expected reaction products. Perhaps these reactions would result in slower ignition. Unfortunately, knowledge of even which reactions occur, not to mention their rate constants, is scant, and it seems of doubtful value to modify an already somewhat questionable reaction scheme by addition of such reactions.

Conclusions

The kinetic analysis for CF_3Br inhibition reproduces the experimental data using either literature values of the rate constants, or "reasonable" values where literature data are unavailable or uncertain. While the reaction scheme is incomplete, in the sense that it does not account for the appearance of COF_2 and other products, such as CO and HF, which would be expected to appear in later stages of the reaction, the fact that only a small fraction of CF_3Br is converted to COF_2 up to the point where 50 to 60% of the CF_3Br has decomposed strongly suggests that these later processes have a small effect on the delay times. Hence, if they were included, the rate constant for Reaction 14 would not have to be changed much (probably less than a factor of two) to accommodate for their effect. It seems possible to state that the reaction scheme for CF_3Br inhibition is basically correct as far as it goes, and only changes in detail would be necessary as additional reactions to account for other products, or updated values of some of the rate constants, are incorporated into it.

Much less confidence can be placed in the proposed schemes for $C_2F_4Br_2$ inhibition, since none of the alternatives proposed is able to reproduce all of the data with "reasonable" rate constants. It seems that one has a choice of an unreasonably high rate constant for Reaction 24, or an unreasonably low one for Reaction 28. From general kinetic comparisons (such as the decomposition of C_2H_5 to C_2H_4 and H) the most likely expectation is that C_2F_4Br decomposition will be fast, with an activation energy of no more than 20 or 25 Kcal, so that the approximation of instantaneous decomposition is realistic, and that some additional reactions which we have not included will result in further inhibition. The importance of later reactions in $C_2F_4Br_2$ inhibition is suggested from the fact that $C_2F_4Br_2$ disappears somewhat faster than CF_3Br at low temperatures, yet ignition delays at higher temperatures are somewhat longer for $C_2F_4Br_2$. While at the moment it is not clear what these additional reactions should be, a kinetic analysis along these lines made in the future when more data are available might well be successful.

Literature Cited

1. Skinner, G.B. and Ringrose, G.H., J. Chem. Phys. (1965), 42, 2190.
2. Skinner, G.B. and Ringrose, G.H., J. Chem. Phys. (1965), 43, 4129.
3. Skinner, G.B., Lifshitz, A., Scheller, K. and Burcat, A., J. Chem. Phys. (1972), 56, 3853.
4. Skinner, G.B., J. Chem. Phys. (1973), 58, 412.
5. Jachimowski, C.J. and Houghton, W.M., Combustion and Flame (1971), 17, 25.
6. Baulch, D.L., Drysdale, D.D., Horne, D.G. and Lloyd, A.C., "Evaluated Kinetic Data for High Temperature Reactions. Vol. 1, Homogeneous Gas Phase Reactions of the $H_2 + O_2$ System", CRC Press, Cleveland, Ohio (1972).
7. Browne, W.G., White, D.R. and Smookler, G.R., Symp. Combustion (1969), 12, 559.
8. Schott, G.L. and Kinsey, J.L., J. Chem. Phys. (1958), 29, 1177.
9. Trotman-Dickenson, A.F. and Milne, G.S., "Tables of Bimolecular Gas Reactions, NSRDS - NBS 9", U.S. Government Printing Office, Washington, D.C. (1967).
10. Warshay, M., J. Chem. Phys. (1971), 54, 4060.
11. Stull, D.R. and Prophet, H., "JANAF Thermochemical Tables, Second Edition, NSRDS-NBS 37", U.S. Government Printing Office, Washington, D.C. (1971).
12. Hartig, R., Troe, J. and Wagner, H.G., Symp. Combustion (1971), 13, 145.
13. Skinner, G.B. and Ruehrwein, R.A., J. Phys. Chem. (1959), 63, 1736.
14. Sehon, A.H. and Szwarc, M., Proc. Roy. Soc. (London)(1951), A209, 110.
15. Benson, S.W. and O'Neal, H.E., "Kinetic Data on Gas Phase Unimolecular Reactions, NSRDS-NBS 21", U.S. Government Printing Office, Washington, D.C. (1970).
16. Westberg, K. and Greene, E.F., J. Chem. Phys. (1972), 56, 2713.
17. Cohen, N., Giedt, R.R. and Jacobs, T.A., Int. J. Chem. Kinet. (1973), 5, 425.
18. Biordi, J.C., Lazzara, C.P. and Papp, J.F., "Abstracts of Papers, Fifteenth Symposium (International) on Combustion", The Combustion Institute, Pittsburgh, Pa. (1974), 171.
19. Wagman, D.D., Evans, W.H., Parker, V.B., Halow, I., Bailey, S.M. and Schumm, R.H., "Selected Values of Chemical Thermodynamic Properties, NBS Technical Note 270-3", U.S. Government Printing Office, Washington, D.C. (1968).

DISCUSSION

A. S. GORDON: Comparison of $H_2 + O_2 \rightarrow 2OH$ (1)

and $H_2 + O_2 \rightarrow HO_2 + H$ (1a)

would seem to favor the latter because, while its endothermicity is greater, the 4 center reaction yielding OH would have a considerable lower (approx. x 10^{-3}) pre-exponential factor.

G. B. SKINNER: I do not feel very strongly that

$$H_2 + O_2 = 2OH \quad (1)$$

is the initiating reaction in hydrogen oxidation. Jachimowski and Houghton (Combustion and Flame (1971), 17, 25) have pointed out that Reaction 1 is experimentally indistinguishable from

$$H_2 + O_2 \rightarrow HO_2 + H \quad (1a)$$

so that the choice between them does not affect calculations of ignition delays or measurable product distributions. They appear to favor Reaction 1 because their calculated activation energy for the initiating reaction was 48.1 Kcal, which is less than the endothermicity (57.7 Kcal) of Reaction 1A, and I agree that their argument has some merit. I suspect that both reactions occur at comparable rates, and that the rate constants calculated by Jachimowski and Houghton are the sum of the two. Scatter of the data is sufficient to obscure curvature of the Arrhenius plot over the temperature range studied.

A. S. GORDON: At flame temperatures, one would have to have very high pressures before $CF_3Br \rightarrow CF_3 + Br$ would be considered as an unimolecular process in its pressure independent regime.

G. B. SKINNER: I do not think that CF_3Br decomposition is a long way from its first-order limit under our conditions. I have not made an RRKM calculation for CF_3Br, but have made a comparison to an earlier calculation (G. B. Skinner and B. S. Rabinovitch,

J. Phys. Chem. (1972), 76, 2418) using Golden, Solly, and Benson's method (J. Phys. Chem. (1971), 75, 1333). The vibrational specific heat, the effective pressure and the activation energy for the reaction are required. From the JANAF tables, C_{vib}/R for CF_3Br at 1300K is 8.5; the effective pressure, assuming a collision efficiency of 10% for argon, is 0.5 atm., and the activation energy is 66 Kcal. Our earlier calculation, for a molecule with C_{vib}/R of 8.6 and activation energy of 61 Kcal, was also made at 1300K and at 0.5 atm., so the parallel is close. The conclusion is that $k/k_\infty > 0.5$ under these conditions. It could also be pointed out that, to the extent that our rate constants for CF_3Br decomposition are based on our pyrolysis results, the fact that both pyrolysis and inhibition were carried out at 5 atmospheres total pressure will lead to a cancellation of errors. That is, the rate constants we used may be considered "effective" first-order rate constants for use at 5 atmospheres total pressure in a gas mixture consisting most of argon.

K. L. WRAY: At times as long as 20 ms behind the reflected shock wave, I would expect that boundary layer effects would have significantly changed the free stream gas temperature from that given by the simple shock equations. Have you corrected for this effect? If not, perhaps this could be responsible for the difficulty you are having in getting reasonable activation energies.

G. B. SKINNER: No corrections for cooling of the gas sample by contact with the walls were made. A few years ago (J. Phys. Chem. (1971), 75, 1) we made calculations on the effects of wall cooling in a 1 1/2-inch diameter shock tube for heating times up to 5 milliseconds. For a typical reaction with an activation energy of 50 Kcal, the effect of wall cooling was to reduce the amount of reaction (and hence the calculated rate constant where conversion is small) by a factor of 0.8 compared to no wall cooling. Since the halocarbon work was done in a 3-inch tube, and since the wall cooling effect increases with time at less than the first power of the time, I would expect about a factor of 0.7 in amount of reaction. Wall cooling will have a substantial (perhaps large) effect on activation energies if conversions at high temperatures are large, but not if they are small (i.e., a kinetic study with conversions ranging from 0.5 to 10%).

R. FRISTROM: Have you considered reactions involving CF_2 and CF_2O, e.g.,

$$CF_3 + H \rightarrow HF + CF_2 \text{ or } CF_2 + O_2 \rightarrow CF_2O + O \quad ?$$

Is reaction 15, $H + CF_3 \rightarrow CF_3H$, reasonable? (See Hunter, Fristrom, Grunfelder, this symposium.)

G. B. SKINNER: Reactions involving CF_2 and CF_2O were not included. Undoubtedly some such reactions must occur, since CF_2O was observed as a reaction product, but they were not included partly because CF_2O appeared toward the end of the induction period, so that it probably did not affect the length of the induction period very much, and partly because kinetic data on such reactions are hard to find. As I suggest in the paper, such reactions may be more important in $C_2F_4Br_2$ inhibition than with CF_3Br. I agree that it would be a good idea to develop a more complete model that would include such reactions and account for all observed products.

The reaction

$$H + CF_3 = CF_3H$$

is "reasonable" in the sense that it could occur, with a collision with (probably) an argon atom to de-energize the CF_3H. In our calculations this reaction was relatively unimportant because in a H_2-rich mixture most of the CF_3 will react with H_2 rather than with H during the greater part of the induction period.

J. BIORDI: Our experience in low pressure methane flames containing CF_3Br suggests that the reaction $CF_3+H\rightarrow CF_3H$ is not very important relative to $CF_3+ H\rightarrow HF+CF_2$. The reasons are these: CF_3H is not detectable in a stoichiometric flame (& ∴ H atom dominated) containing 0.3% CF_3Br, but HF is formed rapidly and early in this flame. The CF_3 radical is also unobservable, although other radical species with mole fraction of the order $10^{-4} - 10^{-5}$ can be detected. The CF_3 radical is also unobservable in flames containing 1% CF_3Br, although in this case we do see small amounts of CF_3H, suggesting that the abstraction reactions do occur at high enough CF_3 concentration. Finally, and most significantly, we do observe the CF_2 radical in substantial amounts in both the 0.3% and the 1% CF_3Br containing flame.

G. B. SKINNER: I would agree that $CF_3 + H = CF_3H$ would be unimportant in low-pressure flames because of the much smaller collision number compared to our shock-tube experiments. Low-pressure flames contain relatively more H atoms than our shock tube reactions, because of slow combination in the hot zone of the flame, and ease of back diffusion of H atoms. I think that $CF_3 + H_2 = CF_3H + H$ is the most important reaction of CF_3 during most of the induction period in the shock tube experiments, since most of our CF_3Br goes to CH_3H. Toward the end of the induction period, the reaction $CF_3 + H = CF_2 + HF$ probably occurs, leading to COF_2, as Dr. Fristrom suggests.

R. C. GANN: There is still some uncertainty in the literature concerning some of the rate constants which, of necessity, you assumed in your model. For instance, your values for reactions 17 and 19 are close to, but not the same as, the values in Nancy Brown's review (this symposium). As you mention, your rate constant expression for reaction 14 is comparable in magnitude with, but still different from, the results of Biordi, Lazzara and Papp. Have you performed a parametric study to put error bars on k_{14} and determine to which of the other reaction rates k_{14} is particularly sensitive?

G. B. SKINNER: Honestly, I did not make note of Nancy Brown's values for Reactions 17 and 19. For Reaction 17 I used the value recommended by Fettis and Knox and reported by Trotman-Dickenson (page 22 of Reference 9 of the paper) for the reverse reaction, and calculated an Arrhenius equation for the forward reaction from the JANAF tables, using log K_f values at 900 and 1300K. For Reaction 19 I chose the value for the reverse reaction recommended by Benson and Buss and reported by Trotman-Dickenson (page 24) and again obtained the forward rate constant from the JANAF tables. In fact, for all of the inhibiting reactions for CF_3Br inhibition, forward and reverse reactions have been related by equations of the Arrhenius type valid for the 900-1300K range.

If all the other rate constants were known exactly, the rate constant for Reaction 14 could probably be given to a factor of 2, but considering other uncertainties the factor is more likely 3. The value of k_{14} is most sensitive to changes in Reaction 16, than to Reactions 17 and 20. The analysis is not sensitive to small changes (1-2 Kcal) in activation energy provided compensating changes are made in A to keep the rate constant the same at 1150K.

E. T. McHALE: Did you perform any parametric-type analysis for the CF_3Br case to determine whether any of the seven assumed inhibition reactions are relatively unimportant? For example, how well can you reproduce your ignition delay times if only the reaction $CF_3Br + H \rightarrow CF_3 + HBr$ is considered?

G. B. SKINNER: We did not vary the rate constants of all the reactions systematically to observe the effect on calculated results. However, by looking at the rates of all the reactions during the calculations it was possible to tell which were contributing most to the overall progress of the reaction. From Table IV of the paper it can be seen that at the half-way point in this calculation, Reaction 16,

$$CF_3 + H_2 = CF_3H + H$$

is the only one proceeding at a rate comparable to that of Reaction 14

$$CF_3Br + H = CF_3 + HBr.$$

Reaction 20, although slow in comparison with the others, is important in the early stages as an initiator. All of the reactions in the scheme were included because they were expected to occur, but it is true that some, such as 15, 18, 19 and 21 have little effect on the results.

One calculation, made to specifically answer your last question, was made with all inhibition reaction rate constants set to 0 except for Reaction 14. This resulted in a delay time of >10 milliseconds, at 1200K, compared to the experimental value of about 3 milliseconds. Probably the most important reason for the change was that, in this calculation, CF_3 radicals "could not" react with H_2 by reaction 16.

11

Initial Reactions in Flame Inhibition by Halogenated Hydrocarbons

RICHARD G. GANN

Chemical Dynamics Branch, Chemistry Division, Naval Research Laboratory, Washington, D. C. 20375

Over the past three decades, many halogenated molecules have been investigated as fire suppressants. Some tests have been carried out on full-scale fires while others have been performed under controlled laboratory conditions. Generally, the effectiveness of the suppressant has been characterized by changes in one or more bulk properties, such as reduction of flame velocity, narrowing of flammability limits, final flame temperature, or simply how much was needed to quench a "standard" fire. All of these allow for the fact that chemical activity of the suppressant is involved, but do not directly examine it.

There are two major considerations in designing laboratory flame studies: fluid mechanics and chemistry. Because of the inherent great difficulties in analyzing combustion systems, we generally simplify the situation by working as nearly as possible in one of two extreme regions. In the first, one assumes "fast kinetics," and thus allows the fluid dynamics to determine the flame propagation. Alternatively, one creates a well-stirred reactor in which the gas mixing is assumed complete and the chemistry determines the system. In the former situation, one must be assured that the chemistry is complete on a time frame which is fast relative to gas motion. In the latter, a more precise knowledge of the rate data is needed to understand the time evolution of the various species that are observed. In all situations a knowledge of the pertinent kinetics is necessary.

Several reviewers in the past decade discussed classical and contemporary studies of flame inhibition chemistry (1, 2, 3, 4, 5). Their analyses of kinetic results was brief for the simple reason that, in general, the reactions involving the halons had not been studied. In the past few years, interest in these processes has increased, and kinetic determinations are beginning to appear in the literature. While their number is still few, the need for the rate data is as great as ever. This Symposium provides a good opportunity to look at what has been done and to point a finger at further needs. In general, the bulk of the data applies to CF_3Br chemistry and this will be reflected in this paper. Where possible, extensions to two other halons which are used as fire suppressants, CF_2ClBr and $C_2F_4Br_2$, will be made.

An attempt to span the full flame history of the inhibitor molecules and their offspring would be fruitless, if only because the full sequence of reactions is barely being guessed at present. Rather, this discussion will review the earliest steps in the scheme. Good kinetic data here is critical, for these reactions may well be determining steps for suppressant effectiveness. Since the available data is still sparse, my approach to each reaction will be to discuss the relevance of the reaction, review the experimental data where possible, discuss the kinetics by analogy if necessary, and guess an approximate value for the rate constant if desperate. The overall discussion can be considered in two sections. The first deals with the reactions which introduce the halon into the inhibition process. The second deals with some avenues for the primary products from these reactions. Throughout the paper, I have converted a melange of units to the following: rate constants in cm, molecule, seconds, activation energies in kcal/mole, and temperatures in Kelvin. Thermodynamic data has been taken from the JANAF Thermochemical Tables (6) and Benson's monograph on thermochemical kinetics (7).

Primary Reactions

Thermal Decomposition. The first question in halon flame inhibition is whether the agent thermally decomposes. If so, halogen-containing species will be generated unimolecularly as well as by bimolecular reations with flame species. There is just one avowed experimental determination of the homogeneous rate constant for

$$CF_3Br \xrightarrow{\Delta} CF_3 + Br \qquad (1)$$

in the literature. Sehon and Szwarc (8) determined a value of $D(CF_3\text{-}Br)$ equal to 64.5 kcal/mole from

1020-1090K using the toluene carrier technique and assumed $E_a = D(CF_3\text{-}Br)$. They also assumed an Arrhenius A factor of 2 x 10^{13} sec^{-1} based on their studies of CH_3Br pyrolysis.

Benson and O'Neal (BO) (9) feel that values for both A and E_a are too low. Based on an RRK calculation they prefer $k_1 = 5 \times 10^{13} e^{-66,300/RT}$. Darwent's (D) compilation (10) recommends a CF_3-Br bond dissociation energy of 68 kcal/mole at 298K and the JANAF (6) thermochemical data (J) produces ΔH = 69.5 kcal/mole, both of which imply a larger A factor is necessary to overlap the Sehon and Szwarc data. Table I shows values for k_1 calculated from the first two references as a function of temperature.

TABLE I

Rate Constants for Thermal Decomposition of CF_3Br

T (K)	SS	BO	D	J
900	0.0043	0.0040	0.0037	0.0033
1020	0.30	0.31	0.32	0.33
1090	2.3	2.5	2.8	3.0
1200	36	42	49	57
1300	290	360	440	534
1400	1700	2230	2900	3670

The two latter columns are obtained by arbitrarily selecting A factors such that the Darwent- and JANAF-derived rates would intersect the experimental data at 983K, the temperature at which the Benson-O'Neal and Sehon-Szwarc formulas intersect. In the narrow temperature range 1020-1090K, the four formulations are quite close to each other. However, upon extrapolation to higher temperatures, still within the flame regime, some divergence occurs.

Looking at the data from another viewpoint provides a better yardstick for assessing the significance of this spread. Let us consider the time frame in which decomposition might be expected. As an example, Table II lists the temperatures at which 1% decomposition occurs for the four models at various heating times.

TABLE II

Temperature Required for 1% Thermal Decomposition of CF_3Br

t(sec)	SS	BO	T(K) D	J
10	864	868	870	873
1	921	923	924	926
0.1	985	985	985	985
10^{-2}	1059	1058	1056	1055
10^{-3}	1145	1141	1136	1133
10^{-4}	1246	1239	1230	1225
10^{-6}	1514	1494	1474	1460

Certainly the local temperature fluctuations in real fires far exceed even the largest difference among the models shown here, and this might lead to the conclusion that k_1 is sufficiently well-known. However, we must keep in mind the fact that the absolute magnitude of all four formulations is based on an assumed A factor, and the relative magnitudes derive from an arbitrary intersection temperature. Over the wide temperature range relevant to flames, different Arrhenius parameters will render greatly varying estimates of the degree of thermal decomposition. At the least, an absolute rate constant determination is needed.

Sehon and Szwarc's experiments were carried out in the presence of a large amount of toluene, an efficient Br scavenger. Skinner and Ringrose (11) (SR) have pyrolyzed 4% CF_3Br - 96% Ar mixtures at 5 atm in a shock tube from 1112-1982K. At their lower values of temperature and fraction dissociated, calculated values of k_1 are close to SS or BO. However, severe differences occur at higher decomposition resulting in an apparent $E_a \sim$ 20kcal/mole. We (12) have performed a few experiments with a 13% CF_3Br-87% N_2 mixture at 26 torr in our flow reactor (13) and found ∿5% decomposition at 1020K and ∿8.5% at 1070K with a residence time of ∿0.8 sec. SS found 12% and 52% decomposition at similar temperatures and ∿1/2 the residence time. In both our and the SR studies, there is no efficient bromine scavenger; and some CF_3Br regeneration, perhaps by

$$CF_3 + Br \rightarrow CF_3Br \qquad (2)$$

$$\text{or}\quad CF_3 + Br_2 \rightarrow CF_3Br + Br, \tag{3}$$

may be occurring.

To affect the apparent rate constant for reaction 1, reaction 2 or 3 must compete with

$$2\ CF_3 \rightarrow C_2F_6 \tag{4}$$

This bimolecular recombination proceeds at close to gas kinetic frequency at these experimental temperatures (14, 15). Also, at these temperatures, the bromine atom concentration is but a small fraction of the molecular bromine concentration (14) and the termolecular bromine atom recombination rate is slow relative to reaction 4 (16). Thus $[CF_3]$ is much lower than [Br], which is in turn lower than $[Br_2]$. (This is, of course, not necessarily valid in actual flames). Amphlett and Whittle (17) have found $k_3 = 3.8 \times 10^{-12} e^{-700/RT}$ cm^3/molecule-s from 451 to 600K, which extrapolates to 3×10^{-12} cm^3/molecule-s at 1000K. Thus, when significant decomposition of CF_3Br has occurred, the rate of reaction 3 is almost certainly competitive with that of 4. This not only perturbs the experimental system but can lead to the re-formation of parent CF_3Br.

There are no reported detailed studies for C_2F_4Br and CF_2ClBr thermal decomposition. However, three less formal studies bear mentioning. Skinner and Ringrose (11) find far more decomposition of $C_2F_4Br_2$ than CF_3Br at roughly equivalent heating times and temperatures. Engibous and Torkelson (18) have reported studies in which a mixture of 2.5% halon 1301, 2402 or 1211 in anhydrous air was passed through a long, thin stainless steel tube which in turn lay inside a furnace. In a quest for a selective analytical technique for hydrocarbons, Stone and Williams (19) passed extremely dilute (∿100 ppm) halons in an O_2-He carrier through a quartz tube filled with quartz beads which also lay inside a furnace. The contact time in the former study is given as 2 sec; in the latter, it was the same magnitude. Both studies were at 1 atm. Neither corrected for a dependence of linear flow velocity and thus residence time on temperature. Each study produced plots of % decomposition *vs*. temperature. A qualitative plot of the data of Stone and Williams appears in Figure 1. The Engibous and Torkelson data is similar. In both studies, the steep slope for CF_2BrCl occurred somewhat parallel to and about 100K lower than that for CF_3Br.

Also in both cases, the C_2F_4Br decomposition begins at a still lower temperature, with the studies disagreeing at higher temperatures and dissociation fractions. The extraction of quantitative kinetic data is not merited due to the presence of oxygen, insufficient temperature control and in the Engibous and Torkelson work, the lack of an efficient scavenger to avoid the aforementioned recombination. What is apparent is that the quartz or stainless steel walls significantly enhance the thermal decomposition of these halons. One can only surmise the effect which would be realized due to the presence of the soot particles prevalent in real fires.

One further observation. In a flame, the halon molecules are being both thermally and chemically decomposed. The relative importance of these two competitive processes will vary with the flame conditions. In a real fire, the halon enters from the air side, and the chemical species and concentrations it encounters first are likely to be those generated late in the combustion sequence. By contrast, in a laboratory study where the halon is premixed with fuel and oxidant, the earliest chemical attack on the halon will come from early flame components, with the intact molecule probably not surviving to meet those conditions mentioned earlier. Thus, even if the thermal conditions were the same in the two cases, different decomposition profiles should be observed.

The conclusion to be drawn is that the homogeneous decomposition data is not sufficient to label the rates for these processes as established. Furthermore, there is also a need for detailed studies of surface effects on the rates of halon decomposition.

Radical Reactions.

$$H + CF_3Br \rightarrow HBr + CF_3 \quad \Delta H = -18 \text{ kcal/mole} \quad (5)$$

This attack on the inhibitor molecule has already been shown to be the primary reactive channel in premixed stoichiometric, low pressure, methane-oxygen flames (20). In those situations, it is also the first step in which the halon depletes active species within the flame. Even though as much attention has been paid to reaction 5 as to any other reaction discussed here, it still exemplifies the uncertain state of the kinetic data. In 1965, Skinner and Ringrose (SR) carried out shock tube studies of H_2-O_2-Ar-CF_3Br mixtures at 5 atm (11).

Based on their numerical modelling of the results, they extracted $k_5 = 7.2 \times 10^{-9} e^{-17,450/RT}$ cm^3/molecule-s from 920-1280K. In 1974, Biordi, Lazzara and Papp (20) reported $k_5 = 3.7 \times 10^{-10} e^{-9460/RT} cm^3$/molecule-s from 700-1550K. Their value is based on derivatives of chemical profiles of CH_4-O_2-Ar-CF_3Br flame at 0.04 atms. In this symposium, Skinner (S) has re-examined the earlier SR results in the light of more recent kinetic data on competing reactions, and a more complete reaction scheme, obtaining an activation energy of 3.4 kcal/mole and an A factor of 2.5×10^{-11} cm^3/molecule-s. Using a fast flow reactor, Frankiewicz, Williams and Gann obtained a lower limit for k_5 of $\sim 1 \times 10^{-12} cm^3$/molecule-s from 750-1200K (21). This was based on the observed effect of trace water in an oxygen discharge on the O + CF_3Br reaction and is only used to establish the magnitude of k_5. All of these results are plotted over the appropriate temperature ranges in Figure 2. It is apparent that further information is needed on this important reaction.

The analogous reaction

$$H + CH_3Br \rightarrow HBr + CF_3 \qquad (6)$$

is thermodynamically and structurally similar to reaction 5 and should go through a similar transition state. Thus it would not be surprising if the rate constant were similar. Seidel (S') (22) has studied reaction 6 using crossed molecular beams. Jones, Macknight and Teng (23), in their recent review of hydrogen atom reactions, have used Seidel's activation energy cross section values to derive $k_6 = 5.9 \times 10^{-10} e^{-6,900/RT} cm^3$/molecule-s. Davies, Thrush and Tuck (DTT) (24) have studied

$$D + CH_3Br \rightarrow DBr + CH_3 \qquad (7)$$

using a discharge-flow apparatus with ESR detection, and have found $k_7 = 8.9 \times 10^{-11} e^{-4,300/RT} cm^3$/molecule-s from 297-480K. It is hardly appropriate to extrapolate these results over a long temperature range, but nevertheless I have done so for the sake of comparison with the CF_3Br data. These data are included in Figure 2 as dashed lines.

There is no data available for H + CF_2ClBr and only one study of H + $C_2F_4Br_2$, that of Skinner and Ringrose in their previously cited work (11).

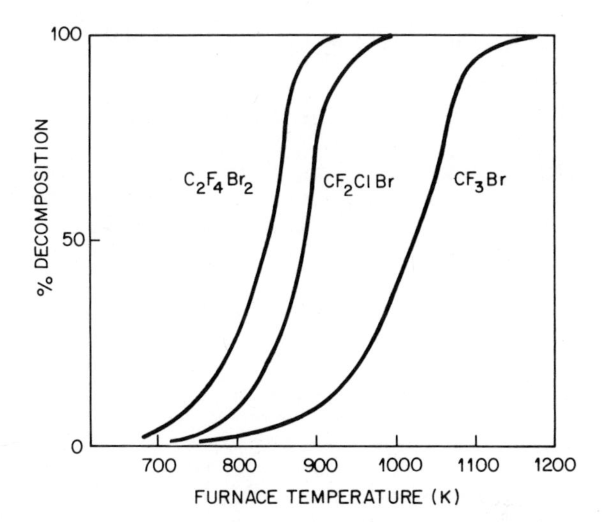

Figure 1. Thermal Decomposition of Halons 2402, 1211, and 1301 in a quartz tube filled with quartz beads (19)

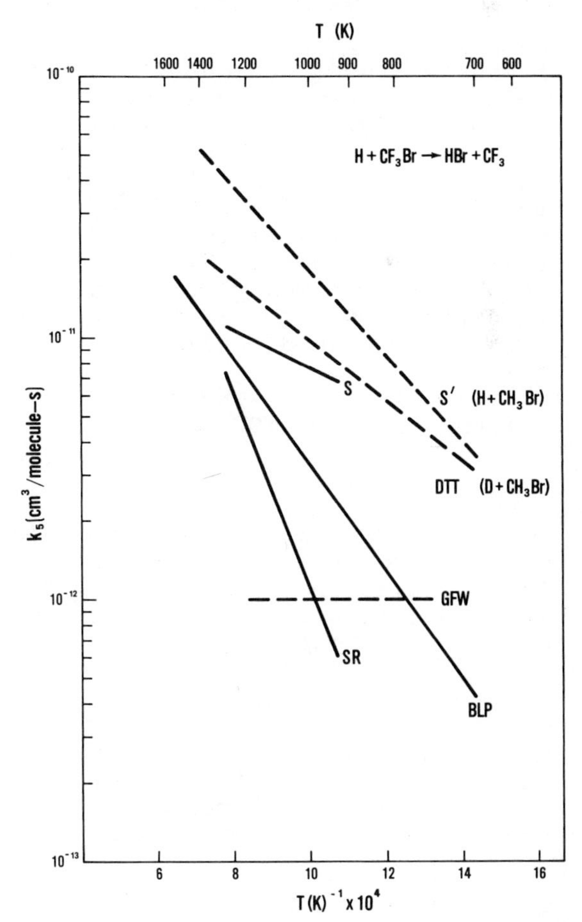

Figure 2. Arrhenius plots for the reaction: $H + CF_3Br \rightarrow HBr + CF_3$. SR: Ref. 11; *BLP: Ref.* 20; *GFW: Ref.* 21 *(lower limit); S′: Ref.* 22; *DTT: Ref.* 24; *S: Skinner, this volume.*

They arrived at $k_8 = 1.7 \times 10^{-8} e^{-14,500/RT} cm^3$/molecule-s from 935-1286K.

$$H + C_2F_4Br_2 \rightarrow C_2F_4Br + HBr \quad (8)$$

Elsewhere in this symposium, Skinner has re-evaluated this data, and cannot obtain a unique formulation for k_8 due to a lack of knowledge about the unimolecular decay,

$$C_2F_4Br \rightarrow C_2F_4 + Br. \quad (9)$$

$$O + CF_3Br \rightarrow OBr + CF_3 \quad \Delta H = +13 \text{ kcal/mole} \quad (10)$$

In 1974, we reported what is still the only direct measurement of this reaction rate (13). Using a discharge-flow apparatus with microprobe sampling and gas chromatographic detection of CF_3Br, we determined $k_{10} = 1.5 \times 10^{-11} e^{-13,500/RT} cm^3$ molecule-s from 800-1200K. Computer modelling predictions of possible effects from secondary reactions were tested against widely-varied experimental conditions, with the conclusion that these reactions did not affect our values for k_{10}. The fact that the experimental activation energy equals the known endothermicity of reaction 10 also lends confidence to the data. The results are shown in Figure 3. Comparison with studies of O + methyl halides is not applicable since the abstraction of a hydrogen atom is exothermic with a low E_a and easily prevails at lower temperatures (25).

A calculation investigating the competition between reaction 10 and

$$O + H_2 \rightarrow H + OH \quad (11)$$

was performed (13), using the Leeds data for reaction 11 (26) and Creitz' propane diffusion flame data (27). The calculation shows that reaction 10 is more efficient than reaction 11 at removing O atoms from the flame at 1000K. Reaction 11 is followed by:

$$H + O_2 \rightarrow OH + O \quad (12)$$

in a principal chain branching sequence in the hot region of flames. Removal of any one radical, H, O, or OH, of necessity reduces the concentrations of the other two and thus their reaction rates with the

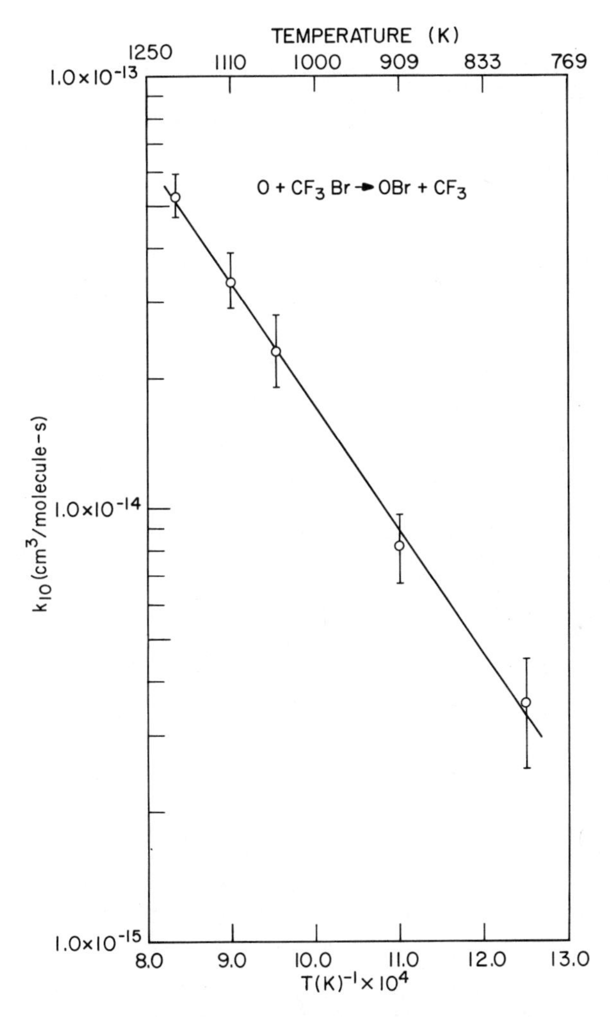

Figure 3. Arrhenius plot for the reaction: $O + CF_3Br \rightarrow OBr + CF_3$. *Data from Ref.* 13.

halon. Competition with the reverse of reaction 12 might also be important, depending on the local OH concentration.

$$Br + CF_3Br \rightarrow Br_2 + CF_3 \quad \Delta H = +23 \text{ kcal/mole} \qquad (13)$$

Consideration of this reaction is included to cover the possibility that atomic bromine, once formed by attack on the halon, might increase the yield of inhibiting species, since reaction 13 would perhaps be followed by

$$H + Br_2 \rightarrow HBr + Br. \qquad (14)$$

The high endothermicity for reaction 13, even if combined with a gas kinetic A factor, makes k_{13} too small to compete with either O or H attack. No experimental determinations of k_{13} or its analogs for the other halons have been reported.

$$CF_3Br + OH \rightarrow HOBr + CF_3 \quad \Delta H = +14 \text{ kcal/mole} \qquad (15)$$

This is the third of the reactions in which CF_3Br could interact directly with the O, OH, H chain branching sequence, reactions 11 and 12. Once again, there is no direct kinetic data for reaction 15. The analogous reaction of CH_3Br with OH leads to H rather than Br abstraction. Since ΔH is about the same as for reaction 10, and the transition states must be similar, it is likely that k_{15} will approximate k_{10} (7). An alternative reaction path:

$$CF_3Br + OH \rightarrow HF + COF_2 + Br \quad \Delta H = -66 \text{ kcal/mole} \qquad (16)$$

is highly exothermic, but the concerted process requires the formation of a strained collision complex with 5 bond rupture-formations, and is thus not likely.

$$CH_3 + CF_3Br \rightarrow CH_3Br + CF_3 \quad \Delta H = -1 \text{ kcal/mole} \qquad (17)$$

This reaction apparently trades one inhibitor molecule for another of similar effectiveness. However, CH_3Br is subject to exothermic attack at the hydrogen positions. Furthermore, CH_3Br appears early as CF_3Br disappears in low pressure CH_4-O_2-Ar-CF_3Br flames (20) and in 1 atm CH_3CHO-air-CF_3Br cool flames (28). Tomkinson and Pritchard (TP) (29) have measured Arrhenius parameters in a static system from 363-518K, deriving $k_{17} = 3.3 \times 10^{-11} e^{-12,500/RT}$ cm^3 molecule-s. Biordi, Lazzara and Papp (BLP) in their low pressure flames find $k_{17} = 1.6 \times 10^{-11}$ $e^{-5,200/RT}$ from 800-1600K. The two curves are shown in Figure 4. The dashed portions of the lines indicate extrapolation for comparison at common temperatures. Tomkinson and Pritchard conclude that $D(CF_3Br) - D(CF_3\text{-}Br) \simeq 4$ kcal/mole, which is at odds with the generally accepted thermochemistry which finds the two bonds equivalent. Biordi, Lazzara and Papp base their results on first and second derivatives of species profiles, and it is not yet certain that the sampling of flames is at a reliable enough level to support rigorous kinetic determination. At any rate, the disagreement between the two reported expressions for k_{17} is significant and indicates a need for further investigation.

$$HO_2 + CF_3Br \rightarrow HOOBr + CF_3 \quad (18)$$

$$ROO + CF_3Br \rightarrow ROOBr + CF_3 \quad (19)$$

$$RO + CF_3Br \rightarrow ROBr + CF_3 \quad (20)$$

These reactions where R is a hydrocarbon radical, are grouped together merely for convenience. They represent attack by oxygenated species which appear early in the flame sequence (30), (31), (32). Both a recent survey (33) and a current literature search reveal no data on reaction 18 or any analogous system, nor was any found on reactions 19 or 20.

Other Intact Halon Molecule Effects. In addition to its chemical interaction with flame species, CF_3Br exhibits a physical presence. The primary effect involves translational cooling of the hot flame gases, or on a bulk scale, bath effects which reduce the flame temperature to a level too low for flame propagation. This is related to the additional heat capacity of the added halon. Larsen, later in this Symposium, discusses this in detail, and it will not be pursued further here.

Relative to N_2, O_2, or the inert gases, the halons, with their increased number of internal (vibrational) degrees of freedom, should quench excited molecules with higher efficiency. It has been

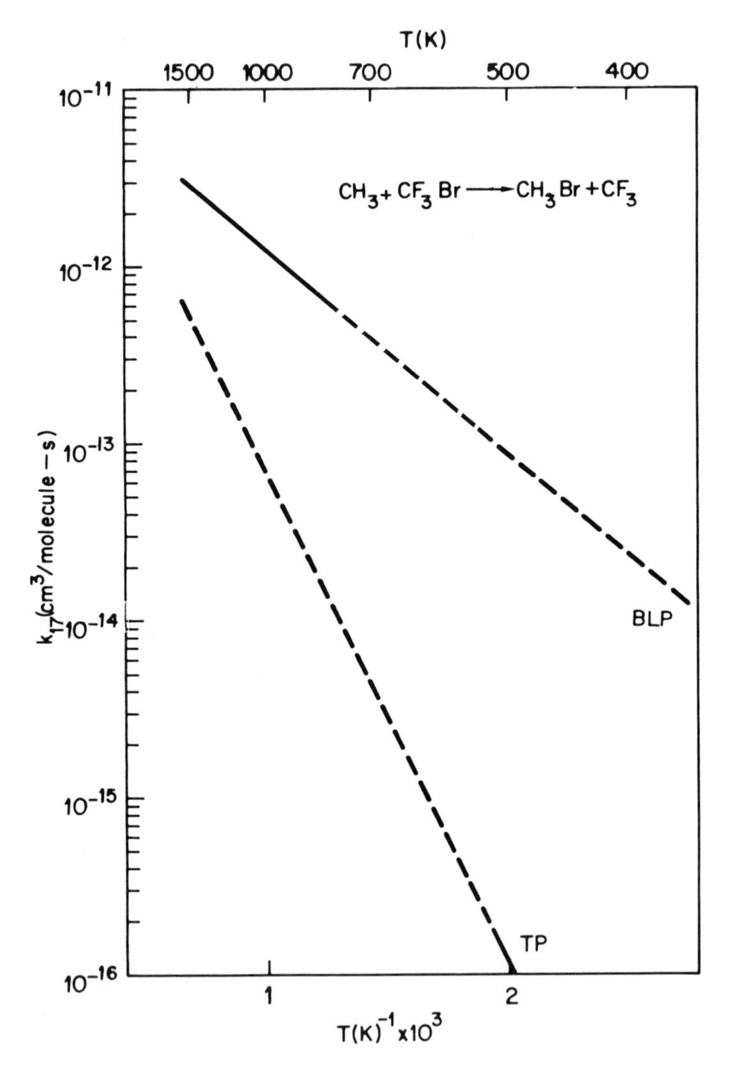

Figure 4. Arrhenius plots for the reaction: $CH_3 + CF_3Br \rightarrow CH_3Br + CF_3$. BLP: Ref. 20; TP: Ref. 29.

known for some time that radicals can be generated either electronically excited and/or with non-Boltzmann rotational and vibrational distributions. If these species have sufficiently long lifetimes and participate in the combustion propagation, then physical quenching by newly introduced, "colder" halons will contribute to flame inhibition. No measurements have been reported on these effects, but since internal energy can significantly increase reaction rates, some investigation of halon quenching rates is necessary.

The halons should also be more efficient than O_2 or N_2 in promoting three-body reactions. This effect is well demonstrated by the work of Walkauskas and Kaufman (34). They measured significant increases in the rate of

$$H + H + M \rightarrow H_2 + M \quad (21)$$

in varying M from N_2 to CH_4 or SF_6. Whether such enhancement, or the equivalent stabilization of excited collision complexes, is important can only be resolved by data which is not now available.

Secondary Fragment Reactions

CF_3 Radicals. Results such as those of Rosser, Wise and Miller (35) indicate that CF_3Br is about 50% more efficient than CH_3Br in reducing flame speed. This probably implies some active participation in the inhibition process by the CF_3 radical. Biordi, Lazzara and Papp (20) were unable to detect CF_3 in their system, implying a rapid conversion to other products. Thus a look at the kinetic data published on its reactivity with flame propagating species is worthwhile. No further comment will be made on the C_2F_4Br or CF_2Cl radicals, which would be produced by analogous reactions with the other two halons, since no data exists on any of their reactions.

$$O + CF_3 \rightarrow OCF_2 + F \quad \Delta H = -81 \text{ kcal/mole} \quad (22)$$

In their review of oxygen atom reactions (36), Huie and Herron conclude that this is probably a major channel for the O-CF_3 reaction. Biordi, Lazzara and Papp, in seeking to explain the rise of [CF_2O] in their flames, conclude that reaction 22 is not fast enough to be the principal explanation (20). There is no reported direct study of k_{22}, the assumption being made that the rate constant is large, comparable to that for the analogous methyl radical reaction. We have, however, performed a few experiments (37) on

$$O + CF_3I \rightarrow CF_3 + OI \quad (23)$$

which is followed by reaction 22. In our fast flow reactor (13), preheated CF_3I was introduced to a 3% O_2 in He mixture at 3 torr and 850K, with or without the microwave discharge on. The principal species, O_2, CF_3I, C_2F_6, OCF_2 were detected by gas chromatography. With the microwave unit off, no CF_3I thermal decomposition (indicated by appearance of COF_2 or C_2F_6) was observed in the approximately 25 ms residence time in the heated section of the flow reactor. With O atoms flowing, all the CF_3I was consumed and both OCF_2 and C_2F_6 appeared, in the rough ratio of 3 : 1. The bimolecular recombination of CF_3 radicals (reaction 4) proceeds at close to gas kinetic frequency at these temperatures (14, 15). Solution of the reaction scheme indicates that for $[OCF_2]$ to be comparable to $[C_2F_6]$, k_{22} must also be of gas kinetic magnitude. Obviously, a more clear-cut determination is desirable.

$$O_2 + CF_3 \rightarrow OCF_2 + OF \quad \Delta H = +1 \text{ kcal/mole} \quad (24)$$

This reaction is also a possible source for the formation of OCF_2 in inhibited flames. Again, no direct rate determination exists. However, in our same series of experiments described in the previous paragraph, the mixture was exposed to 1000K with the microwave unit off. Total thermal decomposition of the CF_3I occurred, with approximately equal amounts of OCF_2 and C_2F_6 formed. Since $[O_2] >> [CF_3]$, it is probable that k_{24} is at least two orders of magnitude lower than k_{22}. We intend to look more closely at this in the future.

$$H + CF_3 \rightarrow HF + CF_2 \quad \Delta H = -45 \text{ kcal/mole} \quad (25)$$

$$H + CF_3 \rightarrow CF_3H \quad \Delta H = -106 \text{ kcal/mole} \quad (26)$$

$$OH + CF_3 \rightarrow OCF_2 + HF \quad \Delta H = -3 \text{ kcal/mole} \quad (27)$$

In each of these three exothermic reactions, a potential chain carrier is bound into a chemically quite stable species. Biordi, Lazzara and Papp (38) attribute the lack of a CF_3 peak and the appearance of CF_2 in their flames to reaction 25. Skinner (39) assigns a value of $k_{26} = 8 \times 10^{-11} e^{-4000/RT} cm^3$/molecule-s, by analogy with his H + CH_3 recombination results, to explain the observation of product CF_3H in his earlier schok tube studies (11). However, while these reactions could well be important in flame inhibition, as yet there has been no kinetic study of them.

OBr Radicals. OBr is formed as a product of reaction 10 and perhaps by other reactions, and its progress to releasing Br atoms into the inhibition scheme is of interest. Until very recently, there had been no report of any kinetic studies of OBr reactions. However, work by Clyne and Watson will soon appear. For reaction 28

$$O + OBr \rightarrow O_2 + Br \qquad \Delta H = -63 \text{ kcal/mole} \qquad (28)$$

these authors quote $k_{28} = 8 \times 10^{-11} cm^3$/molecule-s at 300K (40), and it is reasonable to expect that the reaction proceeds at nearly unit collision efficiency at flame temperatures.

Although no one has yet reported on

$$H + OBr \rightarrow OH + Br \qquad \Delta H = -46 \text{ kcal/mole} \qquad (29)$$

$$H + OBr \rightarrow O + HBr \qquad \Delta H = -31 \text{ kcal/mole} \qquad (30)$$

Clyne and Watson have briefly looked at the analogous OCl reaction and found that the channel which leads to OH formation predominates, probably at a gas kinetic rate (41).

Finally, they have studied the reaction

$$2\, OBr \rightarrow O_2 + 2\, Br \qquad \Delta H = -7 \text{ kcal/mole} \qquad (31)$$

and found $k_{31} = 6.4 \times 10^{-12} cm^3$/molecule-s at 300K (40). This reaction is probably unimportant in inhibited flames since the chemical lifetime of the OBr radical will be far too short for an appreciable density to build up.

HOBr. A literature search reveals no data on the reactions or formation of this molecule. However, it has now been observed in 1 atm, premixed, acetaldehyde - air - CF_3Br cool flames (42), perhaps due to reaction 15 or

Table III

Reactions in CF_3Br Flame Inhibition

(1)	CF_3Br	$\xrightarrow{\Delta}$	$CF_3 + Br$
(5)	$H + CF_3Br$	$\rightarrow$	$HBr + CF_3$
(10)	$O + CF_3Br$	$\rightarrow$	$OBr + CF_3$
(13)	$Br + CF_3Br$	$\rightarrow$	$Br_2 + CF_3$
(15)	$OH + CF_3Br$	$\rightarrow$	$HOBr + CF_3$
(17)	$CH_3 + CF_3Br$	$\rightarrow$	$CH_3Br + CF_3$
(18)	$HO_2 + CF_3Br$	$\rightarrow$	$HOOBr + CF_3$
(19)	$ROO + CF_3Br$	$\rightarrow$	$ROOBr + CF_3$
(20)	$RO + CF_3Br$	$\rightarrow$	$ROBr + CF_3$
(22)	$O + CF_3$	$\rightarrow$	$OCF_2 + F$
(24)	$O_2 + CF_3$	$\rightarrow$	$OCF_2 + OF$
(25)	$H + CF_3$	$\rightarrow$	$HF + CF_2$
(26)	$H + CF_3$	$\rightarrow$	CF_3H
(27)	$OH + CF_3$	$\rightarrow$	$OCF_2 + HF$
(3)	$Br_2 + CF_3$	$\rightarrow$	$Br + CF_3Br$
(28)	$O + OBr$	$\rightarrow$	$O_2 + Br$
(29)	$H + OBr$	$\rightarrow$	$OH + Br$
(30)	$H + OBr$	$\rightarrow$	$O + HBr$
(31)	$2OBr$	$\rightarrow$	$O_2 + 2Br$

$$OH + Br_2 \rightarrow HOBr + Br \qquad \Delta H = -9 \text{ kcal/mole} \quad (37)$$

Summary

Kinetic data for the reactions listed in Table III have been surveyed and discussed. These reactions are those which must be considered in the early stages of flame inhibition by CF_3Br, halon 1301. The equivalents of reaction 1 for CF_2ClBr and $C_2F_4Br_2$ and that of reaction 5 for $C_2F_4Br_2$ have also been included. In all too many cases, no results have been reported. Despite the use of measurement techniques in which the chemistry is often simplified relative to actual flames, for virtually none of the reactions is the rate data sufficiently established to support accurate modelling of inhibited flames. Certainly this situation bears correction. Reliable measurements for reactions 1, 3, 5, 10, 15, 17, 22, 24, 25, and 27 are a minimum requirement. Consideration of reactions 18-20 becomes important in fire-inerting situations. Furthermore, in order to explain on a chemical basis the observed variation in efficiency of the different halons in a given fire suppression procedure, some kinetic data on these reactions is sorely needed for at least halons 2402 and 1211. While this may appear to be a burdensome demand on the rank of kineticists, we should bear in mind the urgency and sublimity of the end result.

Acknowledgements

The author is pleased to thank the following for discussion of recent results or the receipt of pre-publication copies of papers: Dr. G. B. Skinner, Dr. R. E. Huie, Dr. J. T. Herron, Dr. J. C. Biordi, Dr. R. T. Watson, Dr. F. W. Williams, Mr. J. P. Stone, and Dr. J. R. Wyatt. Helpful discussions with Dr. D. J. Bogan and Dr. M. C. Lin are also appreciated. The financial support of the Naval Air Systems Command is gratefully acknowledged.

Literature Cited

1. Fristrom, R. M., Fire Res. Abst. and Rev., (1967) 9, 125.
2. McHale, E. T., Fire Res. Abst. and Rev., (1969) 11 90.
3. Creitz, E. C., J.Res. Nat'l. Bur. Stds., (1970) 74A, 521.

4. Fristrom, R. M. and Sawyer, R. F., 37th AGARD Symposium, Conference Proceedings, No. 84 on Aircraft Fuels, Lubricants and Fire Safety, AGARD-CP-84-71, Section 12, North Atlantic Treaty Organization (1971).
5. Baratov, A. N. in Ryabov, I. V., Baratov, A. N. and Petnor, I. I., eds., "Problems in Combustion and Extinguishment," Amerind Publishing Co., New Delhi, 1974; available as NTIS TT-71-58001.
6. Stull, D. R. and Prophet, H., eds., "JANAF Thermochemical Tables 2nd Edition," NSRDS - NBS 37, U. S. Government Printing Office, Washington, D. C. 20402 (1971).
7. Benson, S. W., "Thermochemical Kinetics," J. Wiley and Sons, New York, 1968.
8. Sehon, A. H. and Szwarc, M., Proc. Roy. Soc., (London) (1951) A209, 110.
9. Benson, S. W. and O'Neal, H. E., "Kinetic Data on Gas Phase Unimolecular Reactions," NSRDS - NBS 21, U. S. Government Printing Office, Washington, D. C. 20402 (1970).
10. Darwent, B. deB., "Bond Dissociation Energies in Simple Molecules," NSRDS - NBS 31 (1970).
11. Skinner, G. B. and Ringrose, G. H., J. Chem. Phys. (1965) 43, 4129.
12. Frankiewicz, T. C., Gann, R. G. and Williams, F. W., Code 6180, Naval Research Laboratory, Washington, D. C. 20375, unpublished results.
13. Frankiewicz, T. C., Williams, F. W. and Gann, R. G., J. Chem. Phys., (1974) 61, 402.
14. Kondratiev, V. N., "Rate Constants of Gas Phase Reactions," National Technical Information Service, Springfield, Va. 22151, 1972.
15. Hiatt, R. and Benson, S. W., Int. J. Chem. Kin., (1972) 4, 479.
16. Brown, N. J., This Symposium.
17. Amphlett, J. C. and Whittle, E., Trans. Faraday Soc., (1966) 62, 1962.
18. Engibous, D. L. and Torkelson, T. R., AD239021, National Technical Information Service, Springfield, Va., 1960.
19. Stone, J. P. and Williams, F. W., Code 6180, Naval Research Laboratory, Washington, D. C. 20375, unpublished results.
20. Biordi, J. C., Lazzara, C. P. and Papp, J. F., "15th Symposium (International) on Combustion," The Combustion Institute, Pittsburgh, Pa., 1975.
21. Gann, R. G., Frankiewicz, T. C. and Williams, F. W., J. Chem. Phys., (1974) 61, 3488.
22. Seidel, W., Z. Phys. Chem. (Frankfurt am Main)

(1969) 65, 95.
23. Jones, W. E., MacKnight, S. D. and Teng, L., Chem. Rev., (1973) 73, 407.
24. Davies, P. B., Thrush, B. A. and Tuck, A. F., Trans. Faraday Soc., (1970) 66, 886.
25. Herron, J. T. and Huie, R. E., J. Phys. and Chem. Ref. Data., (1974) 2, 467.
26. Baulch, D. L., Drysdale, D. D., Horne, D. G. and Lloyd, A. C., "Evaluated Kinetic Data for High Temperature Reactions," Volume 1, Chemical Rubber Co., Cleveland, Ohio, 1972.
27. Creitz, E. C., J. Chromatographic Science, (1972) 10, 168.
28. Wyatt, J. R., DeCorpo, J. J., McDowell, M. V. and Saalfeld, F. E., Int. J. Mass Spec. and Ion Phys., (1975) 16, 33.
29. Tomkinson, D. M. and Pritchard, H. O., J. Phys. Chem., (1966) 70, 1579.
30. Hastie, J. W., Chem. Phys. Lett., (1974) 26, 338.
31. Fish. A., Angew. Chem. (International Edition) (1968) 7, 45.
32. Baker, R. R. and Yorke, D. A., J. Chem. Ed., (1972) 49, 351.
33. Lloyd, A. C., Int. J. Chem. Kin., (1974) 6, 169.
34. Walkauskas, L. P. and Kaufman, F. "15th Symposium (International) on Combustion," The Combustion Institute, Pittsburgh, Pa., 1975.
35. Rosser, W. A., Wise, H. and Miller, J., "7th Symposium (International) on Combustion," p. 175, Butterworths, London, 1959.
36. Huie, R. E., Herron, J. T., "Progress in Reaction Kinetics," in press.
37. Frankiewicz, T. C., Gann, R. G. and Williams, F. W., Code 6180, Naval Research Laboratory, Washington, D. C. 20375, unpublished results.
38. Biordi, J. C., U. S. Bureau of Mines, 4800 Forbes Ave., Pittsburgh, Pa. 15213, private Communication.
39. Skinner, G. B., this Symposium.
40. Clyne, M. A. A. and Watson, R. T., J. Chem. Soc., in press.
41. Watson, R. T., Dept. of Chem., University of Maryland, College Park, Md. 20742, private communication.
42. Wyatt, J. R., Code 6110, Naval Research Laboratory, Washington, D. C. 20375, private communication.

DISCUSSION

J. C. BIORDI: Our experience also raises some questions about the temperature dependence for the thermal decomposition of CF_3Br. Although thermal decomposition, using Benson & O'Neal's value, is negligibly important in a CF_3Br containing methane flame at 1300°K (where the disappearance rate for CF_3Br is at its maximum), it would dominate above ∿1500°K. This in turn, would imply a negative activation energy for the reaction $H+CF_3Br \rightarrow HBr+CF_3$ above about 1400°K, clearly untenable. It may be, though, that at the number density in this flame (P=32 torr), this reaction may be nowhere near its high pressure limit. Studies defining the pressure dependence of the thermal decomposition of CF_3Br would be helpful.

A. S. GORDON: In dealing with the reaction $CF_3+O \rightarrow CF_2O+F$ as an exothermic process, one has to consider the pressure regime, because for $CF_3+O+M \rightarrow CF_3O+M$, followed by $CF_3O+M \rightarrow CF_2O+F+M$, the latter reaction is now controlling for the formation of CF_2O and is endothermic.

R. G. GANN: This is true regarding the thermodynamics. However, if the rate of the bimolecular displacement is extremely fast, as our results indicate, then the termolecular addition reaction will be hard pressed to compete, given the gas densities at flame temperatures and at pressures of the order of 1 atm or less.

J. C. BIORDI: The determination of rate constants in flames, as in any other complicated reacting system, depends upon knowledge of the mechanisms involved and usually, also upon "known" rate constants for some of the reactions. The reaction $CH_3+CF_3Br \rightarrow CH_3Br+CF_3$ that you mentioned is an excellent case in point. What we can determine from our data is the net reaction rate for CH_3Br in the flame, i.e., the sum of the rates all the reactions occurring that form CH_3Br minus the sum of the rates of all the reactions occurring that destroy CH_3Br. Sometimes one is lucky and finds that perhaps two or or even one reaction dominates the behavior of a species. In this case, however, we were required to consider for formation also the reaction $CH_3+Br_2 \rightarrow CH_3Br+Br$; and the abstraction reactions for disappearance, $CH_3Br+H(OH) \rightarrow HBr(HOBr)+CH_3$, could never be neglected. Thus, the quoted rate constant depends upon the rate constants for those other reactions and it is not all easy to assign an uncertainty to a rate constant so determined.

In recent years kineticists have been dealing quantitatively with more and more complicated reacting systems because they have learned to use the computer to test their speculations on mechanisms. Good progress is being made on modelling of flame systems and eventually it will be possible to use these techniques to make more comprehensive analysis of flame profile and rate data than is now practical.

A. W. BENBOW: This comment is in support of your statement that ROO· interactions with flame inhibitors need to be studied. Dr. A. W. Bastow, working in Professor Cullis' combustion research group at the City University, London, has performed some very detailed work on the chemical mechanism whereby hydrogen bromide affects cool flame propagation in n-pentane - oxygen mixtures in the region of 250-350°C. In the absence of the halogen compound, an important reaction is the intramolecular rearrangement of pentylperoxy radicals as a results of the transfer of hydrogen from a point in the C_5 chain to the outer oxygen of the peroxy group, e.g.,

$$\underset{\displaystyle CH_3CHCH_2CH_2CH_3}{\overset{\displaystyle OO\cdot}{|}} \rightarrow \underset{\displaystyle CH_3CHCH_2CHCH_3}{\overset{\displaystyle OO\cdot}{|}}$$

This leads to the formation of C_5 O-heteroxyclic compounds and fission carbonyl compounds and alkenes as the principal cool flame products.

In the presence of even small amounts of HBr (∿2 mole%), there is a striking change in the products of the cool flame reaction, which now consist predominantly of C_5 ketones. It can be shown that this dramatic difference can be ascribed to the replacement of the intramolecular hydrogen abstraction process by an intermolecular reaction involving transfer of hydrogen from the HBr directly to the pentylperoxy radical to form a pentylmonohydroperoxide:

$$CH_3CH(OO\cdot)CH_2CH_2CH_3 + HBr \rightarrow CH_3CH(OOH)CH_2CH_2CH_3 + Br\cdot$$

12

Halogen Kinetics Pertinent to Flame Inhibition: A Review

N. J. BROWN

Department of Mechanical Engineering, University of California, Berkeley, Cal. 94720

The objectives of this paper have been somewhat tempered by what may be termed the total immersion approach to halogen kinetics. There is a vast amount of literature on this subject, and realization of this and the complexity of the field has forced the author to rethink and moderate the original objectives. The objective, simply stated, is to review some of the chemistry believed to play an important role in gas phase halogen flame inhibition and not to review the immense field of flame inhibition chemistry.

Chemical inhibitors are believed to alter the composition profiles of radical species in flames. The detailed mechanism is not understood because of the lack of radical species concentration profile measurements and pertinent kinetic rate data. Radical species represent a small fraction of total flame composition and their profiles change dramatically in the thin reactive zones of flames. Because of the inherent difficulty in their measurement, radical concentrations are often inferred by measurements of stable species and conservation equations and/or by invoking partial equilibrium approximations.

The present state-of-the art is such that the reactions of the hydrogen-halogen chains are believed to play an important role in flame inhibition. The major emphasis will be placed on the kinetics of the inhibitors HCl and HBr. These species were selected because 1) they represent the simplest type of inhibitor, 2) their chemistry has been the object of extensive study, and 3) they play an integral part in the inhibition by more complex chlorine and bromine containing compounds. Reactions with oxygen containing species which may also be important in the inhibition mechanism will not be discussed. The reasons for emphasizing HCl and HBr are that they 1) have more practical importance, 2) are less expensive than iodine compounds, and 3) are more effective inhibitors than fluorine compounds.

Creitz (1), Fristrom and Sawyer (2) and Hastie (3) have written recent review articles on flame inhibition. Fristrom and Sawyer develop a model in their paper which is based on the experiments of Wilson et al. (4,5). The major premise of their model is that the inhibitor molecule reacts by abstracting H atom in competition with the chain branching reaction $H + O_2 \rightarrow OH + O$. The choice of reactions considered in this review which was strongly influenced by their model, are those of the hydrogen-halogen chain:

$$H + HX \rightleftharpoons H_2 + X \qquad (1)$$

$$H + X_2 \rightleftharpoons HX + X \qquad (2)$$

$$H + X + M \rightarrow HX + M \qquad (3)$$

$$X + X + M \rightarrow X_2 + M \qquad (4)$$

Reaction (1) is thought to be competitive with the chain branching reaction and reaction (2) is included since it is competitive with (1). Although the reverse reactions (3) and (4) are slow at flame temperatures, and are not important reaction steps in the halogen inhibition mechanism, we believe that the forward recombination reactions may play an important role. High temperature dissociation kinetics will be discussed since the data is frequently extrapolated to lower temperatures and combined with equilibrium constants to obtain recombination rate coefficients.
Part two of the paper is a review of the chlorine kinetics and part three discusses the bromine kinetics. All activation energies are given in cal/mole and all temperatures are in kelvins. Second order rate coefficients appear in $cm^3 mole^{-1} sec^{-1}$ and the third order recombination rate coefficients are in units of $cm^6 mole^{-2} sec^{-1}$.

Chlorine Reactions

Introduction. Reactions (1) through (4) for chlorine will be discussed in this section. Early workers in the field studied what may best be described as competitive chlorination reactions, and prior to the last decade, chlorine reactions were studied more extensively than any other halogen. An early review by Fettis and Knox (6) and later reviews by Benson et al. (7) and White (8) provide good coverage of the early work. Recent work on the kinetics of ClO_x reactions of atmospheric interest has been reviewed by Watson (9). Here, emphasis will be placed on more recent work.

$H + HCl \rightleftharpoons H_2 + Cl$. This reaction and its reverse are not only important in flame inhibition, but the reverse rate is used as a primary standard in the measurements of rate coefficients of hydrocarbon chlorination. Four review articles discuss the early studies of this reaction 1) Fettis and Knox, (6) 2) Benson et al. (7), 3) White (8)and 4) Watson (9). The analysis and correction of early results with modern thermo data by Benson et al. is excellent. The kinetics of this reaction are still quite controversial and the rate coefficients for the forward and reverse reactions have been the object of recent theoretical and experimental studies. The first measurement of the reverse rate was that of Rodenbush and Klingelhoeffer (10) who studied the reaction at 273 and 298K. The product HCl concentration was measured as a function of time; however k decreased consistently with time during the course of a run. The authors suggest the effect might be due to a general activation of the glass walls of the reaction vessel to the recombination of Cl atoms. Their reported value of k is most likely too low and should not be weighted too heavily due to the large amount of scatter in the data.

Steiner and Rideal (11) measured the forward rate between 900 and 1070K and determined the H concentration from the equilibrium $H_2 \rightleftharpoons 2H$ assuming the normal ortho to para hydrogen ratio of 3:1 obtained in their experiment. Benson et al. point out that the equilibrium o-p H_2 ratio did not exist in the Steiner-Rideal experiments and have recomputed the rate coefficient. They also report Steiner and Rideal used an equilibrium constant for $K_{H_2 \rightarrow 2H}$ that is 55 percent higher than the JANAF (12) value, and conclude that the Steiner-Rideal rate coefficient should be 20 percent higher. Benson et al. further suggest that this corrected rate coefficient should be decreased by 50 percent, since oxygen did not diffuse through the walls of the reaction vessel.

Ashmore and Chanmugam (13) studied the reverse reaction at 523K by determining the rate coefficient relative to that of the reaction, NO + NOCl. They obtained the latter coefficient by extrapolating the 298-328K results of Burns and Dainton (14) to 523K. The method of Ashmore and Chanmugam was based on approximations, and the result of Burns and Dainton which they used was dependent upon the evaluation of other rate coefficients; hence their results could easily be in error. Benson et al. recorrected a numerical error in the Ashmore-Chanmugam results and used more modern thermo data for the Cl_2 + NO equilibrium constant to obtain $k = 4.79 \times 10^{11} cm^3 mole^{-1} sec^{-1}$ at 523K.

Clyne and Stedman (15) studied the forward reaction in a glass flow system by monitoring the H atom concentration indirectly by measuring the emission of HNO formed via H + NO → HNO (the NO was added to the stream at a fixed point after the H + HCl reaction zone). They attempted to make the reverse reaction and the reaction H + Cl_2 negligible by careful choice of experimental parameters; their H decay plots, however, exhibit some curvature at long reaction times which could indicate competition from other reactions. There is also slight curvature in their Arrhenius plots which they fit to a straight line. Their results appear in Table I.

Table I

$HCl + H \rightarrow H_2 + Cl$

Investigators	Reference	Measurement	Temperature (K)	k(cm^3/mole sec)
Clyne and Stedman	15	Flow tube, monitored H atom via HNO emission	195-373	$3.5 \times 10^{11} T^{1/2} exp(-2900/RT)$
Westenberg and de Haas	16	Flow tube, monitored H atom decay with ESR	195-497	$2.3 \times 10^{13} exp(-3500/RT)$

Benson et al. (7) studied the reverse reaction between 476 and 610K using ICl as a thermal source for Cl atoms and measuring Cl_2, I_2, and ICl spectrophotometrically. A mechanism for the $H_2 + Cl_2 + I_2$ system was postulated and an expression for the rate coefficient for the $Cl + H_2$ reaction was derived. They combined their work with the corrected early results and obtained an Arrhenius expression for the temperature range 273 to 1071K; their results are given in Table II.

Westenberg and de Haas (16) studied both forward and reverse reactions in a flow system using ESR detection. Their convention for forward and reverse is opposite ours and the use of "forward and reverse" will apply to the way the reaction is written here, i.e. H + HCl.

The forward rate was studied under pseudo first order conditions in HCl between 195 and 497K by monitoring the H atom concentration via ESR. The H atoms were formed from discharged H_2. In measuring this rate, competition from the reverse reaction was probably not important since $k_{H+HCl} > k_{H_2+Cl}$ and the reaction was studied in excess HCl; however competition from the fast Cl_2 + H reaction would have been a problem if adequate production of Cl_2 had occured via wall recombination. The authors admit this possibility in stating that some Cl_2 was present and that the slope of the H decay plots increased slightly with increasing H_2 concentration. The effect is not due to increased H_2 but to increased initial H (resulting from increased H_2) which reacted to give more Cl which, in turn, could produce more Cl_2. In the evaluation of the rate coefficient, they retained only the data associated with smallest possible initial H_2 concentrations. The results are summarized in Table I.

Measurement of the reverse reaction was accomplished by monitoring the concentration of the ground state ($^2P_{3/2}$) chlorine atom as a function of time with ESR. The reaction was studied under pseudo first order conditions with a large concentration of H_2 and chlorine atom was produced by discharging small amounts of Cl_2. The competitive reactions were the reactions H + HCl and H + Cl_2; however, attempt to minimize these reactions was made by keeping the H_2 concentration as large as possible. The initial slopes of the Cl decay plots had good first order linearity but showed curvature at longer residence times indicating possible competition from other reactions. The Arrhenius parameters are given in Table II.

Westenberg and de Haas calculated the ratio of the measured forward and reverse rate coefficients and the computed ratio was 2 or 3 times larger than the thermodynamic equilibrium constant over the experimental temperature range. This result has been the subject of several current papers. The first question to raise is "does this experiment illustrate a violation of microscopic reversibility" and our opinion is no. Westenberg and de Haas explained their results (in the original paper) as being due to the greater probability of obtaining rotationally excited

Table II

$H_2 + Cl \rightarrow HCl + H$

Investigators	Reference	Measurement	Temperature (K)	k(cm³/mole sec)
Benson *et al.*	7	ICl as source of Cl Measured ICl, I_2 and H_2 spectrophotometrically assumed mechanism for the $H_2 + Cl_2 + I_2$ system.	476-610 evaluation of early results implies expression valid for 476-1071 K.	$4.8 \times 10^{13} exp(-5260/RT)$
Westenberg and de Haas	16	Flow tube, pseudo first order, measured $Cl(^2P_{3/2})$ with ESR.	251-456	$1.2 \times 10^{13} exp(-4300/RT)$
Davis *et al.*	20	Flow tube, measured $Cl(^2P_{3/2})$ with resonance fluorescence.	298	8.4×10^9
Clyne Evaluation	17	I using Clyne and Stedman and Westenberg and de Haas values for reverse reaction and calculating k from equilibrium constant, combined with Davis *et al.* results and Benson *et al.* evaluation.	195-610	$2.23 \times 10^{13} exp(-4260/RT)$
		II - above excluding Benson *et al.* evaluation	195-496	$3.4 \times 10^{13} exp(-4480/RT)$

product HCl in the reverse direction than excited product H_2 in the forward direction.

Clyne and Walker (17) suggest that the reverse rate might have been systematically too low and that the forward rate was correct. They attribute this to Cl wall recombination to Cl_2 and subsequent reaction via $H + Cl_2 \rightarrow HCl + Cl$; hence, the production of Cl via the competitive reaction results in a smaller Cl decay plot slope. This explanation is reasonable and merits further consideration.

Snider (18) gives another plausible explanation for the Westenberg and de Haas results, that is, that the Cl ($^2P_{1/2}$) is more reactive than Cl ($^2P_{3/2}$) which Westenberg and de Haas measured, and this effect could produce a significantly non-Boltzmann distribution of Cl among its electronic states during the forward reaction. Snider supports his point of view with a three step mechanism:

$$Cl(^2P_{3/2}) + H_2 \overset{k_1}{\rightleftharpoons} HCl + H \quad (1)$$

$$Cl(^2P_{1/2}) + H_2 \overset{k_2}{\rightleftharpoons} HCl + H \quad (2)$$

$$Cl(^2P_{1/2}) + H_2 \overset{k_3}{\rightleftharpoons} Cl(^2P_{3/2}) + H_2 \quad (3)$$

If $k_2 > k_1$ and the quenching reaction (3) is not fast enough to restore equilibrium, then Snider's arguments could be correct, however Cadman and Polanyi (19) have shown this not to obtain for the H + HI reaction.

Davis *et al.* (20) have studied the H_2 + Cl reaction at room temperature monitoring the Cl atom concentration using a resonance fluorescence technique capable of detecting Cl ($^2P_{3/2}$ and $^2P_{1/2}$) and they report that the ($^2P_{1/2}$) state made negligible contribution. Their results agree with the Westenberg and de Haas results within 10 per cent at 298 K.

In a further attempt to resolve the controversy, Galante and Gislason (21) argue that the two Cl electronic states are in equilibrium. They reject the proposed two reaction channel mechanism and argue that the measured H + HCl rate is too large rather than the H_2 + Cl rate being too small. They assume the recombination $H + Cl + M \rightarrow HCl + M$ interferes and consumes an additional H atom, and as a consequence the H + HCl rate is large by a factor of two.

Our conclusion with regard to the apparent violation of $k_f/k_r = K$ is that the measured rate for the Cl + H_2 reaction is too low. Whether this can be attributed to Cl_2 production and the competitive reaction Cl_2 + H or to the two channel reaction mechanism via the ($^2P_{1/2}$) and ($^2P_{3/2}$) still remains an unanswered question. The role of the atomic ($^2P_{1/2}$) electronic state in

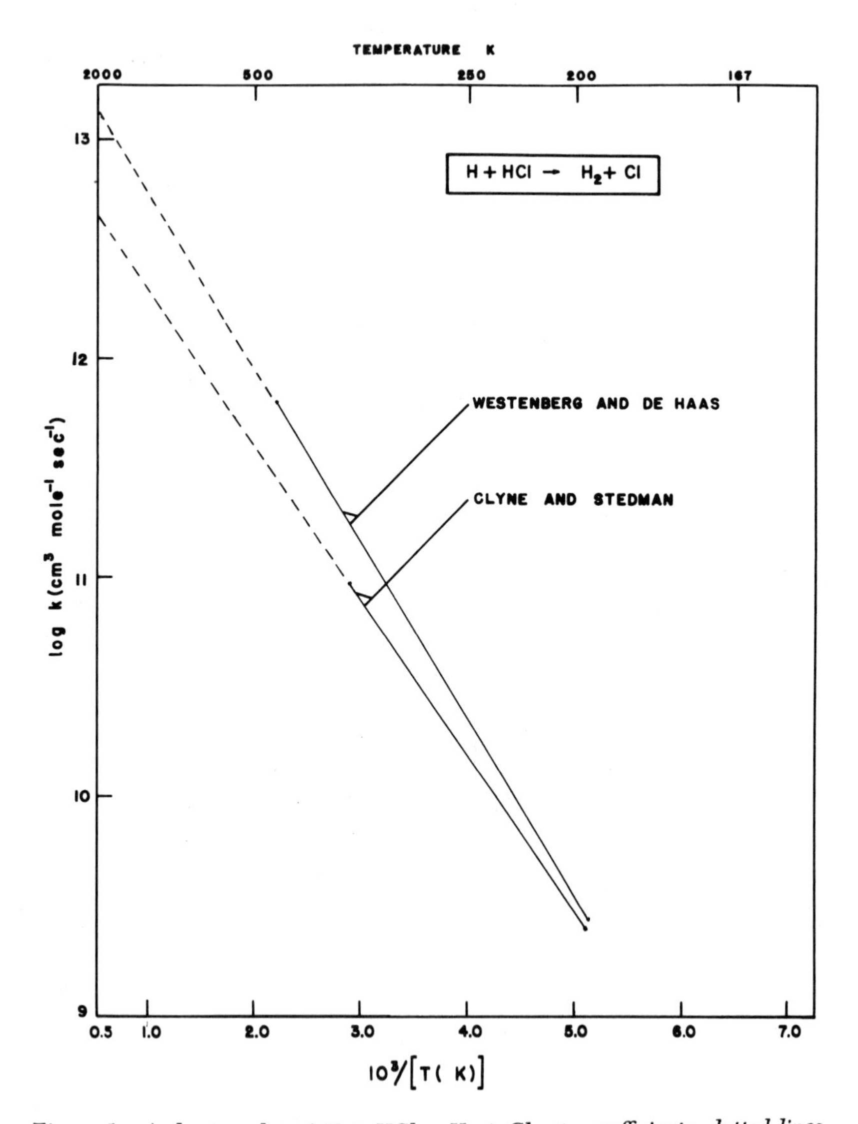

Figure 1. Arrhenius plot of $H + HCl \rightarrow H_2 + Cl$ rate coefficients; dotted lines indicate extrapolation

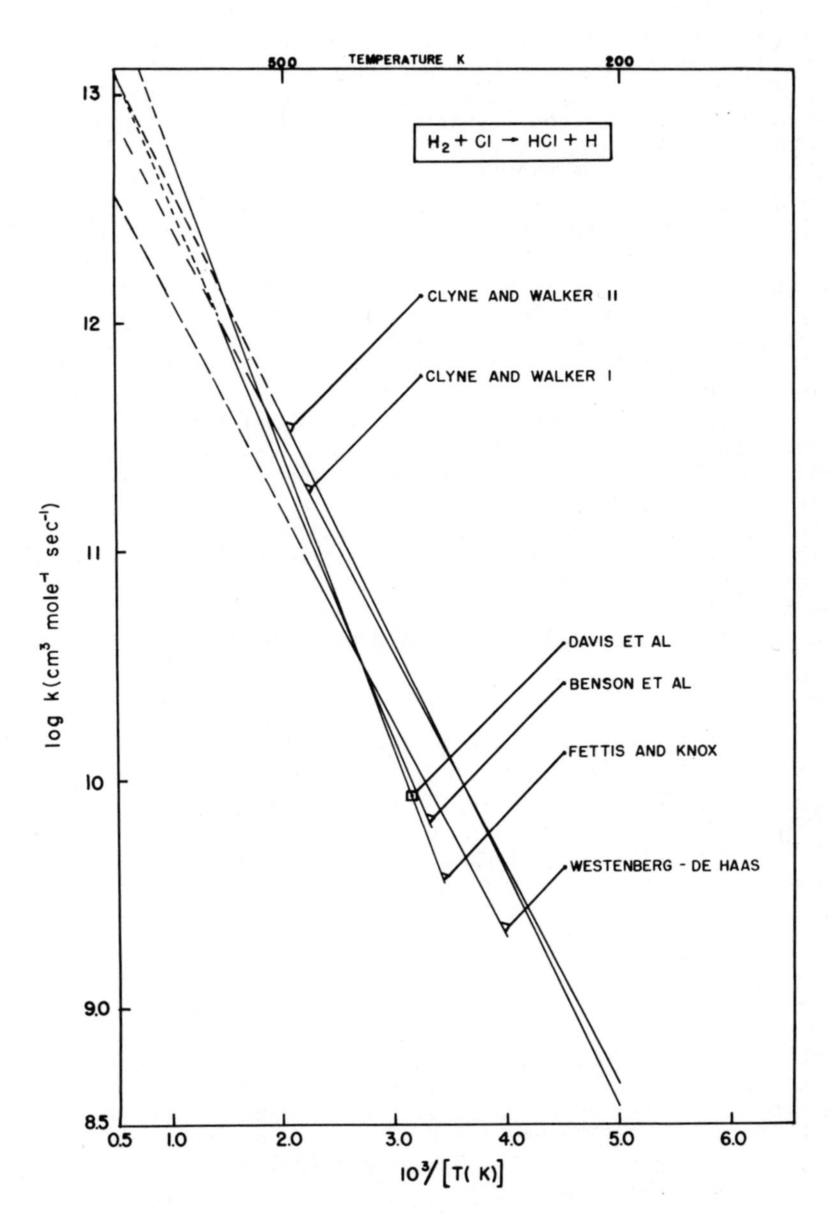

Figure 2. Arrhenius plot of $H_2 + Cl \rightarrow H + HCl$ rate coefficients; dotted lines indicate extrapolation

halogen kinetics is a question often raised by many investigators. Experiments which 1) determine the rate of quenching ($^2P_{1/2}$) atoms at various temperatures, 2) investigate the production of ($^2P_{1/2}$) atoms formed by discharging halogen molecules, and 3) measure the reactivity of the ($^2P_{1/2}$) state related to the ($^2P_{3/2}$) are needed. Careful considerations of these problems by theoreticians would be extremely valuable. More studies of both forward and reverse reactions should be made to investigate whether microscopic reversibility is indeed maintained for systems reacting far from equilibrium.

Figure 1 is an Arrhenius plot for the H + HCl reaction; the agreement between investigators is good; however, the Westenberg-de Haas results are preferable in our opinion, since their measurement was more direct. There are no measurements in the 500 to 2000 K range and the extrapolated results disagree by a factor of 3 at 2000 K.

Figure 2 shows the results of studies of the H_2 + Cl reaction. The Westenberg-de Haas results are low. In our opinion, Benson _et al_. placed too much weight on the low temperature results of Rodenbush and Klingelhoefer. The Clyne I and Clyne II entries are the values recommended by Clyne and Walker. Clyne I, which is the one we favor, is based on the results of Clyne and Stedman and Westenberg and de Haas for the _reverse_ reaction and the data of Davis _et al_. Clyne II is the same as Clyne I with the addition of the results of the Benson _et al_. evaluation.

$Cl_2 + H \rightleftharpoons HCl + Cl$

There has been relatively little work done on the alleged fast reaction, $Cl_2 + H \rightarrow H_2 + Cl$. Klein and Wolfsberg (22) measured the rate relative to the H + HCl reaction using tritium labeled HCl. The temperature interval investigated was 273 - 335 K, and species concentrations were measured with mass spectrometry. The results reported as:

$$\frac{k_{HCl + H}}{k_{Cl_2 + H}} = (0.143 \pm 0.033)\exp(-1540 \pm 130/RT)$$

might be subject to error because of reliance on previously determined rate data. Their paper does not contain enough primary data to verify the above expression with newer kinetic data.

The rate coefficient for H + Cl_2 was measured directly between 294 and 565 K by Dodonov _et al_. (23) with a technique that is best described as molecular beam sampling of a diffusion cloud in a flow. They studied the reaction under pseudo first order conditions, maintaining a nearly constant concentration of H atoms produced from H_2 with a stable discharge. The reactor was a capillary tube containing a large amount of He and a small amount of H atom. Mixtures of Cl_2 and He were introduced at a fixed point and the Cl_2 was measured as a function of distance along the reactor axis. The diffusion coefficient of Cl_2 in He, required for analysis of their results, was measured in their apparatus. Their first order and Arrhenius plots are of good linearity, and the reported rate coefficient is

$$k_{Cl_2+H} + H = 3.7 \times 10^{14} \exp(-1800/RT) \ cm^3 mole^{-1} sec^{-1}.$$

Jacobs <u>et</u> <u>al</u>. (<u>24</u>) measured the same ratio of rates as Klein and Wolfsberg. The reactions were studied in a shock tube between 3500 and 5200 K by measuring HCl concentrations by infrared emission. They indicate that a factor of 2 is the probable error in their experimental results. The ratio

$$\frac{k_{H + HCl}}{k_{Cl_2 + H}} = .2 \exp(-1800/RT)$$

is in reasonable agreement with the value reported by Klein and Wolfsberg.

The Klein and Wolfsberg ratio and the $k_{H + HCl}$ coefficient reported by Westenberg and de Haas (<u>16</u>) can be used to compute $k_{Cl_2 + H}$ and Figure 3 in an Arrhenius plot of this calculated rate coefficient and the Dodonov result. Considering the errors quoted by Klein and Wolfsberg (as a standard deviation) and by assuming the Westenberg and de Haas rate as correct for H + HCl, the agreement is surprisingly good.

<u>H + Cl + Ar ⇋ HCl + Ar</u>

Four shock tube studies of HCl dissociation with Ar as third body spanning the temperature regime 2900 - 5400 K have been conducted. The experimental results are summarized in Table III. Note all the activation energies are less (by several RT) than the HCl bond dissociation energy (102,170 cal/mole), and that they differ considerably. The agreement between the various rate coefficients is reasonably good in the common temperature regime. Of the three experiments employing spectroscopic detection techniques, the absorption measurements of Seery and Bowman (<u>25</u>) are preferable since absorption techniques are generally more reliable. The Breshears and Bird (<u>26</u>) measurement, if their approximations are correct, has the distinct advantage of not assuming anything about the distribution of molecular states.

One difficulty in determining the rate coefficient for HCl dissociation at high temperatures is correctly assessing the contributions of competitive reactions of the $H_2 + Cl_2$ chain. The kinetic data of competitive chain reactions is frequently measured at low temperatures and extrapolation of this data to the high temperatures of shock tube experiments may give rate coefficients

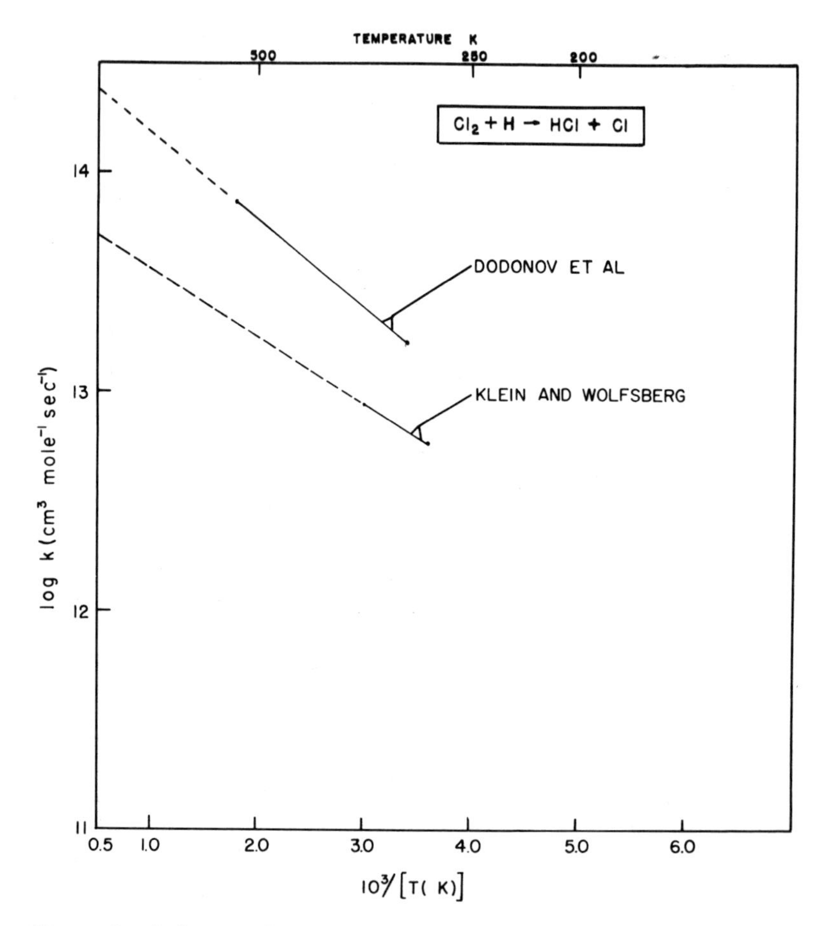

Figure 3. Arrhenius plot of $H + Cl_2 \rightarrow HCl + Cl$ rate coefficients; dotted lines indicate extrapolation

Table III

$HCl + Ar \rightarrow H + Cl + Ar$

Investigators	Reference	Measurement	Temperature (K)	k(cm^3/mole•sec)
Fishburn	28	shock tube, monitored reaction; progress by measuring IR emission from (1-0) band of HCL	3300 → 5400	$1.92 \times 10^{11} T^{1/2} exp(-69,700/RT)$
Jacobs *et al.*	27	shock tube, ir emission of HCl invoking transparent gas approximation	2800 → 4600	$6.6 \times 10^{12} exp(-70,000/RT)$
Seery and Bowman	25	shock tube, measured HCl concentration using vacuum ultraviolet absorption spectroscopy	2900 → 4000	$4.2 \times 10^{13} exp(-81,000/RT)$
Breshears and Bird	26	shock tube, laser schliern technique to monitor post shock density gradient	3000 → 4000	$4.78 \times 10^{13} exp(-82,700/RT)$
Giedt and Jacobs	29	shock tube, monitored radiative recombination of Cl atoms	1600 → 2800	$6.6 \times 10^{12} exp(-70,000/RT)$

which are unreliable. Systematically varying the rate coefficients over the probable error limits is a common method of assessing the effects of uncertainties in the kinetic data and the contributions of competing reactions. The paper of Jacobs *et al.* (27) gives a good discussion of this approach. Jacobs *et al.* found that the activation energy of dissociation reaction was sensitive to the Cl + HCl → Cl_2 + H rate coefficient; however, the rate coefficient of this reaction has only been measured at low temperatures. Increasing $k_{Cl + HCl}$ was found to increase the HCl dissociation activation energy. Fishburn (28) also noted a dependence of the activation energy on the reaction rates for Cl + HCl and H + HCl reactions.

The only measurement of the HCl dissociation rate at temperatures important to combustion was that of Giedt and Jacobs (29). The extended the temperature range to 1600 K monitoring the radiative recombination of Cl atom. They report few experimental details and their measurements raise some doubt since they assumed the intensity of emission varied as $[Cl]^n$ where n = 2; n was subsequently shown by Clyne and Stedman (30) to vary with wavelength. In addition their experimental and calculated reaction profiles did not show good agreement. Their results are shown in Table III.

Figure 4 is an Arrhenius plot of the experimental results. In the common temperature range the agreement is within a factor of 2. Jacobs *et al.* (27) estimate that the accuracy of their rate coefficient is about a factor of 2. Figure 5 is an Arrhenius plot of the HCl dissociation rate coefficients extrapolated to 1300 K. The variation between the rate coefficients at the lower temperature is more than a factor of 10 and the extrapolation introduces errors that are impossible to ascertain.

Little study has been directed toward assessing the relative efficiences of various third bodies in HCl dissociation. Giedt and Jacobs reported no difference in HCl and Ar while Breshears and Bird found HCl 10 times more effective than Ar, and that the temperature dependence of M = HCl departs from the Arrhenius form, with the apparent activation energy increasing at lower temperatures. If one can argue by analogy, using HBr as an example, the Giedt and Jacobs result is incorrect. More work in this area is warranted.

The relatively large difference between bond dissociation energies and activation energies is not an anomaly of the HCl dissociation but a general characteristic of high temperature dissociation reactions. There have been many explanations for the phenomena, 1) impurities, 2) improper consideration of competitive reaction, 3) nonequilibrium distribution of vibrational states, and 4) nonequilibrium distribution of electronic states. Most probably each of the four explanations has some validity in specific experimental environments but the author favors reason (3). Johnston and Birks (31) have written an excellent theoretical paper in which model calculations are presented which

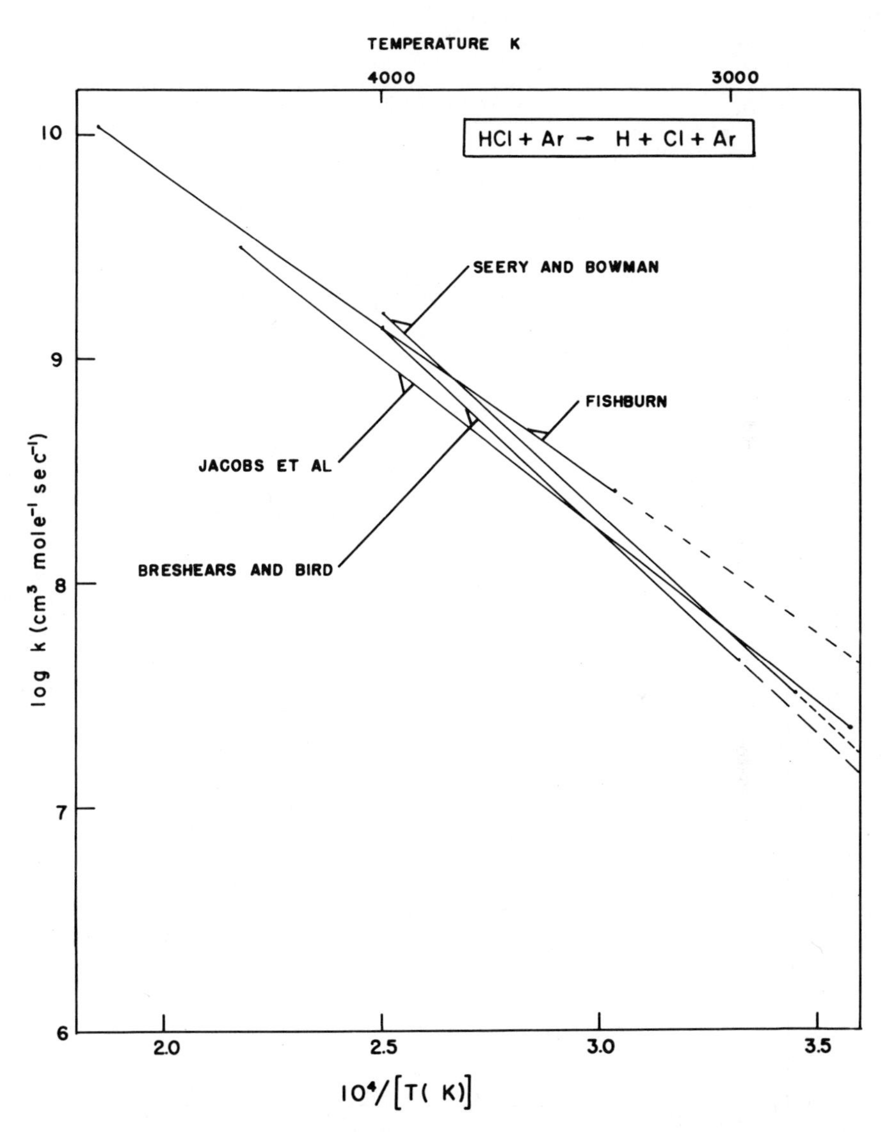

Figure 4. Arrhenius plot of $HCl + Ar \rightarrow H + Cl + Ar$ rate coefficients; dotted lines indicate extrapolation

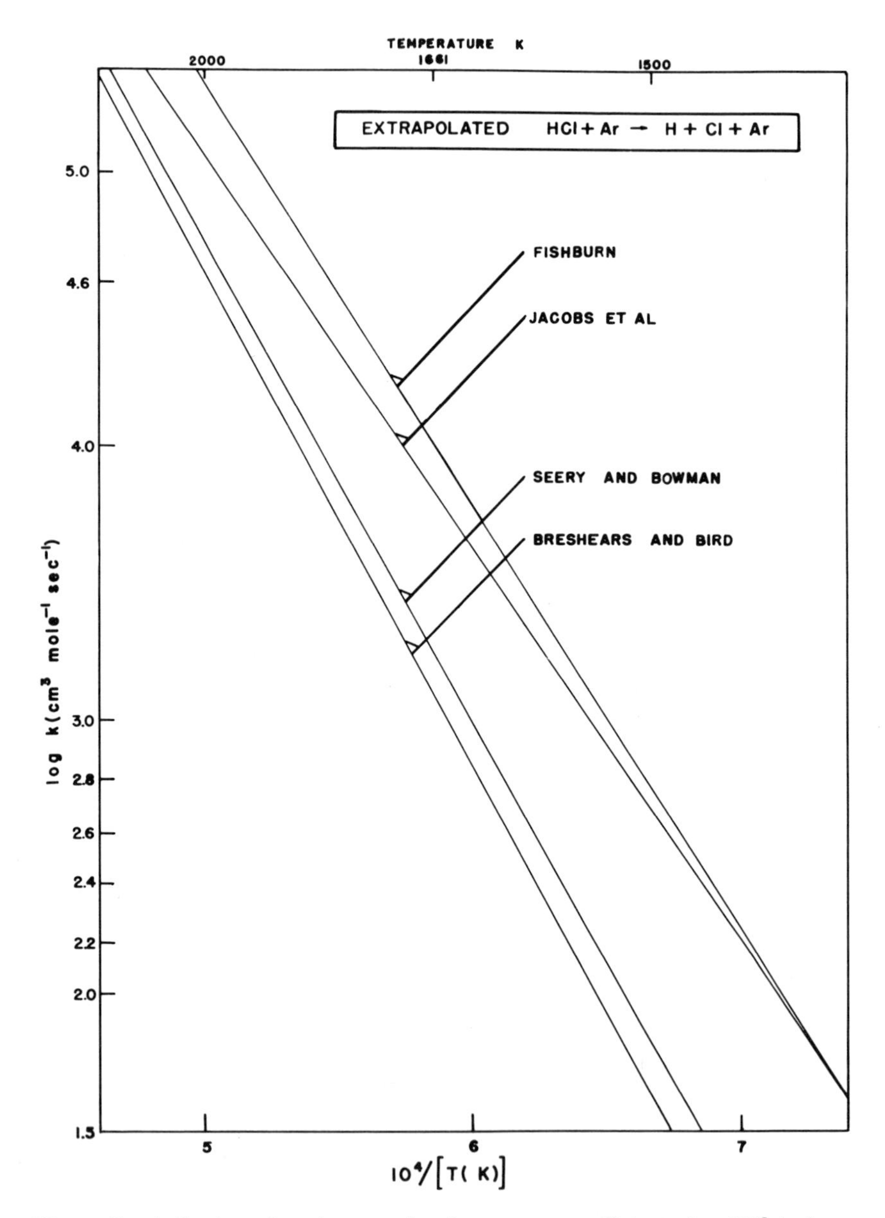

Figure 5. Arrhenius plot of extrapolated rate rate coefficients for $HCl + Ar \rightarrow H + Cl + Ar$

support reason (3). The paper gives a discussion of activation energy concept.

There are no reported measured values of rate coefficients for the reaction, H + Cl + M, and the reaction should be studied in a temperature regime where both forward and reverse reactions could be investigated. Judging from the poor agreement among the extrapolated high temperature dissociation results, calculation of the recombination rate coefficient from the equilibrium constant and the dissociation rate coefficient between 1000 and 2000 K would be unwise.

$Cl + Cl + M \rightleftharpoons Cl_2 + M$

This recombination reaction and the reverse dissociation reaction have been critically reviewed by Lloyd (32). The recommended value for the forward rate coefficient is based largely on the results of Clyne and Stedman (33) who studied the reaction with Ar as third body. The recommended rate coefficient for the temperatures 200 and 500 K is

$$k_{Ar} = 2.19 \pm 1.6 \times 10^{14} \exp(+1800 \pm 500/RT) cm^6 mole^{-2} sec^{-1}.$$

Cl_2 was found 4.5 times more efficient than Ar as a third body between 200 and 500 K; however, the ratio k_{Cl_2}/k_{Ar} is probably temperature dependent.

The "negative" activation energy for recombination arises, in part, from the redissociation of highly excited product Cl_2, an effect that increases with temperature. The product molecule formed via recombination is usually formed in a highly excited vibrational state and it can be thought of as being closer to reactants than products.

The recommended value of the dissociation rate coefficient with Ar as third body in the temperature range 1500 - 3000 K is

$$k_{Cl_2+Ar} = 8.7 \pm 2.0 \times 10^{13} \exp(-48,500 \pm 2000/RT)\ cm^3 mole^{-1} sec^{-1}.$$

This value was based primarily on the results of Jacobs and Giedt (34). The activation energy is approximately 8500 cal/mole less than the bond dissociation energy. Jacobs *et al.* (35) in a later study reported that Cl was 10 times more efficient than Ar in Cl dissociation. The high efficiencies of atoms as third bodies indicates that control of their concentrations in dissociation studies is extremely important. Radicals in flames are also likely to be efficient third bodies and the use of data applicable for M = Ar or N might introduce serious errors in modeling calculations.

Dissociation reactions are slow at temperatures < 2000K and probably are not an important reaction step in halogen flame inhibition. Dissociation rate coefficients are, however, frequently used with equilibrium constants to compute recombination rate coefficients. Since recombination may play an important role in inhibitions it is important to compare the direct recombination rate coefficient with that computed from extrapolated dissociation data. The two rate coefficients for the chlorine recombination reaction are shown in Figure 6. The agreement (indeed a misnomer) is very poor and indicative that more study of recombination and dissociation reactions in a common temperature interval is required. The relationship $K = k_f/k_r$ still has not been confirmed experimentally for reactions measured in shock tubes, and this and the fact that activation energies may well be temperature dependent are further reasons to merit additional study.

HBr Reactions

Introduction. The HBr reactions (1) and (2) have not been studied as extensively as the HCl reactions. There are two reviews of the literature associated with HBr kinetics, an early review by Fettis and Knox (6) and a later review by White (8). Hydrogen-bromine reactions are a classic example of a free radical chain mechanism. It is difficult to experimentally study an isolated reaction of the chain, free from competitive reactions, and the difficulty of assessing the effects of competitive reactions has been largely responsible for the errors in interpretation of early data.

$H + HBr \rightleftharpoons H_2 + Br$. This reaction has been studied simultaneously with the competitive hydrogen bromine chain reactions. There are many errors in the literature due to incorrect evaluation of the rate coefficient ratio k_{H+Br_2}/k_{H+HBr} and inaccurate values of the equilibrium constant $K = [Br]^2/[Br]$. The following discussion will pertain to studies of the chain mechanism and will not always be chronological. Effort will be made to point out rather than correct errors in early data. The $H + Br_2$ reaction will be discussed here, since most of the data associated with it are extracted from studies of the chain mechanism.

Bodenstein *et al.* (36,37,38) studied H + Br reactions between 498 and 577K. Known amounts of H + Br were mixed in a bulb, allowed to react, and the HBr and Br concentrations were measured as a function of time using wet chemical techniques; no rate coefficients were reported. The data were interpreted later by Christiansen (39), Herzfeld (40), and Polanyi (41) who independently showed that the system reacted via a radical chain mechanism. The Br atom concentration was assumed to be governed by the equilibrium

$$Br_2 \rightleftharpoons 2Br.$$

With this interpretation, the now accepted rate law is:

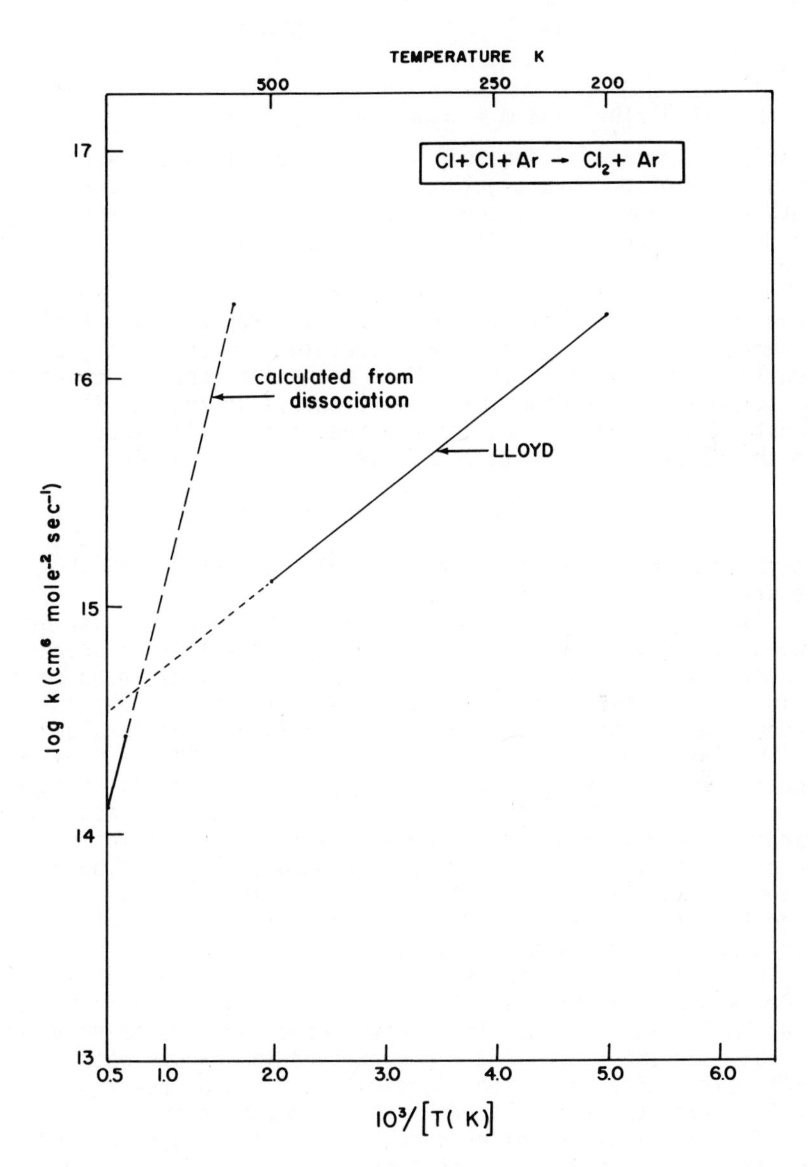

Figure 6. Arrhenius plot of Cl + Cl + Ar rate coefficients; dotted lines indicate extrapolation

$$\frac{d[HBr]}{dt} = \frac{2k_{H_2+Br}K^{1/2}[H_2][Br_2]^{1/2}}{1 + \frac{1}{m}\frac{[HBr]}{[Br_2]}} \quad (1)$$

where $K = [Br]^2/[Br]$ and $m = k_{H+Br_2}/k_{H+HBr}$.
The original data of Bodenstein *et al.* and the rate law have been used by other investigators to determine the rate coefficients k_{H_2+Br} and k_{H+HBr}; these "derived" values have frequently been incorrect because m was not properly evaluated and K was in error.

Bodenstein and Jung (38) performed photochemical experiments in the temperature interval 303 to 575K and reported m = 8.61 at low temperatures and 8.4 at high temperatures. Sullivan (42) has recently shown that the atomic chain mechanism does not obtain in room temperature photochemical experiments. He reports that Bodenstein and Jung did not measure the ratio but instead measured the recombination rate coefficient for the reaction

$$2Br + HBr \rightarrow Br_2 + HBr.$$

If this interpretation is correct, results computed with the Bodenstein-Jung ratio are incorrect.

Bach *et al.* (43) studied reactions of D and H with Br between 549 and 612K by measuring the Br concentration spectrophotometrically. Pease (44) gives a good summary of these results in his text; however m is not evaluated correctly and the value of the equilibrium constant is not the currently accepted value.

Steiner (45) studied the reaction between 820 and 980K using a technique identical to that of Steiner and Rideal (11) discussed earlier. The errors in this study are 1) an incorrect assumption regarding the ortho-para hydrogen equilibrium which makes their rate coefficient about 5 percent too high and 2) use of an equilibrium constant for the reaction $H_2 \rightarrow 2H$ which is 55 percent higher than the JANAF value. The latter error makes their rate coefficient 20 percent too low.

Fass (46) studied the relative rates of the reactions $H + Br_2$ and H + HBr between 300 and 523K by photodissociation of HBr at 1850 Å and 2480 Å. The H atoms produced were translationally hot, and the experiment was contrived to measure both rate coefficient ratios, one associated with hot H atom reactions, the other with the thermalized H atoms. Reaction progress was monitored by using absorption spectroscopy to measure HBr and Br_2 concentrations as a function of time.

An Arrhenius type expression for the rate coefficient ratio, m, was reported as

$$m = \frac{k_{H+Br_2}}{k_{H+HBr}} = 6.8\ \exp(800/RT).$$

The ratio is 26 at room temperature, 15 at 500K and 8.9 at 1500K. Fass indicates that this ratio is valid between 300 and 1400K since his results were combined with the high temperature results of other investigators (47,48,49). The Arrhenius expression for the ratio fits the low temperature data better than the high temperature data.

Fass used He to thermalize the H atoms and he assumed 15 percent remained unthermalized. We believe the Fass results are correct if the assumption of 85 percent thermalization is accurate.

Levy (47) studied H + Br reactions in a flow system between 600 and 1470K. The mixture was thermally quenched after reacting for a specified time interval, then separated by fractional condensation, and the concentrations of HBr and Br were measured as gas volumes. Incomplete quenching and poor separation are two potential sources of error in these data. Levy assumed m = 8.4 in his analysis which, if the Fass results are correct, underestimates the ratio by approximately a factor of 3 at 600K. Levy's Arrhenius plots appear to be reasonably linear; however the scatter in the high temperature data is large. The Arrhenius parameters are given in Table IV and Figure 7 is an Arrhenius plot of the rate coefficient.

The next study was that of Britton and Cole (48) who studied the chain reactions in a shock tube between 1300 and 1700K. Reaction progress was monitored by measuring Br concentration spectrophotometrically. The Br recombination rate coefficient, which was required for data analysis, was computed from dissociation data, and the error in this coefficient affects their results. They reported that their results were most likely uncertain by 25 percent. Their Arrhenius plot was of reasonable linearity, and their rate coefficient is given in Table IV. The ratio m was measured in their experiment and reported as m = 8.3; however the constant value of m is difficult to understand since their tabulated results indicated temperature dependence.

Vidal (50) studied the rate of formation of HBr from various mixtures of H and Br in the temperature range 508-566K. Reaction progress was followed by measuring Br concentrations spectrophotometrically. After quenching the reaction, the HBr concentration was determined by an acid-base titration and the Br , by an iodometric titration. A modern value of $K = [Br]^2/[$ $K = [Br]^2/[Br_2]$ was used in data analysis and the ratio m was determined experimentally and agreed favorably with the Fass results. The results of earlier workers were re-interpreted with the new m and K values and agreement between Vidal's results and those of Bodenstein et al., Bach et al., and Britton and Cole was very good. The Levy data rather than the Levy Arrhenius expression were also compared with the results of others. The agreement at high temperatures between the Levy data and the extrapolated Vidal results was reasonable. Vidal's

Table IV

$Br + H_2 \rightarrow HBr + H$

Investigators	Reference	Measurement	Temperature (K)	k(cm^3/mole·sec)
Levy	47	flow tube, separated gases by fractional condensation and measured HBr and Br_2 pressure	600-1470	$2.04 \times 10^{12} T^{1/2} exp(-17280/RT)$
Britton and Cole	48	Shock tube, measured Br_2 concentration by absorption spectroscopy	1300-1700	$1.79 \times 10^{14} exp(-19200/RT)$
Vidal	50	Flow tube, measured Br_2 concentration by absorption spectroscopy, final HBr and Br_2 concentrations by titration.	508-566 expression valid for 500-1700.	$1.35 \times 10^{14} exp(-18400/RT)$

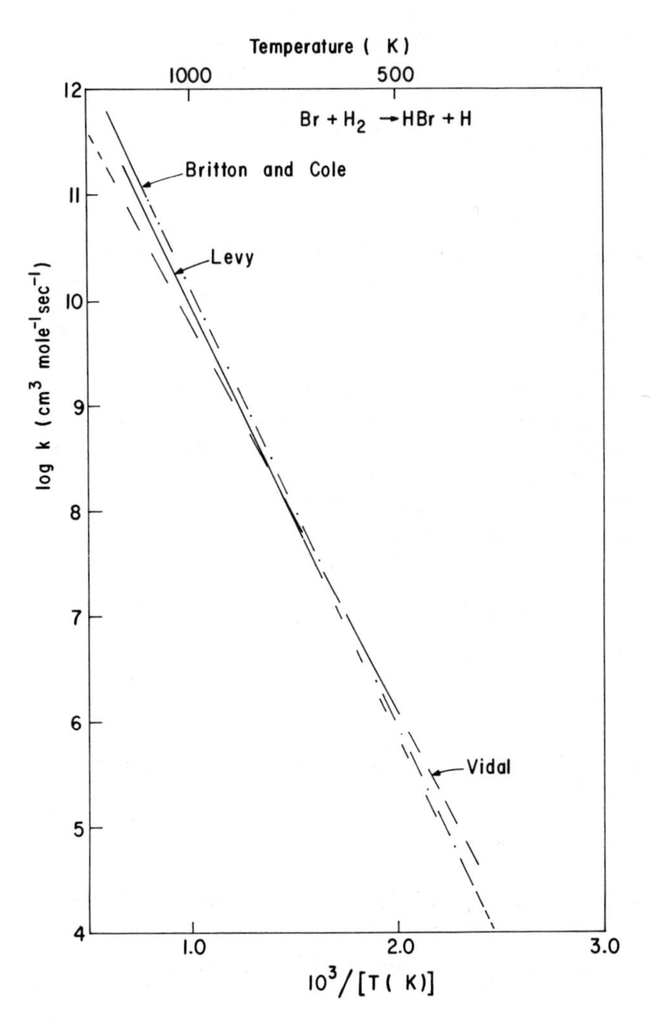

Figure 7. Arrhenius plot of $Br + H_2 \rightarrow HBr + H$ rate coefficients; dotted lines indicate extrapolation

expression for the rate coefficient is given in Table IV. Considering the agreement with the higher temperature results, the Vidal Arrhenius expression is assumed valid for the temperature interval 500 to 1700 K. Figure 7 shows a comparison of the Levy, Britton and Cole and Vidal rate coefficients.

Takacs and Glass (51) report the only direct measurement of the rate coefficient of the H + HBr reaction. The reaction was studied at 295 K in a flow system by monitoring the $Br(^2P_{3/2})$ and $H(^2S_{1/2})$ via EPR spectroscopy; no $Br(^2P_{1/2})$ was detected. The rate coefficient at 295 K was:

$$k_{H+HBr} = 2.05 \times 10^{12} cm^3 mole^{-1} sec^{-1}.$$

If the rate coefficient reported by Vidal is used to compute the k_{H+HBr} coefficient via the equilibrium constant, the value at 295 K is $5 \times 10^{12} cm^3 mole^{-1} sec^{-1}$ which is 2.4 times larger than the Takacs and Glass value.

$Br_2 + H \rightarrow HBr + Br$. No direct measurements of this rate coefficient have been reported. If one assumes the Fass ratio m is correct between 300 and 1400 K and that the Vidal value of k_{H_2+Br} is correct in the same temperature interval, then the rate coefficient for the Br_2 + H reaction can be computed using these results and the equilibrium constant:

$$K = \frac{k_{HBr+H}}{k_{H_2+Br}} = 10^{-.432} \exp(16697/RT). \qquad (2)$$

Assuming

$$k_{H_2+Br} = 1.35 \times 10^{14} \exp(-18400/RT) cm^3 mole^{-1} sec^{-1}$$

the value:

$$k_{H+HBr} = 5.0 \times 10^{13} \exp(-1703/RT) cm^3 mole^{-1} sec^{-1}$$

obtains. When k_{H+HBr} is used with the Fass ratio,

$$\frac{k_{H+Br_2}}{k_{H+HBr}} = 6.8 \exp(800/RT)$$

the rate coefficient is:

$$k_{H+Br_2} = 3.4 \times 10^{14} \exp(-903/RT) cm^3 mole^{-1} sec^{-1}.$$

HBr + M ⇋ H + Br + M. The HBr dissociation has been studied by three investigators in shock tubes between 1500 and 4200 K, and a brief discription of the work appears in the following paragraphs.

The first determination of the rate coefficient was by Giedt et al. (52) who studied the reaction in Ar between 2100 and 4200 K. The initial amount of HBr was varied between 2 and 10 percent and HBr concentrations were measured by monitoring both the HBr, IR emission and UV absorption. The calibration plots exhibited scatter; however, no difference between the results of the emission and absorption measurements was reported. Their first order plots are not of good linearity, indicative of competitive reactions and/or experimental error. Competitive reactions of the H_2 + Br_2 chain were accounted for when shock profiles were computed; however the profiles were most sensitive to the dissociation reaction and less sensitive to the competitive reactions. HBr was reported 15 times more efficient than Ar. The rate coefficient which gave the best fit to their profiles is given in Table V.

Westberg and Green (53) studied HBr dissociation between 1500-2700 K in a shock tube by measuring the emission from Br recombination to $Br_2(^1\pi_u)$. Concentrations were corrected for boundary layer growth; however the calibration measurements showed a great deal of scatter. They list their probable experimental errors as 1) errors in determining shock velocity, 2) uncertainty in determining location of the shock front, and 3) errors due to boundary layer effects, and conclude that the value of the rate coefficient is not seriously jeopardized by these errors. The competitive reactions of the HBr chain were accounted for and it was noted that no realistic variations of the rate coefficients of the competitive reactions produced the rate coefficient reported earlier by Giedt et al. HBr and Ar were found to be equally efficient as third bodies in contrast to the earlier results of Giedt et al. The rate coefficient which was estimated to be accurate within a factor of three is given in Table V.

The most recent study of HBr dissociation is due to Cohen et al. (54) who studied the reaction in pure HBr in the temperature interval 1450-2300 K monitoring the radiative recombination of Br atom. They presented a review of the literature of the HBr chain reactions and were particularly conscientious about assessing the effects of these competitive reactions. Their calibration measurements showed large scatter. The earlier (52) value of activation energy, 50,000 cal/mole, was not valid in the low temperature range and profile matching was only possible if the HBr/Ar efficiency was reduced from the former value of 15 to 3.75. In examining their computed and experimental profiles, one would conclude that no single rate coefficient and choice of rate coefficients for competitive reactions gave an

Table V

$HBr + Ar \rightarrow H + Br + Ar$

Investigators	Reference	Measurement	Temperature (K)	k(cm³/mole·sec)
Giedt *et al.*	52	Shock tube study monitored HBr concentration with both IR emission and UV absorption	2100-4200	$1.5 \times 10^{12} \exp(-50{,}000/RT)$
Westberg and Green	53	Shock tube study, monitored Br radiative recombination emission	1500-2700	$1.2 \times 10^{22} T^{-2} \exp(-88{,}000/RT)$
Cohen *et al.*	54	Shock tube study monitored Br radiative recombination emission	1450-2300	$6.0 \times 10^{21} T^{-2} \exp(-88{,}000/RT)$

equally good fit over the entire temperature interval. The rate coefficient given in Table V is recommended by Cohen et al. for the temperature interval 1500-3800 K. The efficiency of HBr/Ar for the HBr dissociation was approximated at 4:1 and the HBr/Ar efficiency for Br_2 dissociation was approximated as 1:1 while Westberg and Green gave the former ratio the value 1:1 and the latter 3:1.

The results of the three measurements are shown in Figure 8. The agreement is satisfactory and is a bit misleading since there was difficulty in fitting the Cohen et al. results satisfactorily over a large temperature interval and many assumptions were made regarding relative efficiencies and competitive rates. The role of Br atom efficiency was not explored. Our strong conclusion regarding HBr dissociation is that more study is required. The competitive reactions, H + Br and Br + HBr, should also be studied at temperatures above 1000 K using modern kinetic techniques since these competitive reactions play an important role in unraveling the dissociation data. The reverse recombination reaction has not been studied and measurement of this rate in a common temperature range where dissociation studies could be made would contribute to our understanding of these complicated kinetics.

$Br + Br + M \rightleftharpoons Br_2 + M$. This reaction and its reverse have been reviewed by Cohen et al. (54). The dissociation reaction has been studied in shock tubes with Ar as the third body by this investigation; reaction progress was monitored by measurement of Br_2 concentration via absorption spectroscopy. The results are summarized in Table VI which shows that all the reported activation energies are less than the Br_2 bond dissociation energy, 45,500 cal/mole.

The dissociation reaction was studied between 1200 and 2225 K by Palmer and Hornig (55) who observed that Br_2 was more efficient than Ar, and that the relative efficiency was temperature dependent. The next study was by Britton et al. (56, 57, 58); however, the results of their last study, that of Johnson and Britton (58) are considered preferable by the authors. This rate coefficient (58) which is given in Table VI is about 40 percent less than the earlier results (56, 57) with the differbeing attributed to the faulty photomultiplier power supply used in the earlier studies. The Johnson-Britton paper gives a good comparison of the use of reflected and incident shocks in determining dissociation rates. The rate coefficient given here is from an incident shock study which they demonstrated to be the superior technique for Br_2 dissociation studies in the temperature interval 1500 to 1900 K. No attempt was made to correct for the Br_2 efficiency relative to Ar in their 95 percent Ar and five percent Br_2 mixtures.

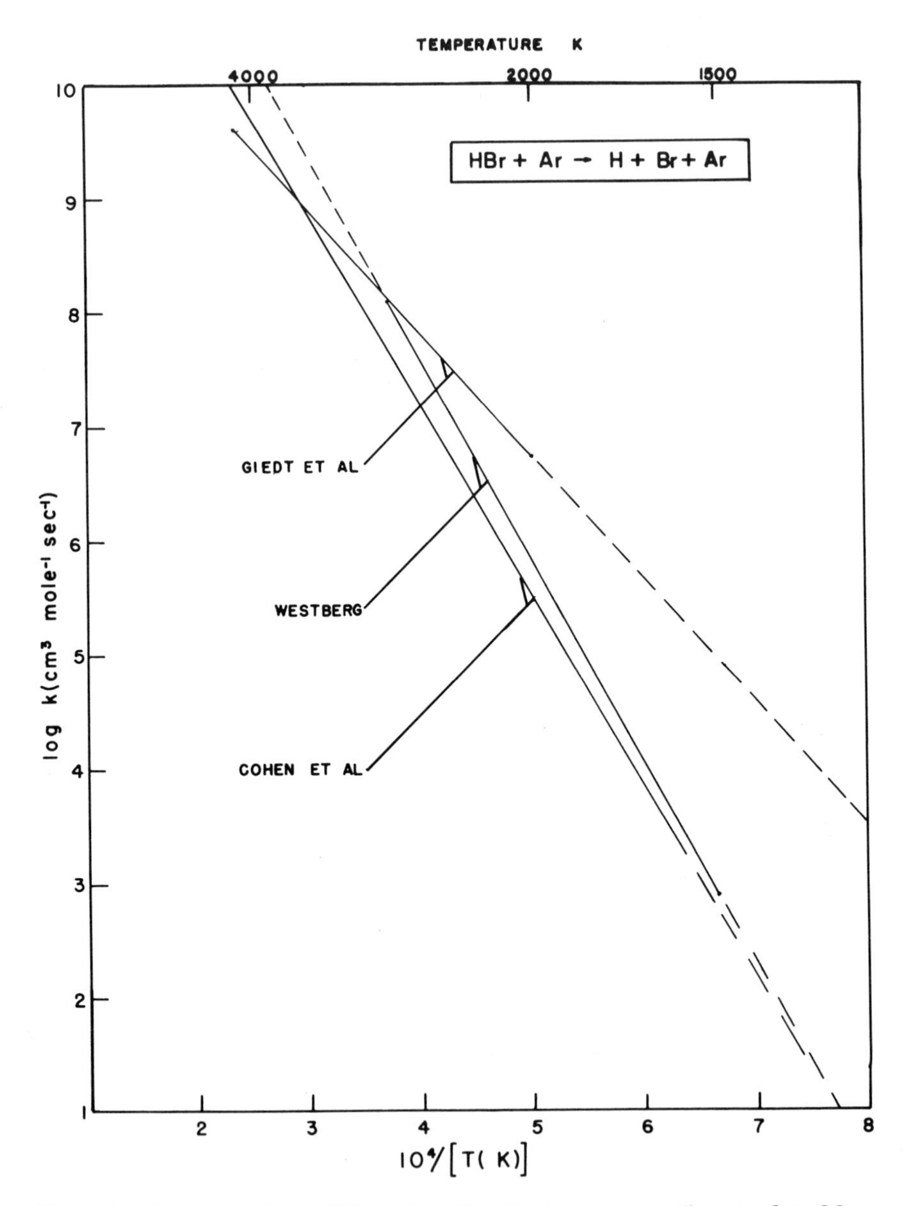

Figure 8. Arrhenius plot of $HBr + Ar \rightarrow H + Br + Ar$ rate coefficients; dotted lines indicate extrapolation

Table VI

$Br_2 + Ar \rightarrow 2Br + Ar$

Investigators	Reference	Measurement	Temperature(K)	k(cm³/mole•sec)
Palmer-Hornig	55	Shock tube, monitored Br_2 concentration by absorption spectroscopy	1200-2225	$2.51 \times 10^{11} T^{1/2} \exp(-30690/RT)$
Johnson-Britton	58	Shock tube, monitored Br_2 concentration by absorption spectroscopy	1500-1900	$5.12 \times 10^{13} \exp(-37700/RT)$
Warshay	59	Shock tube, measured Br_2 concentration by absorption spectroscopy	1200-2000	$2.14 \times 10^{11} T^{1/2} \exp(-31300/RT)$

Warshay has reported a carefully detailed shock tube study of Br_2 dissociation between 1200 and 2000 K in 1:99 percent mixture of the rare gases He, Ne, Ar, Kr and Xe. Boundary layer corrections were made which increased the rate coefficients with the largest corrections found at low temperatures. Mixtures 3:97, Br_2 to Xe were studied to ascertain the Br_2/Xe relative efficiency. The Br_2/Xe relative efficiency was temperature dependent with the values 3.8 and 1.5 at the temperature 1250 and 1900 K, respectively. On this basis, a correction to the 1:99 percent mixtures for Br_2 efficiency was neglected. The relative efficiencies of the rare gases were temperature dependent and at 1500 K the following order obtained: Kr > Xe > Ar > Ne > He. An explanation for the large Kr efficiency is that it is the rare gas whose mass nearly equals the Br atom mass and hence the maximum kinetic energy is transferred in Kr, Br_2 collisions. The Warshay results are summarized in Table VI and plotted in Figure 4.

A fourth study of Br_2 dissociation has been made by Boyd *et al.* (60). The reaction was studied in shock tubes between 1000 and 2985 K monitoring reaction progress via the two body emission of Br atoms. The results of this investigation agree with the high temperature absorption results of Warshay but the rate coefficient determined by emission is about 1/3 that determined by absorption at 1300 K. To add fuel to the fire, the recombination rate coefficient determined via the "emission" dissociation rate coefficient agrees with the recombination rate coefficient determined by flash photolysis by Ip and Burns (61). In the Boyd *et al.* paper, comparison was made with earlier results of Warshay (62) and not the later work (59).

Ip and Burns point out that beyond 1300 K the extrapolated flash photolysis results diverge from the shock tube emission results. Boyd *et al.* used $k_{d,Br_2}/k_{d\ Ar}$ = 2 in their data reduction; however in a later paper (63) they indicate that the low temperature relative efficiency is approximately 5. This would reduce the emission recombination and dissociation rate coefficient and also the agreement between the photolysis experiments and shock experiments. Studying reactions by measuring emission intensities of recombining atoms is still, in the author's opinion, inferior to absorption measurement techniques. With specific reference to Br_2 dissociation, the ($^2P_{3/2}$) Br atoms recombine to the $Br_2(^3\Pi_{0+u})$ state though this state does not correlate with ground state atoms. The amount of quenching of the ($^3\Pi_{0+u}$) state by atoms or molecules is unclear as is the amount of recombination that occurs via the ($^3\Pi_{1u}$) state. Boyd *et al.* believe they measured emission from the $^1\Pi_u \rightarrow {}^1\Sigma g^+$ transition, and, if this is so, are the various molecular electronic states equilibrated? The relative concentration of the atomic ($^2P_{1/2}$) state has not been determined in shock experiments and its role in the radiative process is unknown. As a counter argument one could state that the absorption rate measurements give results which are too high, since absorption from low lying

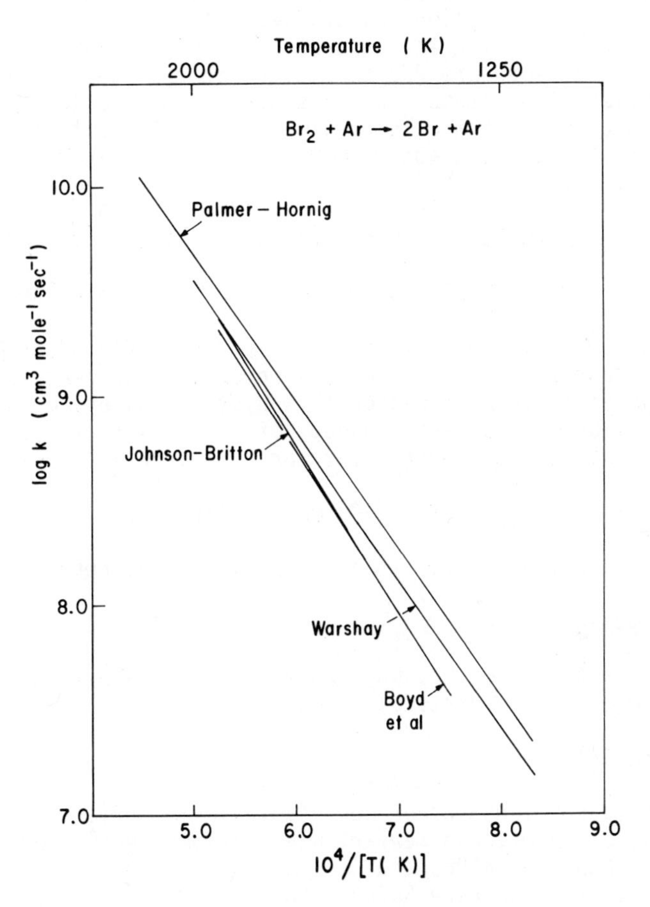

Figure 9. Arrhenius plot of $Br + Ar \rightarrow H + Br + Ar$ rate coefficients; dotted lines indicate extrapolation

states is measured and these states could be underpopulated due to relaxation into higher states. Further studies are required in our opinion before the discrepancy is proven real and is determined not to be an experimental artifact.

The recombination rate of Br atoms in He, Ne, Ar, Kr, N_2 and O_2 was determined in the flash photolysis experiment of Ip and Burns (61). The reaction was studied between room temperature and 1275 K by measuring the absorption of Br_2 at 443 mμ. Flash photolysis reactions are exothermic; hence the equilibrium concentration of Br atoms increases during the course of an experiment because of the temperature rise. Correction for this effect became important above 1000 K. The recombination rate coefficient

$$\log k_{r,Ar} = 15.381 \pm 0.016 - (2.287 \pm 0.125)\log(T/300) + (1.154 \pm 0.194)\log^2(T/300)\ cm^6mole^{-2}sec^{-1}.$$

The value for Ar at toom temperature is $2.4 \times 10^{15} cm^6mole^{-2}sec^{-1}$.

Clyne and Woon-Fat studied the recombination reaction at room temperature in a discharge-flow system monitoring reaction progress by absorption spectroscopy of Br_2 at 415 nm. Their room temperature rate coefficient for $Br + Br + Ar \rightarrow Br_2 + Ar$ is:

$$k_{r,Ar} = 3.4 \times 10^{15} cm^6mole^{-2}sec^{-1},$$

which is 1.4 times greater than the Ip and Burns results.

Acknowledgement

The assistance of Mr. Raymond Yu for his help in the literature search is gratefully noted.

Literature Cited

1. Creitz, E.C., NBS J. of Res. (1970), 74A, 521.
2. Fristrom, R.M., and Sawyer, R.F., "Flame Inhibition Chem.", AGARD Conference Proceedings No. 84 on Aircraft Fuels, Lubricants, and Fire Safety, AGARD-CP-84-71 (1971), Section 12, North Atlantic Treaty Organization.
3. Hastie, J.W., NBS J. of Res. (1973), 77A, 733.
4. Wilson, W.E., Jr.,"10th Symposium (International) on Combustion," 47, Combustion Institute, Pittsburgh, Pa., (1965).
5. Wilson, W.E., Jr., O'Donovan, J.T., and Fristrom, R.M., "12th Symposium (International) on Combustion," 929, Combustion Institute, Pittsburgh, Pa., (1969)

6. Fettis, G.C., and Knox, J.H., "Progress in Reaction Kinetics," (Ed. G. Porter) 2, 1, Pergamon Press, London, (1964)
7. Benson, S.W., Cruickshank, F.R., and Shaw, R., Int. J. Chem. Kinet. (1969), 1, 29.
8. White, J.H., "Comprehensive Chem. Kinetics," (Eds., C.H. Bamford and C.F.H. Tipper) 6, 201, Elsevier, Amsterdam (1972).
9. Watson, R.T., Nat. Bur. Stand., NBSIR 74-516 (1974).
10. Rodenbush, W.H., and Klingelhoefer, W.C., J. Am. Chem. Soc. (1933), 55, 130.
11. Steiner, H. and Rideal, E.K., Proc. Roy. Soc. (1939), A173, 503.
12. JANAF Thermochemical Tables, NSRDS-NBS 37 (1971).
13. Ashmore, P.G. and Chanmugam, J., Trans. Faraday Soc. (1953), 49, 254.
14. Burns, W.G. and Dainton, F.S., Trans. Faraday Soc. (1952), 48, 52.
15. Clyne, M.A.A., and Stedman, D.H., Trans. Faraday Soc. (1966), 62, 2164.
16. Westenberg, A.A., and deHaas, N., J. Chem. Phys. (1968), 48, 4405.
17. Clyne, M.A.A. and Walker, R.F., JCS Faraday Trans. I (1973), 69, 1547.
18. Snider, N.S., J. Chem. Phys. (1970), 53, 4116.
19. Cadman, P. and Polanyi, J.C., J. Phys. Chem. (1968), 72, 3715.
20. Davis, D.D.,, Braun, W., and Bass, A.M., Int. J. Chem. Kinet. (1970), 3, 223.
21. Galante, J.J. and Gislason, E.A., Chem. Phys. Letters (1973) 18, 231.
22. Klein, F.S. and Wolfsberg, M., J. Chem. Phys. (1961), 34, 1494.
23. Dodonov, A.F., Lavrovskaya, G.K. and Morozov, I.I., Kinetika: Kataliz (1970), 11, 821.
24. Jacobs, T.A., Giedt, R.R., and Cohen, N., J. Chem. Phys. (1968), 49, 1271.
25. Seery, D.J. and Bowman, C.T., J. Chem. Phys. (1968), 48, 4314.
26. Breshears, W.D. and Bird, P.E., J. Chem. Phys. (1972), 56, 5347.
27. Jacobs, T.A., Cohen, N. and Giedt, R.R., J. Chem. Phys. (1967), 46, 1958.
28. Fishburn, E.S., J. Chem. Phys. (1966), 45, 4053.
29. Giedt, R.R. and Jacobs, T.A., J. Chem. Phys. (1971), 55, 4144.
30. Clyne, M.A.A. and Stedman, D.H., Trans. Faraday Soc. (1968), 64, 1816.
31. Johnston, H., and Birks, J., Accounts Chem. Res. (1972), 5, 327.
32. Lloyd, A.C., Int. J. Chem. Kinet. (1971), 3, 39.

33. Clyne, M.A.A. and Stedman, D.H., Trans. Faraday Soc., (1968), 64, 2698.
34. Jacobs, T.A. and Giedt, R.R., J. Chem. Phys., (1963), 39, 749.
35. Jacobs, T.A., Giedt, R.R. and Cohen, N., J. Chem. Phys., (1968), 49, 1271.
36. Bodenstein, M. and Lind, S.C., Z. Phys. Chem., (1906), 57, 168.
37. Bodenstein, M. and Lutkemeyer, H., Z. Phys. Chem., (1924), 114, 208.
38. Bodenstein, M. and Jung, Z., Z. Phys. Chem. (1920), 121, 127.
39. Christiansen, J.A., Kgl. Danske Videnskab Selskab, Mat. Fys. Medd., (1919), 1, 14.
40. Herzfeld, K., Ann. Phys., (1919), 59, 635.
41. Polanyi, M., Z. Elektrochem, (1920), 26, 50.
42. Sullivan, J.H., J. Chem. Phys., (1968), 49, 1155.
43. Bach, F., Bonhoeffer, F.K. and Moelwyn-Hughes, E.A., Z. Phys. Chem., (1934), B27, 71.
44. Pease, R.N., "Equilibrium Kinetics of Gas Reactions", 121, Princeton University Press, 1942.
45. Steiner, H., Proc. Roy. Soc., (1939), A173, 531.
46. Fass, R.A., J. Phys. Chem., (1970), 74, 984.
47. Levy, A., J. Phys. Chem., (1958), 62, 570.
48. Britton, D. and Davidson, N., J. Chem. Phys., (1956), 25, 810.
49. Cooley, S.D. and Anderson, R.C., Ind. Eng. Chem., (1952), 44, 1402.
50. Vidal, C., J. Chim. Phys., (1971), 68, 1360.
51. Takacs, G.A. and Glass, G.P., J. Phys. Chem., (1973), 77, 1060.
52. Giedt, R.R., Cohen, N. and Jacobs, T.A., J. Chem. Phys., (1969), 50, 5374.
53. Westberg, K. and Greene, E.F., J. Chem. Phys., (1972), 56, 2713.
54. Cohen, N., Giedt, R.R. and Jacobs, T.A., Int. J. Chem. Kinet., (1973), 5, 425.
55. Palmer, H.B. and Hornig, D.F., J. Chem. Phys., (1957), 26, 98.
56. Britton, D., and Davison, N., J. Chem. Phys., (1956), 25, 810.
57. Britton, D., J. Phys. Chem., (1960), 64, 742.
58. Johnson, C.D. and Britton, D., J. Chem. Phys., (1963), 38, 1455.
59. Warshay, M., J. Chem. Phys., (1971), 54, 4060.
60. Boyd, R.K., Burns, G., Lawrence, T.R. and Lipiatt, J.H., J. Chem. Phys., (1968), 49, 3804.
61. Ip, J.K.K. and Burns, G., J. Chem. Phys., (1969), 51, 3414.
62. Warshay, M., Proc. Intern. Shock Tube Symp. 5th, White Oak, Md., 1965, (1966), 631.

63. Boyd, R.K., Brown, J.D., Burns. G. and Lippiatt, J.H., J. Chem. Phys., (1968), 49, 3822.
64. Clyne, M.A.A. and Woon-Fat, A.R., Dis. Faraday Soc., (1972), 53, 412.

This Work was supported by NSF-RANN under Grant Number GI-43.

DISCUSSION

R. M. FRISTROM: With respect to the problem of extrapolating low temperature kinetic data to high temperature combustion systems, there is one recent hopeful development. The reactions of hydrogen atoms with several methyl halides (CH_3F, CH_3Cl and CH_3Br) have been studied by a new technique called the "point source method" which uses flames as high temperature radical sources (L. Hart, C. Grunfelder, and R. Fristrom, "The Point Source Technique Using Upstream Sampling," Combustion and Flame, (1974), 23, 109). The molecule to be studied is injected at a point and the disappearance is followed as a function of distance from the source. This is an extension of an old method to the flame temperature regime. These same reactions have also been studied by the conventional discharge tube - ESR Methods over a wide temperature range (A. A. Westenberg and N. deHaas, "Rages of H + CH_3X Reactions," J. Chem. Phys., (1975), 62, 3321) and there is an overlap of several hundred degrees in the region around 1000K. Here the absolute rates agree between the two methods within a factor of 2-3; most of the observed discrepancy is probably due to the crude method used for determination of H atom concentrations in the point source method since if relative rates between the halides are compared the two methods agree within 15%. Thus, at least in this one case, it appears that flame kinetics can reliably inferred from lower temperature conventional studies.

13

Halogenated Fire Extinguishants: Flame Suppression by a Physical Mechanism?

ERIC R. LARSEN

Halogens Research Laboratory, The Dow Chemical Co., Midland, Mich. 48640

The mechanism by which halogenated agents act as flame suppressants has been the subject of intensive research for about twenty five years. In the early 1950's Fryburg (1) proposed that the outstanding effectiveness of these agents may be due to a mechanism that involved the chemistry of the combustion processes rather than one or another of the mechanisms which were thought to be applicable to the inert gases. A chemically based mechanism was felt to be attractive since it might explain the fact that certain halohydrocarbons, especially those containing bromine, were from five to eight times as effective as carbon dioxide and nitrogen. Fryburg's proposal led directly to very extensive studies to elucidate the chemistry of flames into which halogenated materials were introduced. These studies, in turn, led to the development of the radical trap mechanism for flame suppression that is in vogue today.

That the halogenated agents enter into the chemistry of the various combustion processes has been adequately demonstrated by studies too numerous to into here. That this chemical reactivity is involved in the primary mechanism by which the halogenated agents act as flame suppressants is, however, open to question.

Several years ago it was reported by the author that an investigation of the flammability of halohydrocarbons showed that these compounds were nonflammable in air only when they contained more than about 70 wt.% halogen (2,3). The type of halogen present did not seem to be important. In other words, when the halogen concentration was expressed on a weight basis, rather than on the commonly employed molar basis, the various halogens were essentially equivalent. This was later shown to also be true for mixtures of halogenated agents and hydrocarbon based fuels when the halogen content of the mixture of agent and fuel was determined at the "flammability peak" (4). The halogen content of peak composition for some sixty agents in heptane-air systems were found to obey the relationship:

$$1. \qquad \frac{[H]}{[A] + [F]} \cdot 100 = 69.8 \pm 3.5\%$$

where [H] is the weight of halogen in the mixture of agent [A] and fuel [F] that is present at the flammability peak when all concentrations are expressed in mg/ℓ. This relationship was found to hold regardless of the type of halogen present in the agent, and showed that the halogens were effective flame suppressants in direct proportion to their atomic weights, i.e. F:Cl:Br:I::1.0:1.9:4.2:6.7. These results are clearly inconsistent with the postulated chemical mechanisms of flame suppression.

Prior to discussing the primary mechanism by which halogenated compounds act as flame suppressants it is necessary to re-examine the role of the inert gases, since we must establish the baseline behavior of these agents before we can hope to study the variant behavior of the halogenated agents. The inert gas agents, e.g. He, Ar, N_2, CO_2, CF_4, etc., are generally conceded to act by a physical mechanism since they do not enter in the chemistry of the various combustion processes to any significant extent.

Figure 1 shows the influence of the inert gases upon the flammability limits of pentane-air mixtures (5). The point at which the upper and lower limit curves meet is commonly called the flammability peak and represents the minimum agent concentration needed to prevent flame propagation in any and all fuel-oxidant mixtures. While the minimum agent concentration, or peak values, of N_2 and CO_2 are relatively high - 42 vol % and 29 vol %, respectively - the halogenated agents generally show peak values of less than 10 vol %. Although the most efficient halogenated agent (CF_2Br_2, 4.4 vol % in heptane) and nitrogen differ in peak concentrations by about a factor of 10, the difference between CF_2Br_2 and C_3F_8 - one of the most effective inert gases - is only about a factor of two. This relatively narrow spread in differences hardly seems to justify a completely new mechanism.

A close inspection of the inert gas agents shows that when the concentrations of pentane, oxygen, and inert gas are replotted on a heat capacity basis, or

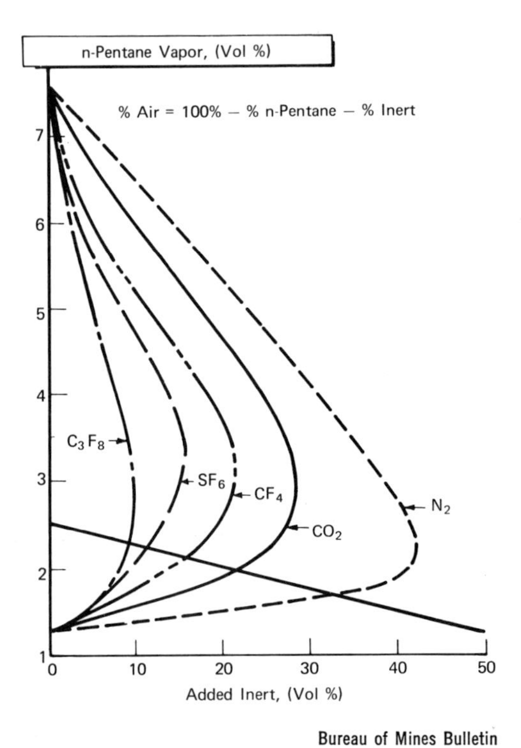

Bureau of Mines Bulletin

Figure 1. Limits of flammability of various n-*pentane–inert gas–air mixtures at 25°C and atmospheric pressure* (5)

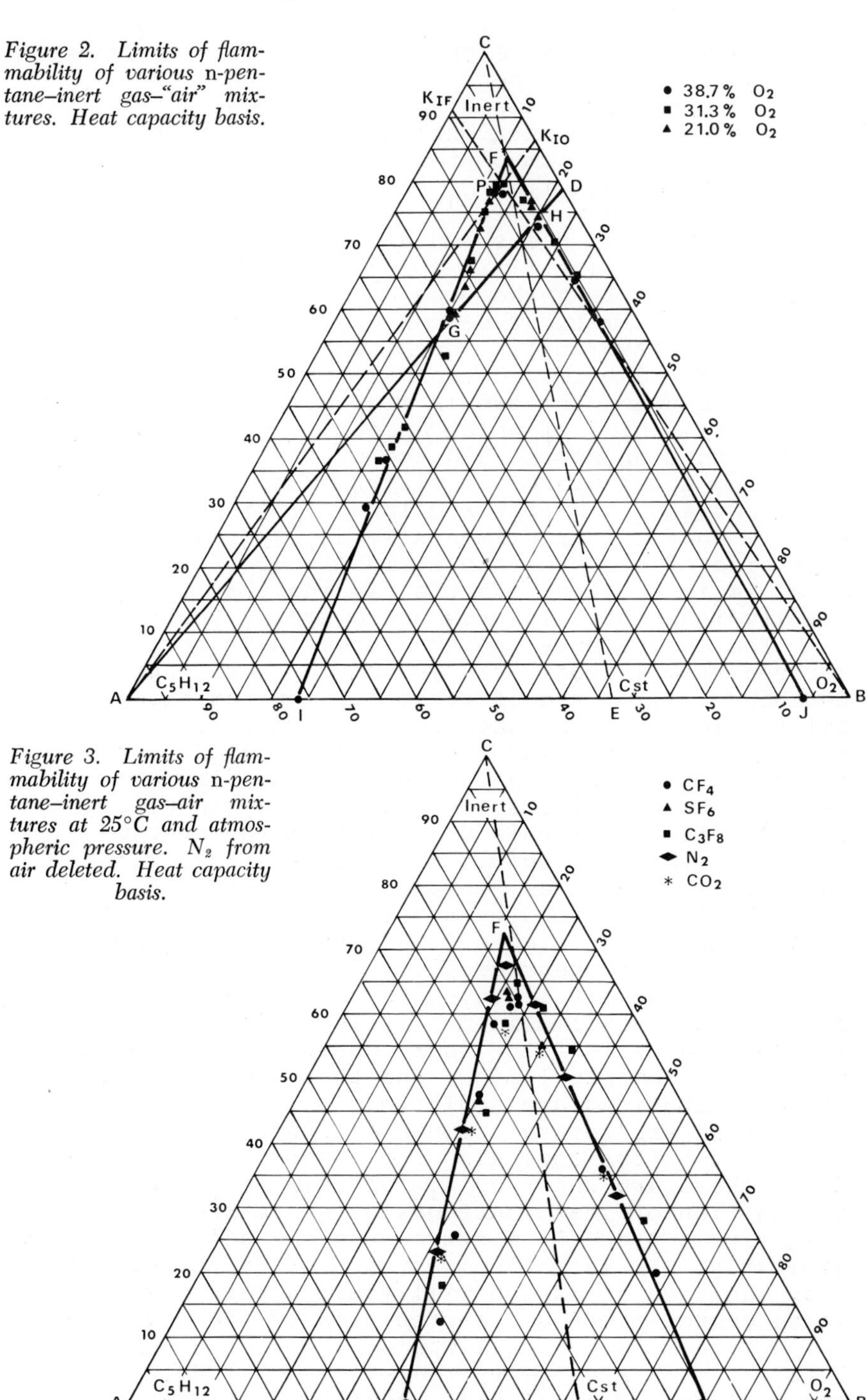

Figure 2. Limits of flammability of various n-*pentane–inert gas–"air" mixtures. Heat capacity basis.*

Figure 3. Limits of flammability of various n-*pentane–inert gas–air mixtures at 25°C and atmospheric pressure.* N_2 *from air deleted. Heat capacity basis.*

for that matter on a weight basis, the differences between the various inert gases disappears (6). Figure 2 shows such a plot using triangular coordinates. The heat capacity contribution of the inert gases is shown here as the sum of the individual contributions of the inert gases i.e. N_2 and CF_4, present in the system. The upper limit (I) and lower limit (J) of pentane in oxygen are shown along the base of the diagram. The upper limit (G) and lower limit (H) in air are shown along the line $\overline{AD}$. Point F, the point at which the limit curves converge, represents the theoretical flammability peak, and point P the experimentally determined flammability peak. The differences between the two is probably due to a variety of factors, chief of which appear to be the strength of the ignition source and the practical difficulty of making up accurate gas mixtures when the differences between the upper and lower limit become very small. That the experimental peak lays on the fuel rich side of the stoichiometric mixture line, i.e. $\overline{C_{st}C}$, is in keeping with the finding that the most readily ignitable mixtures are slightly fuel rich.

The experimental points shown in Figure 2 are derived from a study, carried out by Moran and Bertschy (7), on the effect of CF_4 on the flammability limits of pentane-"air" systems. The "air" employed here contained either 21.0%, 31.3%, or 38.7% oxygen. The solid line represents the flammable limit curves for the system pentane-oxygen-nitrogen. It is obvious from this diagram that the system pentane-oxygen-CF_4 should exhibit essentially the same limit diagram as pentane-oxygen-nitrogen.

The region bound by the triangle ADC is shown in greater detail in Figure 3. In this figure the contribution to the heat capacity due to the nitrogen content of the air employed as the oxidant was deleted, and the data renormalized. It is obvious from these figures that all of the inert gases show essentially the same flammable limits, and that their heat capacity contributions are additive.

There are several points in Figure 2 that are important to the analysis of flammability peak (P) data; these are shown as K_{IO} and K_{IF}. K_{IO} is the composition of that inert gas-oxygen mixture which will give a peak (P) mixture upon the addition of fuel. Similarly, K_{IF} is the composition of inert gas which upon addition of oxygen will yield a peak mixture. These compositions are represented by the equations:

$$2. \qquad K_{IO} = \frac{[I]Cp_I}{[I]Cp_I + [O_2]Cp_{O_2}} \cdot 100$$

and

3. $$K_{IF} = \frac{[I]Cp_I}{[I]Cp_I + [Fuel]Cp_F} \cdot 100$$

where $[I]Cp_I$, $[O_2]Cp_{O_2}$ and $[Fuel]Cp_F$ represent the individual heat capacity contributions of the inert gases, oxygen, and fuel, respectively. The ratio of the heat capacity contributions of all of the inert gases, $\Sigma[I]Cp_I$, at the peak (P) to the total heat capacity of the system can be represented by the equation:

4. $$KCp = \frac{\Sigma[I]Cp_I}{\Sigma[C]Cp_C} \cdot 100$$

where $\Sigma[C]Cp_C$ is the sum of the individual heat capacity contributions of all components of the system (3,6).

Table I shows the flammability peak compositions for various methane-oxygen-agent and pentane-oxygen-agent systems. Those systems marked with an asterisk contain only a single inert gas while the other systems contain N_2 along with the indicated agent. This data shows quite clearly the additivity of the heat capacity contributions of the inert gases to the peak composition, and confirms the fact that all of the inert gases behave as flame suppressants by virtue of their contributing to the total heat capacity of the system without contributing to the heat producing capacity of the system. The data for Ar and He also suggests that neither dilution of the fuel or oxygen, nor the thermal conductivity of the agent are particularly important to the flame suppression process.

The mean value of K_{IO} (86.9 ± 1.2) for the inert gases is particularly germane to the mechanism by which the halogenated agents act as flame suppressants. Recently, Creitz on the basis of an extensive study of the flame inhibiting ability of CF_3Br determined the concentration of CF_3Br required to prevent flame propagation in various fuel-oxygen mixtures (8). The "limiting concentration of oxygen-agent atmosphere" determined by Creitz for the fuels CH_4, C_2H_6, C_3H_8, and C_4H_{10} are shown in Table II along with the equivalent values for various other oxygen-agent atmospheres. When CF_3Br is compared to the inert gases, i.e. Ar, He, and N_2, on a volume % basis it would appear that the data might support a chemical mechanism of flame suppression. Such support is less clear when CF_3Br is compared with CF_4. When the various agents are compared on a heat capacity basis it would appear that such support is entirely absent.

Table I

Flammability Peak Compositions (7,11,12)

Fuel	Agent	Peak Fuel Vol%	Peak Agent Vol%	KCp	K_{IO}	K_{IF}
CH_4	N_2*	6.3	81.5	80.3	86.9	91.3
	CO_2	7.5	22.3	77.8	84.7	89.9
	He	5.8	36.3	78.8	85.5	91.0
	He*	5.0	85.0	79.0	85.8	92.6
	Ar	5.0	48.4	81.7	88.0	92.2
	Ar*	4.0	88.2	83.2	88.8	93.9
C_5H_{12}	N_2*	2.1	86.2	80.8	88.0	90.9
	CO_2	2.8	28.0	77.5	86.1	88.7
	CF_4	2.5	26.0	79.7	86.5	91.0
	SF_6	3.0	15.7	79.7	87.1	90.4
	C_3F_8	2.8	10.6	80.6	87.1	91.5

Mean Values KCp = 80.2 ± 1.7

K_{IO} = 86.9 ± 1.2

K_{IF} = 91.0 ± 1.5

*Sole Agent

Experimental Conditions: Fuel-O_2-N_2- Agent Mixtures with Flame Propagation Vertically through Tube

Table II

Limiting Concentrations of Oxygen-Agent Atmospheres

Fuel	Agent	O_2 Vol%	Agent Vol%	O_2 Cp%	Agent Cp%
CH_4	Ar	8.2	91.8	11.2	88.8
CH_4	He	10.5	85.9	14.2	85.8
CH_4	N_2	13.0	87.0	13.0	87.0
C_3H_8	N_2	11.9	88.1	12.0	88.0
C_5H_{12}	N_2	12.0	88.0	12.0	88.0
C_5H_{12}	CF_4	25.0	75.0	13.7	86.3
CH_4*	CF_3Br	34.0	66.0	17.9	82.1
C_2H_6*	CF_3Br	32.0	68.0	16.6	83.4
C_3H_8*	CF_3Br	33.3	66.7	17.5	82.5
C_4H_{10}*	CF_2Br	33.8	65.3	18.0	82.0

*Diffusion Flame - E. C. Creitz, Fire Tech., 8 131 (1972) (8)

The value which appears in the last column (Agent Cp) of Table II is the value for K_{IO} discussed previously. The difference between the value of K_{IO} for the inert gases (86.9 ± 1.2) and for CF_3Br (82.5 ± 0.6) is small enough to be attributable to the fact that the inert gas data is derived from premixed combustion tube determinations which the data for CF_3Br is derived from a study on the agent concentrations required to extinguish small diffusion flames.

Table III shows the flammability peak compositions for a number of halogenated agents that were studied by the Purdue Research Foundation (9). These agents are representative of a total of 46 agents studied with respect to their ability to suppress flame propagation in heptane-air mixtures. This study by Purdue has been frequently employed to support the premise that the halogenated agent act as flame suppressants by a chemical, rather than physical, mechanism (1). The variety of agents shown are representative of the different halogens, singlely and in combination. The validity of this premise is questionable when the agents are compared on a heat capacity basis rather than on the usual volume basis. When compared on a heat capacity basis no one type of halogen stands out as clearly superior to any other. It is axiomatic of the radical trap mechanism that bromine containing agents, which enter into the combustion chemistry, should be superior to the fluorine containing compounds and CO_2 which behave as inert gases.

The mean value of K_{IO} for these systems is 84.0 ± 2.9 which lays midway between the values of K_{IO} found for the inert gas systems (86.9 ± 1.2, Table I) and for CF_3Br (82.5 ± 0.6, Table II) determined from Creitz's data. The value of K_{IF} found for these halogenated agents compares closely with that found for the inert gases; 87.9 ± 2.5 and 91.0 ± 1.5, respectively.

In a more recent study by Bajpai and Wagner (10) the effect of the agents CF_3Br and CF_2ClBr upon the flammability limits of several hydrocarbon fuels were determined in a modified combustion tube. The flammability peak values derived from this study are shown in Table IV, along with the corresponding values of KCp, K_{IO}, and K_{IF}. This study tends to bear out the overall findings cited above, and again fails to support any clear superiority of the halogenated agents over the inert gases. The value of K_{IF} here corresponds even more closely with that found for the inert gases than does the Purdue study (9). The value of K_{IO} which in this case is the sum of the heat

Table III

Flammability Peak Compositions

Experimental Conditions: n-Heptane-Air Mixtures with Flame Propagation Vertically through Tube, Spark Ignition

Purdue Res. Found. and Dept. of Chem.: Sum. Rep't Sept. 1, 1947 to Aug. 31, 1948, Purdue Univ. (Contract W-44-009 eng-507 with Army Eng. R&D Labs) (9)

Agent	Peak Agent Vol%	Peak Heptane Vol%	KC_p	K_{IO}	K_{IF}
CF_2Br_2	4.2	4.4	67.2	81.2	79.5
CH_2Br_2	5.2	2.1	73.5	80.9	89.0
C_2H_5I	5.6	3.2	70.5	81.4	84.7
CH_3Br	6.1	2.1	73.4	80.7	88.8
CH_3I	6.1	2.1	73.1	80.6	88.8
CF_3Br	6.1	3.0	71.8	81.8	85.4
C_2H_5Br	6.2	2.3	73.5	81.6	88.4
CCl_4	11.2	3.2	74.6	84.6	86.4
$CClF_3$	12.3	3.3	74.0	84.4	85.4
C_2F_6	13.4	3.0	77.8	86.6	88.6
CF_2Cl_2	14.9	3.9	73.8	85.2	84.7
$CHCl_3$	17.5	3.6	74.6	85.9	85.2
CHF_3	17.8	2.2	77.3	84.8	89.7
$c\text{-}C_4F_8$	18.1	2.3	85.3	90.5	89.7
CF_4	26.0	1.8	82.1	87.7	92.7
CO_2	26.0	1.8	78.8	85.5	91.1

Mean Values $KC_p = 75.1 \pm 4.4$

$K_{IO} = 84.0 \pm 2.9$

$K_{IF} = 87.9 \pm 2.5$

Table IV

Flammability Peak Compositions (10)

Fuel	Agent	Peak Agent Vol%	Peak Fuel Vol%	KCp	K_{IO}	K_{IF}
CH_4	CF_3Br	6.8	9.2	74	82	88
	CF_2ClBr	4.3	9.9	72	81	87
C_3H_8	CF_3Br	8.0	4.3	75	83	89
	CF_2ClBr	7.0	5.2	73	82	87
n-C_4H_{10}	CF_3Br	7.2	3.0	75	82	90
	CF_2ClBr	6.3	2.8	75	82	90
i-C_4H_{10}	CF_3Br	7.2	4.0	73	82	87
	CF_2ClBr	6.3	2.9	75	82	90

Mean Values $KCp = 74.0 \pm 1.2$

$K_{IO} = 82.0 \pm 0.5$

$K_{IF} = 88.5 \pm 1.4$

capacity contributions of the agent and the nitrogen from the air employed as the oxidant agrees exactly with the value for CF_3Br derived from Creitz's data(8).

On the basis of the data developed here and from the previous finding that the halogens are effective as flame suppressants in direct proportion to their atomic weight it is felt that the halogenated agents probably play a duel role in the suppression of flames. Their primary role is similar to that of the inert gases, that is, they act as heat sinks without contributing significantly to the heat producing potential of the system. In this respect it is anticipated that they act in the region ahead of the moving flame front rather than within the flame per se.

In addition to this primary mechanism the halogenated agents, with a few exceptions, also appear to act as fuels once they are heated to a high enough temperature. This ability to act as a fuel is consistent with the relationship shown in equation 1 and it is probably in this role that these agents participate in the chemistry of flames. In this role the halogenated agents would be flame suppressants only within the context that the addition of excess fuel to a flammable mixtures produces a non-flammable mixture.

In order to make any kind of a strong case for a more complex mechanism of flame inhibition than is found for the inert gases it would appear to be necessary to show that the agent produces a definite effect upon the overall flammable envelope. This will be difficult to accomplish using air as the oxidant as has been done with the great bulk of the work carried out to date, and it is strongly recommended that future studies on the mechanism of flame inhibition be carried out using air that is significantly enriched with oxygen, eg. >50% O_2.

Experimental Data

The data relied upon in this paper were developed by a number of different groups, including Coward and coworkers (11,12), Zabetakis (5), Moran and Bertschy (7), Creitz (8), Bajpai and Wagner (10) and others (9). The "flammability peak" values derived from these references are given in the various tables. The reader is referred to the original articles for the complete data.

Only those studies were evaluated in which a number of inert gases on halogenated agents were evaluated for their ability to prevent flame propagation by determining the effect of the added inert gases upon the limits of flammability of various fuel-oxidant mixtures. In most cases the limit data was obtained by upward flame propagation in an apparatus similar to that described by the Bureau of Mines (12), the one exception being Creitz's study (8) which employed a diffusion flame. In the combustion tube studies the tubes employed were sufficiently large so that increasing the diameter of the tube would have little effect upon the values obtained.

For the method employed to convert the composition of the systems from the normal volume percent basis to the heat capacity basis employed here the reader is referred to reference (6).

Literature Cited

1. Fryburg, G., NACA Tech. Note 2102, p. 27 (1950).
2. Larsen, E. R., in Handbook of Experimental Pharmacology, New Series, Volume XXX, (M. B. Chenoweth, ed.), Springer-Verlag, New York, 1972, p. 28.
3. Larsen, E. R., Abs. 166th National ACS Meeting, Chicago, Illinois, Aug. 26-31 (1973), Paper No. INDE 054.
4. Larsen, E. R., JFF/Fire Retardant Chemistry, 1 4 (1974).
5. Zabetakis, M. G., "Flammability Characteristics of Combustible Gases and Vapors," Bull. 672, Bur. Mines, 1965.
6. Larsen, E. R., JFF/Fire Retardant Chemistry, 2 5 (1975).
7. Moran, Jr., H. E., and Bertschy, A. W., "Flammability Limits for Mixtures of Hydrocarbon Fuels, Air, and Halogen Compounds", NRL Report 4121,(1953).
8. Creitz, E. C., Fire Tech., 8 131 (1972).
9. Purdue Research Foundation and Department of Chemistry: Fire Extinguishing Agents, Summary Report Sept. 1, 1947 to Aug. 31, 1948, Purdue University (Contract W-44-009 eng. 507 with Army Eng. Res. and Dev. Labs.).
10. Bajpai, S. N. and Wagner, J. P., I&EC Product R&D, 14 54 (1975).
11. Coward, H. F. and Hartwell, F. J., J. Chem. Soc., 1926, 1522.
12. Coward, H. F. and Jones, G. W.: "Limits of Flammability of Gases and Vapors", Bull. 503, Bur. Mines, 1952.

DISCUSSION

D. CHAMBERLAIN: Did you include bromine and hydrogen bromide in your correlation?

E. R. LARSEN: Data on the effect of HBr was not available for the fuels which were discussed in this paper. Data that is available for the system propane-air-hydrogen bromide indicates that HBr is not significantly different in its behavior.

R. ALTMAN: Don't you need to use the $H_T - H_{298}$ data instead of C_p (298) to demonstrate the veracity of your argument, and isn't the T to use the flame temperature or some fraction thereof instead of 298K?

E. R. LARSEN: As I pointed out in reference 6, the method used here which employs C_p (298) allows a good approximation of the flammability behavior of these systems. In a more rigorous treatment the specific heat of each component should be integrated over the temperature range between the temperature of the cold gas and the limit flame temperature. This is more important with rich mixtures than with lean mixtures since, while the light gases show a slight linear increase in C_p with increasing temperature, the heavier more complex fuels and agents, e.g., C_5H_{12}, CO_2, CH_4, CF_4, etc. show a large non-linear increase in C_p over the range of 300-1000°K. This factor is not particularly important in systems containing large amounts of light components such as N_2 and O_2, i.e., studies employing air as the oxidant, and, consequently, was for the most part ignored in evaluating the "peak" data. In some cases, especially those involving extinction, or flame velocity reduction, in pre-mixed gas burners the use of C_p (700°K) seems to give better results. There are, however, not enough examples of such systems where a large variety of agents were employed to determine whether this approach offers any distinct advantage.

A. D. LEVINE: Your correlations based on heat capacity are valid in a practical sense for predicting flammability limits. Will this correlation work when agents are added to a hot flame where chemical kinetics may be important?

E. R. LARSEN: The correlation appears to hold in those cases where the agent is injected into either the fuel side or the air side of a diffusion flame, or for that matter into a burner which is fueled with a premixed fuel-air mixture. Some adjustment obviously has to be made since the burner itself may contribute to the heat losses incurred by the system.

A. S. GORDON: If we consider heat capacity on a molar basis, then CF_4 and CF_3Br should have the same heat capacity/mole to a good first approximation. Now if we add the same quencher/fuel ratio to a given fuel/oxygen mix, the two systems should have the same heat release potential and total heat capacity. Yet at a CF_3Br/fuel ratio where the flame is extinguished, full replacement of CF_3Br by CF_4 does not extinguish the flame.

E. R. LARSEN: I agree that CF_4 and CF_3Br have essentially the same heat capacity. I do not, however, have data that either supports or refutes your second statement and I am not sure that a set of circumstances such as you postulate is in fact possible. CF_4 is essentially a true inert gas which contributes nothing to the energy producing ability of the system and causes minimal, if any, change in the fuel-oxygen stoichiometry. CF_3Br, on the other hand, does enter the combustion process and cause a marked change in the fuel-oxygen stoichiometry. CF_3Br, in fact, becomes part of the fuel content of the system. Within this context CF_3Br would also act as a flame suppressant in the same manner that the addition of excess fuel to a flammable mixture can result in the production of a non-flammable mixture. Consequently, it would not surprise me to find that full replacement of CF_3Br by CF_4 did not extinguish the flame.

F. A. WILLIAMS: In figure 17 of my paper is shown the dependence of measured flame temperature at extinction on the nitrogen - CF_3Br ratio for a heptane diffusion flame burning in a mixture of oxygen, nitrogen and CF_3Br. The temperature at extinction in CF_3Br is more than 500°C above that in nitrogen. This tells me that thhe cooling caused by CF_3Br at extinction is considerably less than that caused by nitrogen. The chemical inhibition, i.e., slowing of the overall reaction, by the suppressant at any given flame temperature, therefore must be important. In a paper to be published in the Fifteenth International Symposium on Combustion, Kent and I amplify this point further. Our conclusion is that your correlation can work well because of two competing effects that cancel.

E. R. LARSEN: That the temperature at extinction in CF_3Br is more than 500°C above that in nitrogen comes as no great surprise to me, in fact it is what I would have expected if my theory is correct. As I have pointed out N_2 serves strictly as a heat sink, but CF_3Br can play a duel role, i.e., heat sink and fuel. In order to evaluate you comments it is necessary to know the concentrations of all components of both systems. However, in view of the fact that the ignition energy of a CF_3Br-O_2-heptane mixture near the flammability peak is several orders of magnitude higher than is the ignition energy of a corresponding N_2-O_2-heptane mixture I would assume that a much hotter flame must occur in order to obtain a self-propagating flame.

W. D. WEATHERFORD, JR.: Southwest Research Institute has obtained data which could be considered to support a physical inhibition mechanism under certain conditions. I will mention it at this time to further illustrate the complexities of halogen compound flame inhibition mechanisms. We have confirmed that 5 liquid volume percent halon 1011 (The halon 1011 composition was 85% (wt) 1011, 10% 1020, and 5% 1002.) in diesel engine is satisfied with such a fuel. In fact, the total ignition delay in a single-cylinder research diesel engine is not influenced by the presence of this concentration of halon 1011 in the fuel. In addition, minimum autoignition temperature measurements yield AIT values of about 480°F for halon 1011 cooncentrations in diesel fuel from 0 to 30 liquid volume percent, whereas the value is about 920°F for the fuel-free halon 1011.

The observed mist flammability, both at ambient conditions and in a diesel engine, probably stems from the vapor flammability characteristics shown in the following illustration. These data were measured by Southwest Research Institute Army Fuels and Lubricants Laboratory (Wimer, W. W., et al, "Ignition and Flammability Properties of Fire-Safe Fuels," NTIS, AD 784281, 1974). The results clearly indicate that 5 liquid volume percent tech grade halon 1011 can be flammable under certain conditions

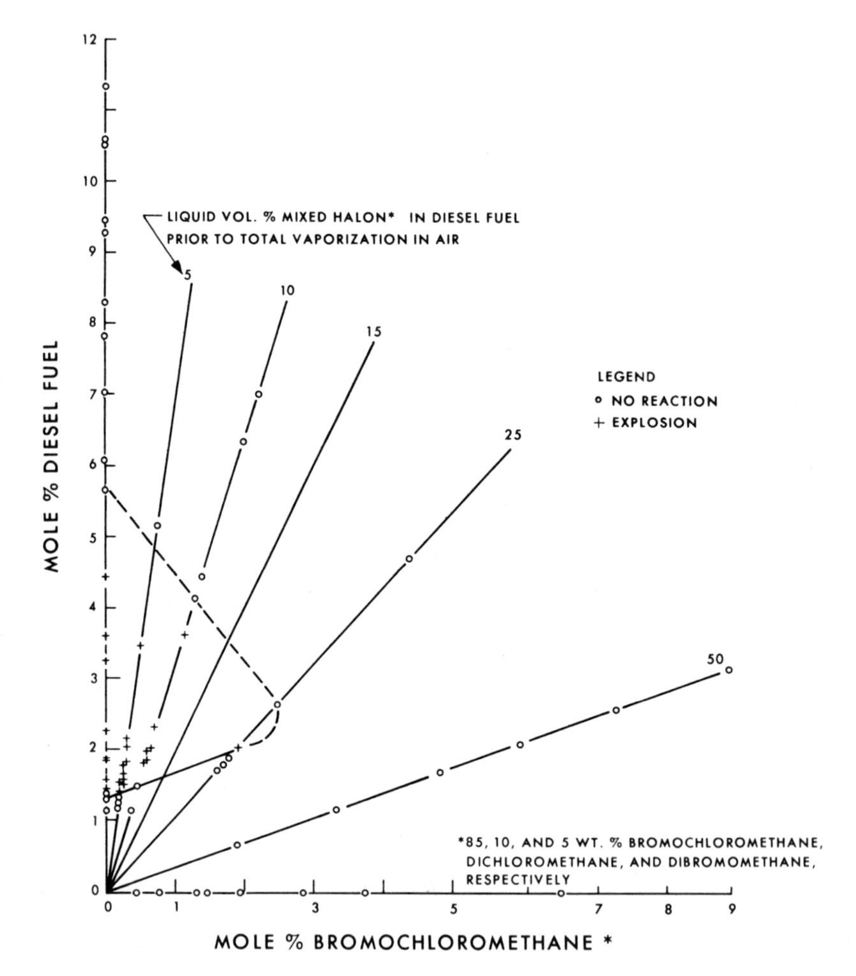

Flammability characteristics of diesel fuel bromochloromethane mixtures in air at 150 ± 10°C and 1 atm*

and nonflammable under others.

J. DEHN: The author gives as his most successful analysis of Purdue University results the approximate constancy of the ratio (heat capacity of inerts)/(total heat capacity) (Egn. 4). For n-heptane+air systems he finds that when the nitrogen+halon inerts contribute 75% to the heat capacity of the mixture then no flame can propagate. I submit that for the cases he has treated this ratio is basically (heat capacity N_2)/(heat capacity air) with observed variations introduced by the other components of the system and experimental errors in the data. For example, the specific heats (cal/mole-°K) of nitrogen and oxygen (and so of air) are equal to 7.0, while those of heptane and CF_3Br are 39.0 and 16.5 respectively at the initial temperature of the gas mixture. For peak values of 6.1% CF_3Br, 3% n-heptane and the rest air (author's Table III) we obtain

$$\frac{.061(16.5) + .73(7.0)}{.03(39) + .91(7.0) + .061(16.5)} \times 100 = 72\%$$

Actually the Purdue report (pp 65 and 146) gives 2.5% fuel instead of 3% which raises the above slightly to 73%. If we neglect the halon and the fuel we obtain

$$\frac{.73(7.0)}{.91(7.0)} \times 100 = 80\%,$$

the approximation we are using for the percentage of nitrogen in air. Similar calculations for the other compounds cited reveal the same basic reason for the cluster around 75%.

If only weight fractions are used without heat capacities

$$\frac{.061(149) + .73(28)}{0.3(100) + .73(28) \_ .18(32) + .061(149)} \times 100 = 77.1\%$$

If we neglect the halon and fuel we have

$$\frac{.73(28)}{.73(28) + .18(32)} \times 100 = 78\%, \text{ etc.}$$

The author observes another approximately constant weight ratio (agent)/(agent+fuel) and claims that somehow this demonstrates a purely physical mechanism. Why this should be so is not clear. At any rate, I submit that this ratio is basically unity (halon weight divided by itself) with deviations introduced mainly by the weight of the fuel and with other variations due to the fuel percentage found at a peak and experimental uncertainties. For example, all ratios involving n-pentane center around 0.89 while those involving the heavier n-heptane center around the value 0.83.

The author also claims that halons or hydrocarbon+halon mixtures become non-flammable in air when halogen atoms account for more than about 70% of the fuel+halon weight. The examples shown in the author's reference 4 do not convince one that this figure is a practical guide for avoiding fires since values as low as 63% and as high as 80% appear there. Moreover, CO_2 and N_2 which contain no halogen atoms fit weight fraction schemes elsewhere in the paper why shouldn't they also fit here? As for pure halons in air being non-flammable when halogen atoms make up at least 70% of their weight, we might refer to the Zabetakis report cited by the author (Bureau of Mines Bull 627, pp 102-5). Here data is given to show that CH_3Br (84% Br by weight), CH_2Cl_2 (84% Cl by weight) and C_2HCl_3 (81% Cl by weight) can be flammable in air at one atmosphere pressure or below, given a proper ignition source and container size. Similar matters are discussed in Bur. Mines Bull. 503, pp 101 ff. 70% by weight halon is a poor practical guide.

We may ask why that fraction of the total mixture heat capacity contributed by the inerts should have any particular significance. It would seem more reasonable to look for a critical heat capacity value of the total mixture if heat capacity is indeed the dominant factor. In purely thermal theories of flame propagation in pre-mixed gases at least two physical factors are usually taken to be important. These are the specific heat, c_p, and the thermal conductivity, λ. Simple theories lead to a flame velocity proportional to some power of the ratio λ/c_p, since a flame should propagate well when it is easy to transmit heat and raise the temperature of the unburned gas (at least for typical hydrocarbon+air mixtures in two-inch tubes). A good fire suppressant should decrease λ and increase c_p, leading to a decrease in flame velocity and eventual extinction. The Table below gives the average c_p and λ values for a stoichiometric n-heptane+air mixture and for Purdue peak percentage mixtures with various compounds. The N_2 peak mixture however was taken from the Zabetakis report already cited (p. 36) since the

Purdue work did not include N_2.

	c_p	λ	λ/c_p
n-heptane+air	7.6	56.5	7.43
+ 4.2% CF_2Br_2	8.1	54.9	6.78
+ 5.2% CH_2Br_2	8.1	54.5	6.73
+ 6.1% CF_3Br	8.6	54.0	6.28
+ 7.6% CH_2ClBr	8.1	53.4	6.59
+ 17.5% $CHCl_3$	9.1	49.6	5.45
+ 9.3% HBr	7.56	52.4	6.93
+25.5% HCl	7.47	50.3	6.73
+29.5% CO_2	8.1	51.0	6.30
+ 33% H_2O	8.2	50.7	6.18
+ 42% N_2	7.3	56.7	7.77

NOTE: λ units: 10^{-6}cal/cm-sec-deg
c_p units: cal/mole-deg

With three exceptions (N_2, HBr and HCl) c_p is increased compared to the fuel+air reference, while all except N_2 lower λ in this sample. The first thing we note is the fact that there is no critical value of c_p or λ or of the ratio λ/c_p which can be correlated with the phenomenon of a peak percentage. If we simply look at averages without paying attention to the type of deviations we might deceive ourselves into thinking that we have found critical values and proceed to argue that no chemical explanation is needed. This should be avoided. The brominated methane examples given have relatively small peak percentages in spite of the relatively high λ values of the mixtures and the contributions which some of these compounds might be expected to make to the heat of combustion. This suggests that for such compounds where dilution is much less important than for other compounds in the Table factors besides λ and c_p are at work. HBr and HCl are particularly interesting since they actually lower the average heat capacity of the mixture yet are more effective on a volume basis. Even on a weight basis HCl is more effective than CO_2. The fact that N_2 increases the λ/c_p ratio partially explains the jump between peak values of H_2O and N_2. We particularly note that N_2 is heavier than H_2O yet more is required. If we insert a water curve on Figure I of the paper under discussion, we see that a simple correlation of effectiveness with weight breaks down immediately. Finally we note that $CHCl_3$ is not especially effective in spite of its favorable λ/c_p ratio in the mixture. All of this leaves us without a correlation between effectiveness in this particular experiment and these two thermal parameters. We can either elaborate our physical explanation by including other factors or can turn to a partial chemical explanation (if indeed it is profitable to make such distinctions).

While I dispute the author's claim of showing there is only one physical mechanism at work in halon flame suppression, I do not necessarily deny his assertion. He may be right in some cases in spite of the fact that he has not proved his case here. My personal prejudice is toward the view that some combination of physical and chemical mechanisms is probably at work in most inhibition or promotion phenomena, with one or the other dominant perhaps in particular cases, although no one has yet established the conditions which must hold. While strong indications of at least partial chemical mechanisms have been presented at this symposium and elsewhere, there is still a lingering doubt since little attempt is usually made to evaluate possible physical factors which must also be at work. Similarly the present paper and my comments make no attempt to evaluate the chemistry of explosion burette work. It might be worthwhile to employ chemical analysis with isotopic labeling in such pre-mixed gas flame-propagation experiments as has been done in static or flow reactor systems which propagation distance is not a factor.

The chemistry of combustion inhibition is a difficult field which has opened up only in the past few years. Kineticists are to be congratulated for the factual light they have already shed on old theories. While the gulf between our present chemical knowledge and the goal of improved fire suppression is still quite wide, the practical importance of the goal is certainly worth the effort.

E. R. LARSEN: Dr. Dehn's critique of my paper is quite lengthy and detailed and requires a similarily detailed answer. It should be noted that Dr. Dehn's critique is strictly applicable to a paper (ref. 2) that I gave several years ago. In that earlier paper I had dealt in length upon the use of mass fraction as a means of expressing limit data, with only a cursory reference to the fact that expressing the composition on a heat capacity basis appeared to be superior. In the present paper the opposite tack was taken. In spite of this change Dr. Dehn, after hearing my paper, decided to stand by his critique.

Dr. Dehn's points are well taken and were considered during the analysis of the flammability limit data that is available in the literature. Unfortunately, I can not agree with his reasoning, nor with his conclusions.

Flammability limit, and peak, data which appears in the literature has been determined in a variety of systems and with a variety of ignition sources. As I previously stressed the Purdue study was carried out twenty, or so, years ago and employed a spark ignition source of unknown strength. From our experience spark sources employed during that period were not intense enough to ignite all flammable mixtures. It is probable that the limits of ignitability were being measured in the Purdue study; rather than the limits of flammability. This would result in an experimental flammability peak that would be somewhat lower than it should be. Such an effect is apparent when a comparison is made of the Purdue peak value for CF_3Br (6.1 vol %) and the CF_3Br peak (7.2 vol %) cited in Table IV. For this reason I have shied away from laying undue emphasis upon the peak value of any single agent. In spite of this problem the Purdue study is internally self-consistent and is large enough (46 agents) so that the trends drawn from it should be valid even though the individual peak values may be unreliable. It is partly for this reason that the differences between the values of KC_p for the inert gases and for the halons do not overly disturb me.

The reasoning employed by Dr. Dehn in critizing my analysis of the peak data is obviously invalid; both when the compositions are expressed on a mass fraction basis or on a heat capacity basis. To deny the validity of a direct conversion of vol % data to these other units, a simply mathematical manipulation, is to deny the validity of the volume % data itself. Dr. Dehn submits that for the cases in which air is employed as the oxidant, as in the Purdue study, KC_p is "basically (heat capacity N_2)/(heat capacity air) with observed variations introduced by other components of the system and experimental errors in the data." Dr. Dehn's ratio (heat capacity N_2)/(heat capacity air) is literally point D in figure 2. If the "observed variation introduced by the other components of the system and the experimental error in the data" are really this large then I submit that all of our studies are worthless, since the same argument can be employed when volume % data is employed.

Dr. Dehn's comments must also be viewed as invalid since they are applicable to all points bound by the triangle ADC in figure 2, and apply equally well to the inert gases and to the halons, including points laying near, but not on, side $\overline{AC}$, and certainly to all points laying along line $\overline{AD}$.

The weight ratio (agent)/(agent + fuel) was expressed in the current paper on a heat capacity basis rather than on a mass fraction basis as was done in the paper referred to by Dr. Dehn. The importance of this ratio is again obvious in figure 2, since this ratio simply represents any composition consisting solely of agent and fuel: those compositions laying along side $\overline{AC}$. As is apparent in figure 2 any mixture of inert gas and fuel which has a composition laying between points A and K_{IF} will, upon the addition of oxygen, eventually yield a flammable mixture, i.e., pass through the flammable envelope (IPJ).

One would anticipate that if the halons were disproportionately more effective than the inert gases they would show values for this ratio which are much smaller than is found for the inert gases. It might also be expected that this ratio would be much smaller with the bromine containing halons than for the chlorine or fluorine containing halons. That this is not the case is shown by the values of K_{IF}, i.e., the value of this ratio for peak compositions, shown in Table III. The constancy of this ratio is strong evidence in support of an argument that the primary mechanism of flame suppression is the same for both the inert gases and the halons, and, conversely, is difficult to rationalize away with a chemical mechanism.

The observation that the halons and halon-hydrocarbon mixtures generally become non-flammable in air when the halogen atoms account for more than 70% of the fuel = halon weight is dealt with in detail in reference 4. As pointed out in reference 4 the flammability peak composition is most interesting from a mechanistic standpoint since it represents that composition occurring at the point of convergence of the upper and lower limit curves. The peak composition represents the minimum amount of agent required to inert any and all fuel-oxidant mixtures. Therefore, the composition of the system at this point is unique with respect to the agent and fuel in each agent-fuel-oxidant system, and combinations of fuel and agents that occur at the peak can be treated as single entities. The halogen content of the mixture is fixed, and may be calculated, in much the same way that the halogen contant of a finite compound is fixed.

The halogen content for some 60 peak compositions were calculated and the mean halogen content of the mixtures was found to be 69.8 ± 3.5 wt %, i.e.,

$$\kappa = \frac{\text{wt \% Halogen in}}{\text{the mixture}} = \frac{\text{wt Halogen x 100}}{\text{wt Agent + wt fuel}}$$

to determine whether or not any one species of halogen was disproportionately more effective than any other, as demanded by the radical trap theory of flame suppression, the peak data for the 46 compounds included in the Purdue study were broken down by halogen type, and the mean values of κ for each group determined. One group (9 agents) contained only fluorine (κ=68.0), and a fourth and final group (5 agents) contained iodine or iodine and fluorine (κ=73). On the basis of this analysis, which is somewhat more complete in ref. 4, no evidence was found to support the contention that any one halogen was more effective than any other, and the relative effectiveness of the halogen is shown to be directly proportional to their atomic weights, i.e., F : Cl : Br : I = 1.0 : 1.9 : 4.2 : 6.7. The fact that one bromine atom is 4.2 times as effective as a fluorine atom on a molar basis is not supportive of the contention that they function by different mechanisms.

The fact that those halons and halon-hydrocarbon systems studied showed a mean value of 70 wt % halogen does not mean that there are no exceptions: three being cited by Dr. Dehn. It should be pointed out that while CH_3Br is flammable in air even though it contains 84 wt % Br, the Purdue study shows a peak percentage for the heptane-CH_3Br system than contains 73 wt % bromine. I would be remiss here if I did not turn Dr. Dehn's question about and ask how the chemical mechanism explains the flammability of CH_3Br in air in view of its high molar concentration of bromine.

I do agree with Dr. Dehn that employing 70 wt % halogen as a practical guide is poor policy, except for saying if the compound, or mixture of fuel and halon, contains less than 70 wt % halogen it should be rigorously tested before assuming that is non-flammable. In general, any flammability limit data obtained using an unknown spark source is potentially a booby-trap (ref. 2).

Dr. Dehn asks why the fraction of the total mixture heat capacity contributed by the inerts should have any particular significance. I do not have a good answer for this question at this time other than it seems to work. If we examine the peak data for the system pentane-air-agent we find that the agent concentration decreases as the total heat capacity of the mixture increases, but the ratio of the heat capacity contribution of the agent and the total heat capacity of the mixture remains reasonably constant. The N_2 in the air is properly part of the agent system and its heat capacity contribution is included in $\Sigma[I]Cp_I$.

	Agent (vol %)	$\Sigma[I]Cp_I$	$\Sigma[C]Cp_C$	KC_p %
Pentane	N_2 (42.2)	6.00	7.42	80.8
	CO_2 (28.0)	6.29	8.11	77.5
	CF_4 (20.6)	7.25	9.10	79.5
	SF_6 (15.7	8.12	10.19	79.7
	C_3F_8 (10.6)	8.62	10.69	80.6
	CF_4 (41.2)*	8.97	11.45	78.7
	CF_4 (46.8)**	8.67	11.02	78.3

* Oxygen enriched "air" - 31.3% O_2
** Oxygen enriched "air" - 38.7% O_2

On the basis of the behavior of the inert gases I see little reason to expect that a critical value of Cp will be found, unless this ratio is that value.

I am aware of the fact that in purely thermal theories of flame propagation in pre-mixed gases at least two physical factors are usually taken to be important: the specific heat (Cp) and the thermal conductivity (λ). As far as I can ascertain λ was originally proposed as being important by Coward and Hartwell in an attempt to justify the fact that helium appeared to be more effective than they thought it should be in view of its low heat capacity: especially when compared to argon which has the same heat capacity but a much lower λ value. They very high thermal conductivity of helium then was employed to rationalize the efficiency of helium.

In a previous paper (ref. 6), I noted that neither He nor Ar were significantly different than N_2, H_2O, or CO_2, when compared to a pure heat capacity basis. If the present thermal theories require that λ play an important role in flame propagation then I suggest that these theories might require rethinking.

If Dr. Dehn had extended his table to include more inert gases he would have observed that Cp for "peak" compositions can extend from at least 5.3 to about 13.5, $\lambda(10^{-6})$ can extend from about 42 to 295, and his "constant" ratio λ/Cp from about 5.5 to 55.

I agree with Dr. Dehn that looking at averages, especially with a limited data base such as employed by Dr. Dehn, can cause one to be decieved into thinking that we have found critical values and proceed to argue that no chemical explanation is needed. Conversely, one can be deceived by observing that because the halogenated agents undergo thermal cracking and enter into the flame chemistry this chemistry must be inhibitory in nature.

Proposed mechanisms of flame suppression in which the halogen behaves as a radical trap give rise to a dichotomy which is not generally dealt with. On the one hand, the halogens and their compounds are well known as effective flame suppressants with bromine occupying a prominent position with respect to efficiency. On the other hand, it is also known that bromine is especially effective as an effective oxidation, and hence combustion, promoter.

Is it not possible that the free radical chemistry that is reputed to result in flame inhibition might be the chemistry of flame promotion, or the initiation of the combustion at a temperature lower than it might otherwise occur in a given system.

It should be pointed out here that the data for KCp in Table III does have a bias built into it, and that the simple use of averages can be misleading. As mentioned in my oral presentation, but overlooked in the written manuscript, I ran a regression analysis of this data and found it to fit the equation:

$$\Sigma[I]Cp_I = 1.0\ \Sigma[C]Cp_C - 2.0$$

which tends to account for the fact that the value of KCp tends to increase as we descend the column.

This bias is associated with the fact that each agent-fuel-oxidant system has its own unique flammability diagram. If we look at the diagrams for a series of agents, such as was done in figure 3, we find that they form a family of diagrams having a common base (line $\overline{AB}$). For each different agent the apex C of the diagram will be evaluated above the plane of the page, so that our collection of diagrams open somewhat like the leaves of a book.

The diagram which occupies the page is that of N_2, and the points for the other agents are shown as projections upon the N_2 diagram.

I submit that the burden of proof for the support of a new and novel mechanism lies upon those who propose it. I do not deny that the halons enter into the flame chemistry and in most cases appear to act as fuels. Their behavior in premixed gas systems, such as employed in combustion tubes, is essentially indistinguishable from that of the inert gases.

It is for the reasons set down in this paper and in previous papers that I maintain that the primary role of the halons in flame suppression is as heat sinks, and that they have a common mechanism with the inert gases.

As a further note in proof, I would like to say that just prior to this meeting I requested Dr. R. V. Petrella of our laboratory to determine whether or not the composition 48.9 vol % CF_3Br, 7.2 vol % C_5H_{12}, and 43.8 vol % O_2, the whole at one atmosphere pressure, was flammable. Indeed, it proved to be. On a weight fraction basis the ratio

$$\frac{[CF_3Br] \times 100}{[C_5H_{12}]+[O_2]+[CF_3Br]} = 79.2$$

is reasonably close to the ratio found in the Purdue study for the system Heptane-Air-CF_3Br, which was criticized by Dr. Dehn above, i.e.,

$$\frac{[CF_3Br]+[N_2]}{[C_7H_{16}]+[O_2]+[N_2]+[CF_3Br]} \times 100 = 77.1$$

14

The Role of Ions and Electrons in Flame Inhibition by Halogenated Hydrocarbons: Two Views

EDWARD T. McHALE

Atlantic Research Corp., Alexandria, Va. 22314

ALEXANDER MANDL

Avco Everett Research Laboratory, Inc., Everett, Mass. 02149

(Editor's note: The question of whether charged species, known to exist in flames, are important in flame inhibition is a continuing matter of controversy. Dr. Edward T. McHale addressed the subject at the Symposium. The entire text of his presentation is not given here because essentially the same material, co-authored with Dr. D. Spence, appears elsewhere [Combustion and Flame (1975), 24, 211]. The following brief summary contains the essential points of the paper. Following it is a reply to Dr. McHale's paper by Dr. Alex Mandl.)

E. T. McHALE: Experimental measurements of the kinetic cross-sections were made as a function of temperature for seven alkyl halides (CH_2Br_2, CCl_4, CH_3Br, CF_3Br, $CHCl_3$, CH_3I, $CFCl_3$) and SF_6 for the electron dissociative attachment reactions

$$RX + e^- \rightarrow R + X^- \qquad (a)$$

The results of these measurements were compared with "peak percentage" values for the same compounds. Since peak percentages represent a measure of inhibiting effectiveness, one would expect to find some correlation between these and the kinetic rate constants if ions or electrons had a significant role in flame suppression. The complete lack of correlation that was found represents substantial evidence that charged species are not important in inhibition processes involving halogenated hydrocarbons.

Further evidence for the absence of ionic reactions in inhibition was also presented in the form of calculations of the relative rates of reaction (a) and reaction (b):

$$RX + H \rightarrow R + HX \qquad (b)$$

In a sample calculation for a stoichiometric methane-air flame with CF_3Br as the RX species, the ratio (b)/(a) was shown to be

$$\frac{\text{Rate Reaction (b)}}{\text{Rate Reaction (a)}} = 3 \times 10^2$$

The calculation was based almost entirely on experimentally measured values of the necessary parameters reported in the literature. This large calculated value of the ratio of neutral to ionic inhibition reaction rates reinforces the previous conclusion that combustion suppression is not a result of charged species in flames inhibited with alkyl halide compounds.

A. MANDL: Dr. McHale has commented on some work he has recently completed with Dr. Spence from which he has concluded that ions are not important in flame inhibition. Whereas under the assumptions and conditions of his mechanism one could conclude that ions do not play an important role, I think that a general statement implying that ion chemistry is unimportant in flame inhibition cannot be made.

In particular, if one starts with the same set of equations that Dr. McHale has presented,

$$RX + e^- \rightarrow X^- + R \qquad (1)$$

$$X^- + H \rightarrow HX + e^- \qquad (2)$$

$$RX + H \rightarrow HX + R \qquad (3)$$

where RX = inhibitor and X = halogen, and compares rates (2) and (3) instead of (1) and (3), one has a direct comparison between the effect of ions and inhibitors on the free radical, atomic hydrogen.

Thus

$$\frac{\text{Rate 2}}{\text{Rate 3}} = \frac{k_2[X^-][H]}{k_3[RX][H]} \qquad (4)$$

and since at flame temperature, typical rates for reactions (2) and (3) might be $k_2 \sim 2 \times 10^{-9}$ cm^3/molecule-s and $k_3 \sim 3 \times 10^{-12}$ cm^3/molecule-s.

$$\frac{\text{Rate 2}}{\text{Rate 3}} \simeq 10^3 \frac{[X^-]}{[RX]} \tag{5}$$

Typical inhibition concentrations reported at this meeting indicate that $[RX] \sim 10^{-2}$ mole fraction. Thus, for systems in which $[X^-] > 10^{-5}$ is a high equilibrium density, it should be pointed out that there are nonequilibrium paths for ion formation.

For example, my own shock tube work (1) on the cesium halides has shown that they collisionally dissociated almost completely into the nonequilibrium ionic branch. R. S. Berry and his coworkers (2) have shown that in general many of the alkali halides dissociate preferentially into the nonequilibrium ionic branch. This is due to the fact that the ionic and neutral potential curves cross a large internuclear separation for a large number of these salts.

In cases for which a salt such as KCℓ is injected into a flame its dissociation will be mainly into K^+ and $Cℓ^-$. If a percent of KCℓ is added to a flame it is not unreasonable to assume that this will result in $[X^-] > 10^{-5}$. This would imply that the inhibition in this case would be through an ionic mechanism.

J. C. Biordi et al have shown at this meeting that nonequilibrium chemistry is quite important in flames. Additionally, large nonequilibrium electron densities have been observed (3) in flames; and in experiments (4) in which alkali metal salts were added to flames, large increases in flame current were observed for many of the effective flame inhibitors. It is therefore not unreasonable to assume that ion chemistry can play an important role in the explanation of how some quenchers work.

Literature Cited

1. Mandl, A., J. Chem. Phys., (1971), 55, 2918.
2. Ewing, J. J., Milstein, R. and Berry, R. S., J. Chem. Phys., (1971), 54, 1752.
3. Fenimore, C. P., "Int. Encyclopedia of Phys. Chem. and Chem. Phys.," Topic 19, Vol. 5, "Chemistry of Premixed Flames," McMillan, New York, (1964).
4. Birchall, J. D., Combustion and Flame, (1970),

15

Theoretical Investigation of Inhibition Phenomena in Halogenated Flames

SERGE GALANT and J. P. APPLETON

Mechanical Engineering Department, Massachusetts Institute of Technology, Cambridge, Mass. 02139

Literature reviews on chemical inhibition (1),(2),(3) reveal that a great deal of experimental work has been carried out in an attempt to gain a better understanding of inhibition phenomena in flames. Since strong experimental evidence (2),(3),(4) supports the view that a basic understanding of inhibition phenomena can be obtained by studying inhibition effects in premixed laminar flames, researchers have studied inhibition effects in such systems. Yet, two somewhat different approaches have prevailed. One crude but efficient way of screening potential inhibiting agents has been to measure the decrease in nominal laminar flame speed as a function of the inhibitor concentration. (5),(6),(7), (8),(9),(10). Another related type of experiment has been concerned with the modifications of the ignition limits of a given mixture when inhibiting agents are introduced. (3),(6),(9). On the basis of such experimental evidence, a plethora of speculative kinetic models have been proposed (for exhaustive listings, see References (1),(2),(3)). A feature common to almost all these proposals is their lack of conclusive agreements when it comes to making "*ab initio*" predictions of the reduction of the laminar flame speed.

On the other hand, it can be argued that a better insight into inhibition can be gained by looking at flames on a molecular level (2). In particular, spectroscopic measurements (11),(12), (13),(14) have yielded a wealth of information by providing actual nonequilibrium concentration and temperature profiles. Most important, speculations have been quantified into models which identified some of the critical kinetic steps. However, the uncertainty in those measurements and the lack of detailed knowledge about the various nonlinear couplings between transport phenomena and chemical reactions which occur in flames, have impaired any conclusive description of the actual inhibiting effects.

Our theoretical approach is aimed at reconciling these two experimental approaches: using the rate coefficients and diffusion parameters as obtained from "molecular experiments" (11),

(12),(15),(16),(17), we intend to use a new numerical technique to solve the general one dimensional unsteady flame equations and, in turn, to predict the local and global behavior of inhibited flames, viz. temperature and concentration profiles but also, flame velocities and ignition limits. Such an approach is not new *per se*. Day *et al*. (6) solved approximations to the flame equations to study inhibition of rich hydrogen-air flames. In doing so, they were able to pinpoint the fundamental kinetic steps needed to account for experimental observations. Yet a proper adjustment of unknown reaction rates was required to match to the experimentally observed decrease of the laminar flame speed.

In the present investigation, a computing procedure which had been successfully applied to the ozone-oxygen flame (18) has been improved and generalized to be used for an automatic computer code which describes both ignition and steady propagation of deflagrations in planar and spherical coordinates. The calculations reported here have been obtained with a mathematical model which includes all effects that are believed to be important, i.e., molecular diffusion, heat conduction and thermal diffusion, together with a general reaction kinetic scheme and fully realistic thermodynamic data. A short overview of the equations and of the computational method to solve them is given in the first section of this paper.

As noted earlier, a first logical approach is to model inhibition phenomena in flames for which the kinetic, transport and thermodynamic data are fairly well established. The archetype is evidently the hydrogen-oxygen (or hydrogen-air) flame which is abundantly documented in the literature. In the second section we consider inhibition effects for a low pressure rich H_2-O_2 flame (19) supported on a porous plug burner. In so doing, we hope to demonstrate the capability of the numerical method to describe actual experiments. These calculations are the first in a series of simulations dealing with inhibition phenomena in hydrogen-oxygen air flames. Further work on ignition limits and flame velocity computations will be reported later.

Theory

The Mathematical Model The equations which describe the unsteady one-dimensional propagation of a premixed flame in a frame of reference fixed with respect to the moving flame front are (20)

mass continuity: $$\frac{\partial \rho}{\partial t} + r^{-\gamma}\frac{\partial}{\partial r}(r^{\gamma}\rho v) = 0$$

species continuity: $$\rho\frac{\partial n_j}{\partial t} + \rho v\frac{\partial n_j}{\partial r} = \frac{r^{-\gamma}\partial}{\partial r}(r^{\gamma}\rho G_j) + \frac{w_j}{m_j}$$

where $$G_j = D_j\frac{\partial n_j}{\partial r} - \frac{D_j n_j}{\bar{n}}\frac{\partial \bar{n}}{\partial r} - \frac{D_j^T}{\rho T m_j}\frac{\partial T}{\partial r}$$

$$\bar{n} = \sum_{j=1}^{m} n_j$$

energy equation:

$$\rho C_p \left\{\frac{\partial T}{\partial T} + v\frac{\partial T}{\partial r}\right\} = r^{-\gamma}\frac{\partial}{\partial r}\left(r^{\gamma}\rho\lambda\frac{\partial T}{\partial r}\right) + \frac{\partial T}{\partial r}\sum_{j=1}^{m} C^{*}_{pj}C_j - \sum_{j=1}^{m} H^{*}_j \frac{w_j}{m_j} \quad (1)$$

equation of state: $\rho = P/(RT\bar{n})$

overall mass conservation: $\sum_{j=1}^{m} w_j m_j = 0$

$$\sum_{j=1}^{m} n_j m_j = 1$$

$$\sum_{j=1}^{m} m_j G_j = 0$$

Here t is the time, r is the coordinate along the dimension of interest, v is the velocity component in the r direction, ρ is the local density of the flowing gas, n_j is the concentration of the j-th species in moles per gram of mixture, m_j its molecular weight, C^{*}_{pj} its molar heat capacity at constant pressure, H^{*}_j its molar enthalpy, D_j its molecular diffusion coefficient, $D_j T$ its thermal diffusion coefficient and w_j is the net mass production (or destruction) rate of the j-th species per unit volume. It has been tacitly assumed that species diffuse according to Fick's law, that heat transfer by conduction within the system obeys Fourier's law, λ being the heat conduction coefficient, and that the mixture consists of ideal gases at constant pressure P. Finally, by setting $\gamma = 0$, 1 or 2 the equations can be obtained in terms of cartesian, polar or spherical coordinates respectively.

Together with the set of equations (1) there must be a set of adequate initial and boundary conditions which complete the description of the system.

Two types of initial conditions have been dealt with in the past. For steady state flame structure calculations, guessed S-shaped profiles are assumed to describe the variations of the dependent variables (concentration, temperature) through the flame region initially, If enough energy is stored in these initial profiles, a flame will propagate and the profiles ultimately assume their steady-state shapes. Physically, this means that steady state flame propagation does not depend upon the way in which the system was ignited. Yet, such arbitrary initial conditions have no physical basis. In order to consider the ignition phase, it would be preferable to use initial conditions which were compatible with a more physical situation such as exposing "pockets" of combustion products of different dimensions to a large body of combustible mixture (18). More generally, initial conditions which are square (discontinuous) waves in concentrations and/or temperature might be thought of as representing the early phases of ignition (extinction) phenomena: indeed, they are approximate models of the large concentration and temperature gradients which initially do exist in, say, spark or laser ignited systems. Boundary conditions for steady flame propagation have been discussed at length. The problem is to specify internally consistent boundary conditions for an actual non-adiabatic multi-dimensional flame. For the calculations presented in the second part, we use the concept of a porous plug burner (20),(21) which is valid only for planar coordinates, i.e., $\gamma=0$. At the flameholder a temperature gradient is assumed to exist which tends to stabilize the position of the flame. Because of diffusion of the product species from the flame all the way back to the flameholder, the chemical composition of the gas at the flameholder is not specified. However, since the mass flux of the j-th species must be continuous at the cold boundary (r=0), we obtain

$$(n_j \rho^0 u^0 - \rho G_j)\big|_{r=0} = \rho^0 u^0 n_j^0$$

where the superscript ^{0} denotes conditions in the mixing chamber, and r = 0 at the cold boundary. Using the definition of G_j in (1) we get

$$\rho \left\{D_j \frac{\partial n_j}{\partial r} - \frac{D_j n_j}{\bar{n}} \frac{\partial \bar{n}}{\partial r} + \frac{D_j^T}{\rho T m_j} \frac{\partial T}{\partial r}\right\} = \rho^0 u^0 (n_j - n_j^0) \quad r=0,\ t > 0$$

Also, the temperature at the flameholder is supposed to remain constant with time. Finally, at the hot boundary ($r\to\infty$), it is assumed that the species diffusion fluxes and heat diffusion flux are zero. Therefore, the set of boundary conditions reads:

$$r = 0 : \rho\left\{D_j \frac{\partial n_j}{\partial r} - \frac{D_j n_j}{\bar{n}} \frac{\partial \bar{n}}{\partial r} + \frac{D_j^T}{\rho T m_j} \frac{\partial T}{\partial r}\right\} + \rho^0 u^0 (n_j^0 - n_j) = 0, \quad t>0$$

$$T(0,t) = T_o \qquad , t \geqslant 0$$

$$r\to\infty \quad \frac{\partial n_j}{\partial r} = 0, \quad \frac{\partial T}{\partial r} = 0, \quad t > 0$$

Before searching for solutions of set (1), further information is needed about the functional form of the diffusion coefficients, heat capacities, enthalpies, and chemical production terms. These modelling questions are considered in the Appendix.

Computational Method. It is only recently that numerical techniques to describe one-dimensional premixed laminar flames have become available. Spalding (21) and Dixon-Lewis (22) developed similar relaxation methods where the unsteady flow equations are solved for arbitrary initial conditions. Although computation times seem quite reasonable, trial and error must be used to determine an economical stable step size, because the equations are linearized locally in time. Alternatively, the steady state equations and their corresponding eigenvalue--the adiabatic flame velocity--may be solved via shooting methods quasi-linearization or finite difference techniques (23), (24). The latter method can become very time consuming, whereas, the former two present difficult instability problems in both the backward and forward integration because of the sensitivity of the numerical solutions to the guessed boundary conditions (19).

From this brief glance at already existing methods to solve the flame equations, it can be recognized that the choice of a numerical technique is governed by trade-offs between computation time and problems of stability, convergence and accuracy. Since one of our objectives was to develop a generalized computer program to provide the user with a fast and reliable technique which could analyze a wide range of physical situations, it was our belief that none of the methods discussed above fulfilled all the requirements for a general user package. The technique presented now is the so-called "method of lines" which was first applied to one dimensional flame propagation by Bledjian (18). Basically, instead of approximating the partial derivatives with respect to all the independent variables by finite different expressions, one can convert the original set of parabolic partial differential equations (1) to a system of ordinary differential equations by discretizing only the space variable, the time being left continuous. In our case, for a system of m reacting species, the system of m parabolic nonlinear differential equations, m-1 species continuity equations and 1 energy equation , is transformed into a set of m N nonlinear differtial equations where N is the number of grid points in the spatial direction: the initial conditions to solve this system are provided by the initial conditions applied to the original partial differential equations.

Basic numerical advantages characterize this method

- once the problem is reduced to an initial value problem, very efficient numerical methods of integration are available, with a computation cost proportional to the dimension of the system (mN)
- unsteady problems can be dealt with in a general fashion in particular, the modelling of ignition and/or extinction phenomena becomes feasible.
- flames in spherical or cylindrical coordinates can be treated with no major inherent difficulty. The immediate

shortcomings of the technique stem from the particular nature of the resulting set of differential equations.

- Since rapid changes within the flame occur over a very limited spatial domain, a large number of discretized points N is needed to accurately describe those changes. In case of an insufficient number of grid points, strong instabilities which most often yield negative concentrations appear readily.
- On the other hand, it has been shown with the help of simple linear parabolic equations (25) that, as the number of grid points N is increased, the "stiffness" of the differential equations becomes very critical, i.e., smaller and smaller time step sizes are needed to obtain stable and accurate results.
- Moreover, since the original equations are already "stiff" due to the presence of nonlinear chemical source terms, this immediately suggests the use of an implicit "stiff" method of integration such as Gear's code (26). But then the cost of computation is no longer proportional to the dimension of the system, m N: it varies as $(mN)^2$ or even $(mN)^3$ because the Jacobian matrix of the nonlinear system must be inverted when a loss of accuracy is detected.

Most of the numerical problems inherent to the solution of the flame equations in their original form were overcome by transforming both the dependent and independent variable. The use of the von Mises transformation (21) together with a mapping (27) bringing the infinite or semi-infinite domains $]-\infty, +\infty[$, $[0,\infty]$ into $[0,1]$ allowed for a non-uniform grid size in the r-direction and improved the stability characteristics of the resulting set of equations. The stiffness of the equations was further reduced by using the new dependent variables

$$\tilde{\tilde{n}}_j = \log n_j \qquad j=1,\ldots m$$

$$\tilde{\tilde{T}} = \log T$$

which also eliminates the problem of dealing with negative concentrations. The discretization of the spatial derivatives was accomplished by using a non-central five-point difference scheme (28). Such a scheme allows for a sensible reduction in the number of spatial grid points needed to follow the flame history within a given accuracy. Finally, a slightly modified version of the variable step size variable order integration method devised by Krogh (29) was implemented: it leads to stable, accurate and relatively inexpensive calculations for the set of equations considered above.

Numerical Simulations

Motivations. The flame which was chosen is a low pressure, rich, H_2-O_2 flame, for which temperature and concentration measurements exist (19). Although such a flame is not a very good approximation to the burning of inhibited polymers in an atmospheric environment, it was selected to

i) demonstrate the ability of the computational scheme to reliably describe an actual burner type experiment,

ii) pinpoint eventual improvements to be made in the modelling (kinetics, transport, properties,...),

iii) obtain a general description of inhibition phenomena when the chosen flame is seeded with halogenated compounds (HBr, Br_2 ...),

iv) check the influence of thermal diffusion which had been considered as a secondary effect but which is claimed to be important under some experimental conditions (14).

Within this context, no attempt was made to critically evaluate the parameters which are needed to describe both uninhibited and inhibited flames. Rather, the results shown here should be considered as a check for the validity of our flame model and a first step toward a basic understanding of inhibition phenomena in gaseous mixtures.

Results

H_2-O_2 flame. The present predictions have been obtained using a set of rate constants listed in Table I (from there on, referred to as Flame 1 : 75% H_2, 25% O_2).

TABLE I

Reaction Mechanism and Reaction Rate Data
(a) Reaction Mechanism and Forward Rate Constants
Reaction Rates are expressed as $K = AT^c \exp(-E/RT)$ with k in $cm^3\ mole^{-1} sec^{-1}$ ($cm^6\ mole^{-2}\ sec^{-1}$ for three body reactions), T in K and E in kilocalories

Flame 1

Reaction No.	Reaction	A	c	E	Ref.
1	$OH+H_2 = H + H_2O$	2.9×10^{13}	0.	5.155	(30)
2	$H+O_2 = OH+ O$	2.74×10^{13}	0.	16.815	(30)
3	$H_2+O = H + OH$	1.74×10^{13}	0.	9.459	(30)
4	$O + H_2O = OH+ OH$	5.75×10^{13}	0.	18.016	(30)
5	$H+OH+M = H_2O+ M$	3.0×10^{15}	0.	-1.49	(31)
6	$H+H +M = H_2+ M$	2.04×10^{16}	-0.31	0.0	(31)
7	$H+O +M = OH+ M$	4.6×10^{15}	-1.75	-0.777	(31)
8	$H_2+O_2 = OH+ OH$	2.1×10^{14}	0.	58.133	(30)
9	$H+O_2+M = HO_2+ M$	1.5×10^{15}	0.	-0.993	(31)
10	$H + HO_2 = OH+OH$	$7. \times 10^{13}$	0.	0.	(30)
11	$HO_2 + O = OH+O_2$	$1. \times 10^{13}$	0.	0.	(30)
12	$OH + HO_2 = O_2+H_2O$	$1. \times 10^{13}$	0.	0.	(30)
13	$H + HO_2 = H_2+O_2$	6.3×10^{12}	0.	0.	(30)

(b) Third Body Efficiencies. For Reactions (6) and (7) all third bodies are taken to be equally efficient. For reactions (5) and (9), the following relative efficiencies are used, all others being set equal to 1.

Reaction No.	H_2O	H	References
5	5.	1.	(30),(31)
9	16.25	2.5	(30),(31)

The species considered were H_2, O_2, H_2O, H, O, OH, HO_2. H_2O_2 was set aside because of its numerically proven minor importance, and the lack of thermodynamic data above 1500°K. The computation of the transport coefficients is detailed in Appendix A. The predicted and experimental values of temperature are compared in Figure 1a whereas in Figure 1b the heat release rate along the burner is plotted. The agreement is fair at low temperature: the deviations from experimental values might be due to modelling inaccuracies when computing the thermal conductivity and the thermal diffusion coefficients. The heat loss by conduction to the burner is equal to 0.062 $cal.cm^{-2}\ s^{-1}$, whereas, Eberius et al. found it equal to 0.15 $cal.cm^{-2}\ s^{-1}$ (19). At higher temperature the agreement is very poor; this can be explained by experimental heat losses along the burner which tend to flatten the temperature profile. However, since this is a low pressure flame, the recombination of radicals in the hotter regions is extremely slow: hence, the high temperature profiles which are displayed have not reached their steady state values (19).

In Fig. 2, the predicted and measured concentrations profiles for H_2, O_2, H_2O, H and OH are compared, together with the theoretical values displayed in Ref. 19 for a model without thermal diffusion effects. The agreement is improved and emphasis should be put on the influence of thermal diffusion over most of the reaction zone. (See Figure 4a--dotted lines--to compare the relative magnitudes of the various H atom fluxes.) The importance of those effects is illustrated by the fairly good agreement between experimental and predicted values for the H and OH concentration at low temperature. This also demonstrates that the contribution from thermal diffusion effects cannot be correctly estimated on the basis of steady-state flux calculations alone, as it has been done before (22). In fact, the preflame region is characterized by strongly nonlinear feedback effects which are hardly predictable without an adequate numerical treatment.

Finally, it should also be mentioned that polarity effects of the water vapor molecule, which have not been considered in those calculations, can account for the discrepancy in the concentration profiles over the cold part of the flame. They may change,

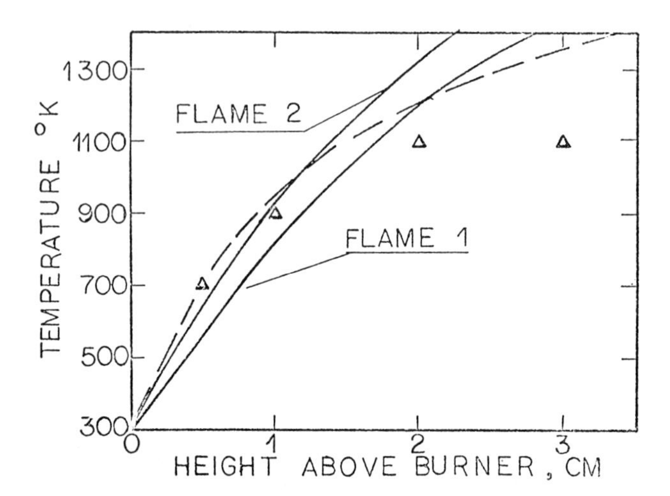

Figure 1a. Temperature distribution above the burner

△ *Experimental points taken from Ref.* 19

——— *Solution of the flame equations taken from Ref.* 19

——— *Solution of the flame equations for Flames 1 and 2, including thermal diffusion*

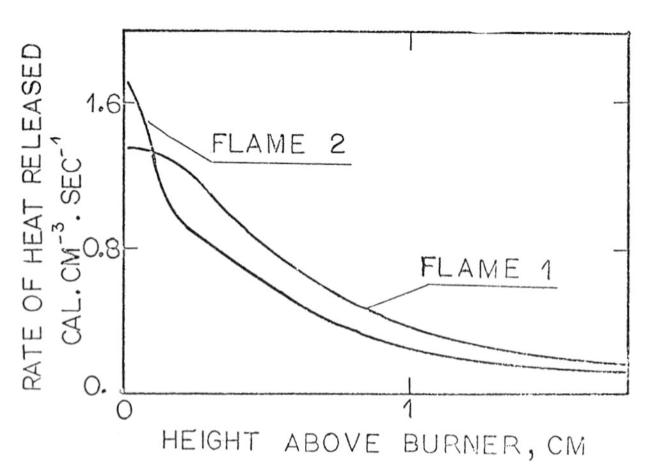

Figure 1b. Variation of heat release rate through Flames 1 and 2

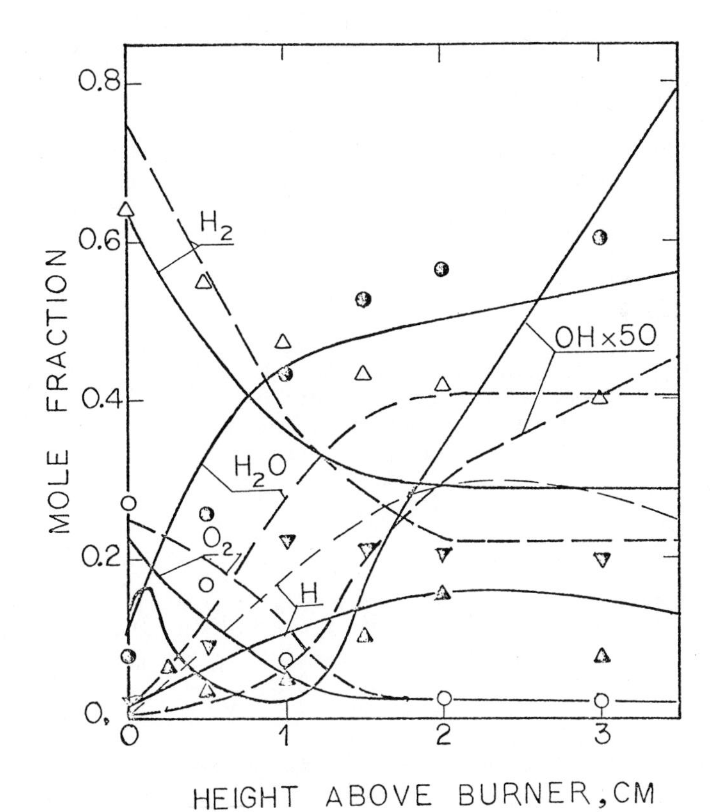

Figure 2. Species concentration distribution above the burner
△, H_2; ○, O_2; ●, H_2O; ▼, H; ▲, OH: Experimental values taken from Ref. 19
——— *Solution of the flame equation taken from Ref. 19*
——— *Solution of the flame equations for Flame 1 including thermal diffusion*

for instance, the location of the inversion temperature of the H_2O thermal diffusion coefficient. Of course, errors at high temperature arise from the artificially high temperature distribution (see OH and H_2O profiles) and the fact that the experimental H_2 concentration profile was not corrected for recombination of H atoms in the sampling probe.

$H_2 - O_2$ - HBr flame. Considering again the same low pressure $H_2 - O_2$ flame, 5% HBr per volume of the original mixture was added to the premixed cold gas. Although some reaction time calculations about an equivalent system have been published (32), no experimental data is available for comparison. This numerical experiment was therefore intended to provide a basis for a preliminary investigation of typical effects characterizing inhibition. To the original kinetic scheme describing Flame 1 was added a set of 5 reactions involving Br, Br_2, HBr as listed in Table II. (From thereon referred to as Flame 2)

TABLE II

Reaction Mechanism and Rate Data for Bromine Kinetics. Notation as in Table I. All third bodies are taken to be equally efficient.

Flame 2

Reaction No.	Reaction	A	c	E	Ref.
14	Br + Br + M = BR_2 + M	5×10^{14}	0	0	(6),(32)
15	H_2 + Br = HBr + H	1.4×10^{14}	0	19.573	(6),(32)
16	H + Br_2 = Br + HBr	4.5×10^{14}	0	2.980	(6),(32)
17	H + Br + M = HBr + M	3.2×10^{15}	0	0	(6),(32)
18	HO_2 + Br = O_2 + HBr	1.7×10^{12}	0	0	(6),(32)

In Figure 1b, the heat release rates for Flame 1 and Flame 2 are compared. In both flames, the major contributions to the heat release rate in the preflame region are due to reactions involving H and HO_2. For Flame 1, Reactions 9, 10, and 1, respectively, play major roles, whereas in the early part of Flame 2 the Reaction H + HBr $\rightarrow$ H_2 + Br contribute about 30% of the total heat release rate. Hence, for the same cold gas flow rate a higher heat transfer rate by conduction to the porous plus is needed to stabilize the flame (0.75 $cal.cm^{-2}$ sec^{-1}). If Figures 1a and 1b are combined to show the heat release rate of various temperatures a delayed start of the important heat-releasing reactions can be observed, although this effect is not as marked as it could be expected from similar computations (6). Such a feature has already been noticed experimentally in halogen inhibited flames (12),(33),(34). In Figure 4, the H concentration

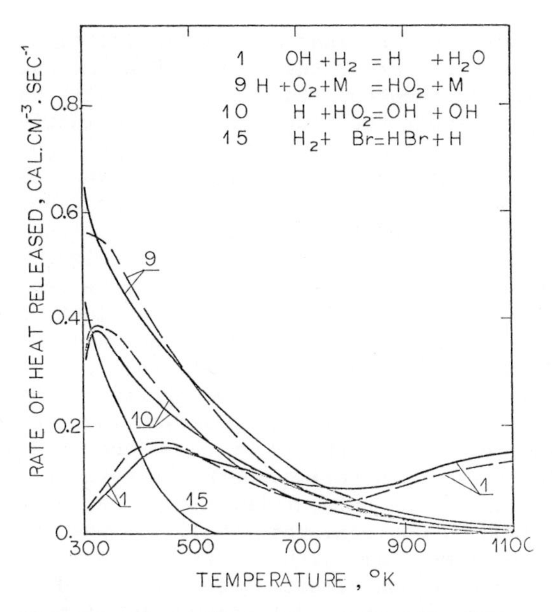

Figure 3. Heat release rate as a function of temperature along the burner for some important reactions

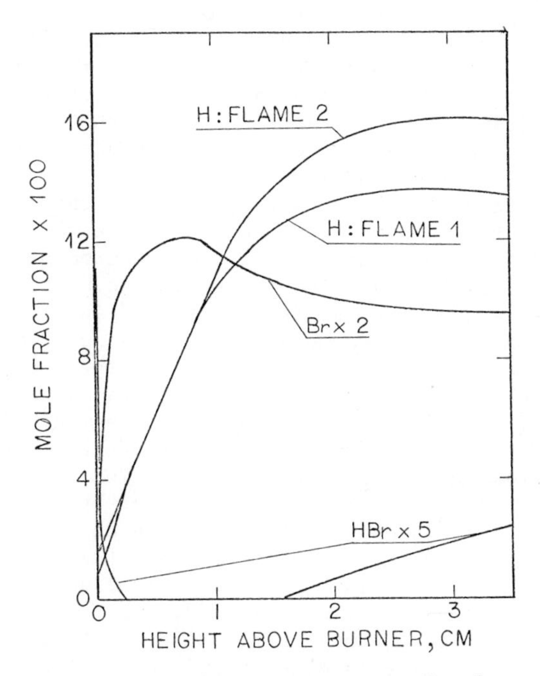

Figure 4. H and Br concentration profiles above the burner for Flames 1 and 2

profiles for both the original and the inhibited H_2-O_2 flames are compared. As expected, the maximum H atom concentration is markedly depressed when the flame is seeded with HBr. In the very early region near the plug, HBr is immediately attacked by the H atoms diffusing back to create Br atoms. The maximum Br concentration is reached around 900°K where the reactions become branching. The competition between reactions involving bromine containing compounds and the preponderant reactions describing the hydrogen-oxygen flame is illustrated in Figure 5. The overall rates of the major reactions listed in Table I and the reactions listed in Table II are plotted against distance along the burner. Two kinetically distinct zones are revelated which have already been discussed by Fristrom and Sawyer (3). At low temperature H atoms are scavenged by the inhibiting species HBr which is rapidly depleted via Reaction 15. The depletion of the H atom level is furthered by three-body reactions such as 6 and 9. The transition region where the rate of the branching reaction 2 balances the rate of the recombination reaction is reached at T=900°K, whereas, this condition was attained at T=800°K for the same flame without HBr. This increase in the reaction temperature has already been considered as characteristic of inhibition phenomena (3). Above this temperature, chain carrying reactions such as 1 dominate to set an almost constant level of H atoms which will diffuse back into the colder parts of the flame. In this region, the level of Br atoms is also almost constant and this favors scavenging of H atoms via Reaction 17 which regenerates HBr. (See Figure 4).

However, the high concentration of Br atoms--and therefore of all radical species--over a large portion of the flame, is a direct consequence of the artificially high temperature in this region. A more realistic solution should accentuate the decay of those radical species.

It is worthwhile noting that all 5 reactions of Table II contribute to the maximum Br concentration level. This is clearly indicated when the overall bromine atom production rate is considered (Figure 5). The change in the hydrogen flux when the flame is seeded with HBr is illustrated in Figure 6. The presence of brominated scavengers reduces significantly the overall flux of H atom diffusing back into the preflame region, which in turn leads to a decrease of the reaction zone thickness. Also, another important conclusion is the negligible influence of thermal diffusion on the overall flux of Br atoms for this particular premixed low pressure flame. The level of Br atoms along almost the whole burner length is set by chemical reactions alone. However, this cannot be considered as a refutation of early experimentally supported assumptions favoring thermal diffusion effects in laminar diffusion flame as a basic contribution to chemical inhibition. Indeed, for hydrogen-air or hydrocarbon air flames, the distribution of the thermal diffusion fluxes is drastically changed and the extension of these

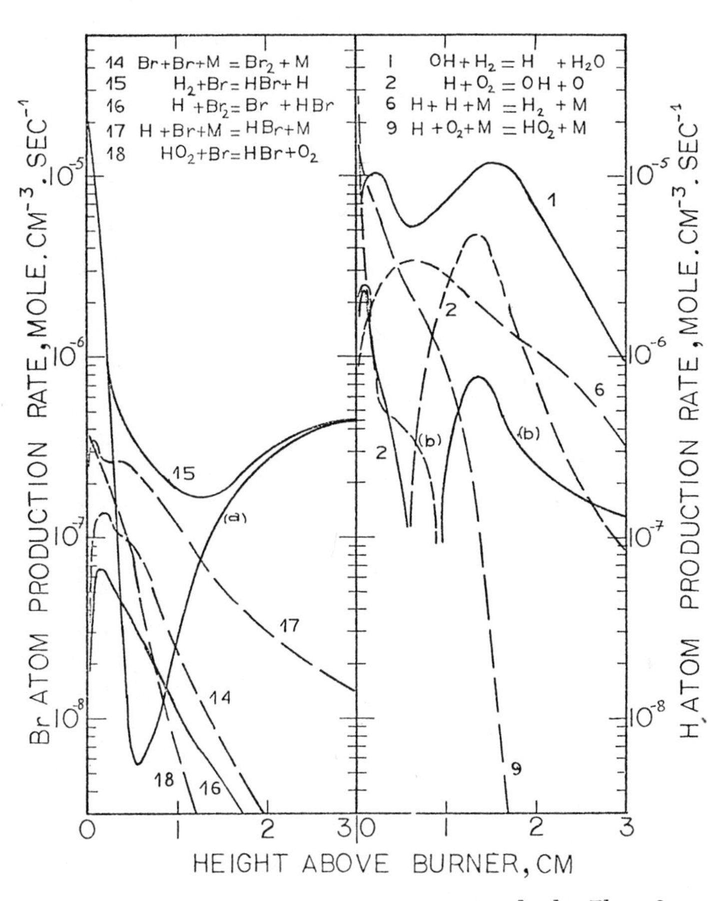

Figure 5. Net rate of important reactions involved in Flame 2

————	*destruction of Br or H atoms*
————	*production of Br or H atoms*
(a)	*overall production of Br atoms*
(b)	*overall production and destruction of H atoms*

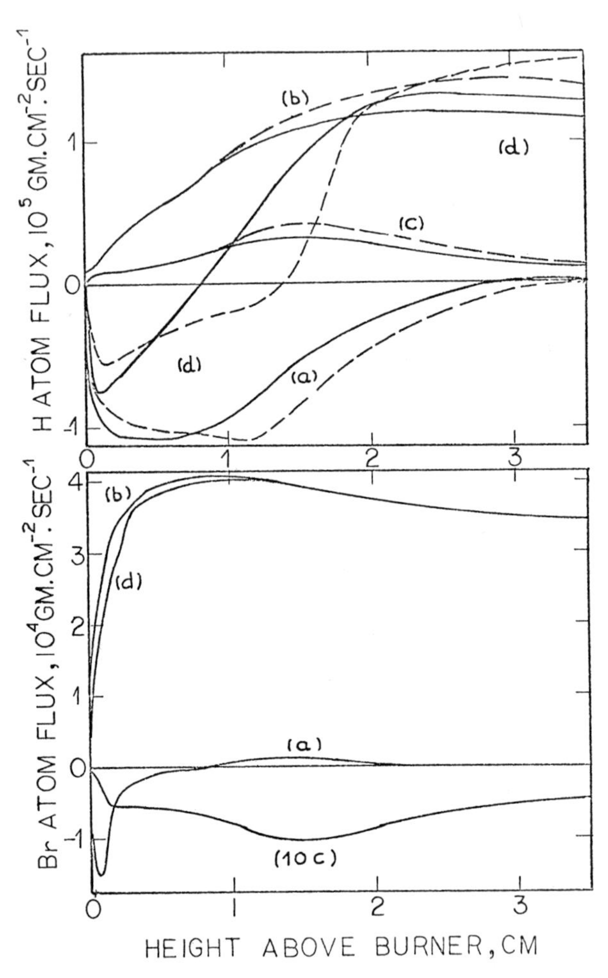

Figure 6. Computed fluxes of atomic hydrogen and atomic bromine for Flames 1 and 2

————	*Flame 1*
————	*Flame 2*
(a)	*Molecular diffusion flux*
(b)	*convective flux*
(c)	*Thermal diffusion flux*
(d)	*overall flux = (a) + (b) + (c)*

results to laminar diffusion flames is not clear, since higher temperature gradients are expected in such flames.

Conclusion

A new computational technique which is able to solve the one-dimensional unsteady propagation of a premixed flame under a wide variety of boundary and initial conditions has been successfully applied to the description of uninhibited, rich, H_2-O_2 flames supported on a burner. It is believed that further improvement of the transport property modelling should lead to good agreement with experimental data. Moreover, the introduction of thermal diffusion effects yields a satisfactory description of the interplay between chemistry and transport phenomena in the preflame region.

For inhibited flames, some basic experimental observations characteristic of inhibition phenomena in halogenated flames have been reproduced at least qualitatively, i.e., reduction of the reaction zone thickness, increase of the reaction temperature and a delayed start of the heat releasing reactions. However, the assumption of thermal diffusion effects has been ruled out, at least for low pressure laminar pre-mixed rich H_2-O_2 flames.

Subsequent modelling work will be geared toward the matching of laminar flame speed measurements and some quantitative explanation of the modification of ignition limits by halogenated species.

Appendix: Thermodynamic, Transport and Chemical Kinetic Model

In order to accurately describe the flame structure, the use of thermodynamic, transport and kinetic data which describe as realistically as possible the experimental system are needed.

The thermodynamic data were taken from the JANAF Tables (35) using our own least-square fit as a function of temperature to compute varying enthalpies and heat capacities at constant pressure throughout the flame region. The thermal conductivity of the mixture was computed via the Mason-Saxena approximation (36) as a function of the local temperature and composition. The thermal conductivities for each of the pure monatomic substances were evaluated as in Ref. 14, whereas, in the case of polyatomic gases, Svehla's semi-empirical formula (38) was used. Pure gas thermal conductivities calculated in this way agree with experimental data to within 10 to 20 percent over the whole temperature range, for a Lennard-Jones 6-12 potential, except for strongly polar species such as H_2O. Similarly, the multi-components diffusion model employed is that suggested by Hirschfelder *et al.* (37). The validity of such a model has been discussed for mixtures of gases at high temperature by Samuilov and Tsitelauri (40). A comparison of the thermal diffusion ratios for binary mixtures found in this way, with very limited low temperature experimental data listed by Hirschfelder *et al.* shows agreement within 20%.

The final piece of information needed to describe a specific flame structure is the set of rate constants for the chemical reactions which are considered to occur. For simultaneous reactions involving m species of the form

$$\sum_{j=1}^{m} \nu_{\ell j}^{+} n_j \underset{K_\ell^-}{\overset{K_\ell^+}{\rightleftarrows}} \sum_{j=1}^{m} \nu_{\ell j}^{-} n_j$$

the net mass production rate of species j, for unit volume, is given by

$$w_j = m_j \sum_{\ell=1}^{n} (\nu_{\ell j}^{-} - \nu_{\ell j}^{+}) K_\ell \rho^{\sum_{j=1}^{m} \nu_{\ell j}^{+}} \prod_{j=1}^{m} n_j^{\nu_{\ell j}^{+}} - K_\ell \rho^{\sum_{j=1}^{m} \nu_{j\ell}^{-}} \prod_{j=1}^{m} n_j^{\nu_{j\ell}^{-}}$$

where the forward and backward rate K_ℓ^+, K_ℓ^-, for the ℓ-th reaction are related through the detailed balancing relationship

$$\frac{K_\ell^+}{K_\ell^-} = k_\ell$$

k_ℓ being the equilibrium constant for the ℓ-th reaction. The rate constants are usually given by equations of the form (see Tables I and II).

$$K_\ell^{\pm} = A_\ell^{\pm} T^{c_\ell^{\pm}} \exp(-E_\ell^{\pm}/RT)$$

Acknowledgment

This research has been supported by the National Bureau of Standards through Grant Number NBS3-9006.

Literature Cited

1. Creitz, E.C., Journal of Research NBS (1970), 74A, p. 521.

2. Hastie, J.W., Journal of Research NBS (1973), 77A, p. 733.

3. Fristrom, R.M. and Sawyer, R.F., AGARD Conference Proceedings #84 on Aircraft Fuels, Lubricants and Fire Safety (1971).

4. Friedman, R., Fire Research Abstracts and Review, (1971), 13, p. 187.

5. Miller, D.R., Evers, R.L. and Skinner, G.B., Comb. and Flame (1963), 7, p. 137.
6. Day, M.J., Stamp, D.V., Thompson, K. and Dixon-Lewis, G., "Thirteenth Symposium (International) on Combustion," The Combustion Institute, Pittsburgh (1971), p. 705.
7. Homann, K.H. and Poss, R. Comb. and Flame, (1972), 18, p. 300.
8. LeBras, G. and Combourieu, J., J. Chimie Phys. (1974), 71 #74, p. 465.
9. Butlin, R.N. and Simmons, R.F., Comb. and Flame (1968), 12, p. 447.
10. Fiala, R. AGARD Conference Proceedings No. 84 on Aircraft Fuels, Lubricants and Fire Safety, (1971).
11. Hastie, J.W., Comb. and Flame (1973), 21, p. 49.
12. Wilson, W.E., O'Donovan, J.T. and Fristrom, R.M., "Twelfth Symposium (International) on Combustion," The Combustion Institute, Pittsburgh (1969) p. 929.
13. Pownall, C. and Simmons, R.F., "Thirteenth Symposium (International) on Combustion," The Combustion Institute, Pittsburgh (1971), p. 585.
14. Lerner, N.R., and Cagliostro, D.E., Comb. and Flame (1973), 21, p. 315.
15. Hart, L.W., Grunfelder, C. and Fristrom, R.M., Comb. and Flame, (1974), 23, p. 109.
16. Poulet, G., Barassin, J., LeBras, G. and Combourieu, J., Bull. Soc. Chim. France (1973) #1, p.1.
17. Barassin, J. and Combourieu, J., Bull. Soc. Chim. France (1973), #376, p. 3173.
18. Bledjian, L., Comb. and Flame (1973), 20, p. 5.
19. Eberius, K.H., Hoyermann, K. and Wagner, H. GG. "Thirteenth Symposium (International) on Combustion," The Combustion Institute, Pittsburgh (1971), p. 713.
20. Williams, F.A., "Combustion Theory," Addison-Wesley, Reading, Mass. (1965).
21. Spalding, D.B. and Stephenson, P.L., Proc. Roy. Soc. (London) (1971) A324, p. 315.
22. Dixon, Lewis, G., Proc. Roy. Soc. (London) (1967), A298, p. 495.
23. Wilde, K.A., Comb. and Flame, (1972), 18, p. 43.
24. Dixon-Lewis, G., Proc. Roy. Soc. (London)(1970), A317, p.235.
25. Lambert, J.D., "Computational Methods in Ordinary Differential Equations," John Wiley & Sons, New York (1973).
26. Gear, C.W., "Numerical Initial Value Problems in Ordinary Differential Equations," Prentice-Hall (1971).
27. Lavan, Z., Nielsen, H. and Fejer, A.A., Phys. of Fluids (1969), 12, p. 1747.
28. Hicks, J.S. and Wei, J., JACM (1967), 14, p. 549.
29. Krogh, F.T., Paper presented at the Conference on the Numerical Solution of Ordinary Differential Equations, the University of Texas at Austin (1972). Also in lecture notes in Mathematics, (1973), 363, p. 22.

30. Stephenson, P.L, and Taylor, R.G., Comb. and Flame (1973), 20, p. 231-244.
31. Dixon-Lewis, G., Greenberg, J.B., and Goldsworthy, F.A. "Fifteenth Symposium (International) on Combustion," The Combustion Institute, Pittsburgh (1974), (In press).
32. Lovachev, L.A., Babkin, V.S., Bunev, V.A., V'Yun, A.V., Krivulin, V.N., and Baratov, A.N., Comb. and Flame (1973), 20, 259-289.
33. Wilson, W.E. "Tenth Symposium (International) on Combustion," The Combustion Institute, Pittsburgh (1965), p. 47.
34. Fenimore, C.P. and Jones, G.W., Comb. and Flame (1963), 7, p. 323.
35. Stull, D.R. and Prophet, H., "JANAF Thermochemical Tables," Second Edition, NSRDS-NBS37 (1971).
36. Mason, E.A. and Saxena, S.C. Phys. of Fluids (1958), 1, p. 361.
37. Hirschfelder, J.O., Curtiss, C.F. and Bird, R.B., "Molecular Theory of Gases and Liquids," 2nd Ed., John Wiley & Sons, New York (1964).
38. Svehla, R.A., NASA Technical Report R-132 (1962).
39. Fristrom, R.M. and Westenberg, A.A., "Flame Structure," McGraw-Hill (1965).
40. Samiulov, E.V. and Tsitelauri, N.N. Teplofizika Vysokikh Temperatur (1969), 8, p. 1174.

DISCUSSION

N. J. BROWN: The OH minimum has been observed for other hydrogen-rich flames. This can be explained in terms of two competitive reactions: one which generates OH: $H + HO_2 \rightarrow 2OH$, the other which consumes OH: $H_2 + OH \rightarrow H_2O + H$.

S. GALANT: The rate constants used by Eberius *et al.* (19) are those recommended by K. Schofield, Planet. Sp. Sci. (1967), 15, p. 643, viz:

$$(10)\ H + HO_2 \rightarrow OH + OH \quad K_{10} = 6.62\ 10^{13}$$

$$(1)\ H_2 + OH \rightarrow H_2O + H \quad K_1 = 3.79\ 10^{13}\ e^{-5490/RT}$$

both in $cm^3\ mole^{-1}\ sec^{-1}$. They compare very well with those listed in Table I. Since Eberius *et al.* did not observe any OH minimum using a model in which thermal diffusion effects were excluded, it is believed that thermal diffusion effects might explain the occurrence of the OH minimum in the cold region of the flame. Indeed, thermal diffusion tends to markedly depress the H_2 concentration which in turn slows the removal of OH radicals and the creation of H atoms via Reactions (1) and (10). However, a sensitivity analysis for the rates of the above reactions has not been carried out and we agree that they are critical when it comes to describing the OH kinetics over the entire flame region together with the chain branching reactions

$$H + O_2 = OH + O$$

$$H_2 + O = H + OH.$$

N. J. BROWN: In the Eberius *et al.* paper, they demonstrated that the flux (for H and H_2) due to thermal diffusion was nearly equivalent to that due to molecular diffusion. This flux calculation is very dependent on one's choice for the thermal diffusion coefficient and, in addition, one's choice for its temperature dependence. Could this have affected your results?

S. GALANT: The thermal diffusion model used in our calculations is detailed in Hirschfelder et al.(37). The thermal diffusion coefficient for water vapor exhibits an inversion temperature around T = 350°K. This clearly indicates that the thermal diffusion coefficients in the cool part of the flame are very sensitive to one's choice for the potential parameters which describe the various interactions among species. We are currently improving the model by including polarity effects due to the H_2O molecule. This should result in a more realistic description of all transport properties for the species involved.

R. M. FRISTROM: Did you take into account the heat transfer to the burner surface by H atom recombination?

S. GALANT: No surface recombination reactions were considered in our calculations although provisions have been made in the computer code to include them.

R. M. FRISTROM: What are the computational requirements for the H_2-O_2 flame illustrated?

S. GALANT: 20 grid points were taken to cover the spatial domain. About 300 Kilobytes of core space on an IBM 370/168 were needed to carry out the integration. Since the numerical technique is of a relaxation type, the computation times are very much dependent upon one's choice for the inital profiles. For the H_2-O_2 flame, the CPU time amounted to about 178 seconds whereas 425 seconds were needed for the inhibited flame. Such a difference is explained by the fact that the stable step size for the H_2-O_2 -HBr flame is about 4 times smaller (0.687 μsec) than that for the clean H_2-O_2 flame (2.76 μsec). Also, 10 species including Br, HBr, Br_2 are needed to describe the inhibited flame against 7 species for the clean H_2-O_2 flame.

J. C. BIORDI: Do you predict formation of Br_2 close to the burner?

S. GALANT: Our model predicts a maximum in the Br_2 concentration (2.1×10^{-6} mole fraction) which is located 0.1 cm away from the burner. Such a maximum can be kinetically explained by the competition between the two reactions (see Figure 3).

(14) $Br + Br + M = Br_2 + M$

(16) $H + Br_2 = Br + HBr$

J. C. BIORDI: How do your results compare with Dixon-Lewis for $H_2/O_2/N_2$/HBR flames? i.e., would you also conclude from your studies that termolecular reactions are necessary to account for the observed reduction in burning velocity with HBr?

S. GALANT: Figure 5 clearly shows that the Br concentration overshoot is adequately described only when termolecular reactions such as

$$Br + Br + M = Br_2 + M$$

$$H + Br + M = HBr + M$$

are taken into consideration. Such a conclusion was reached by Day _et al_. (_6_) and also by Lovachev _et al_. (_32_). More recent work dealing with H_2-O_2-N_2-HBr flame (_6_) seems to indicate that reactions such as

$$HO_2 + Br = HBr + O_2$$

$$OH + HBr = H_2O + Br$$

might also play a significant role in the inhibition mechanism.

GENERAL DISCUSSION

(Editor's Note: Following the formal presentations and the discussion specific to each, an open end forum was conducted. The approximately one hour period was taped and transcribed. The participants were permitted to edit their comments for structure and clarity, and relatively little was excised by them or the editor.)

R. G. GANN: This is, I hope, the time slot in which we earn our keep. I would like to get comments relevant to the questions: "What do we know now that we didn't know yesterday?" and "Where Do we go from here?" The whole idea has been to bring together not only the people who are presenting the papers, but other people who are either interested in the field, who are getting into the field, or who have worked in the field. Now we have to tie together the wide variety of information that we've heard in the last couple of days. I will entertain comments from the floor.

R. L. ALTMAN: Based upon the paper that Joan Biordi gave, in which she observed very little change in the concentration of stable species in low pressure flames upon addition of the halon, her work suggests that however the halon acts, it must be by some recombination reaction which is pressure sensitive. The very fact that halons put out fires in ambient air and she doesn't observe any great changes in concentration of the same stuff at low pressure further supports this argument that the basic reaction must be pressure sensitive. That's probably the most important thing that their study has shown: that halon flame inhibition is concentration dependent.

J. C. BIORDI: I think you've said the right thing for the wrong reason. In any general mechanism of flame inhibition, the recombination reactions will have to be included in the model. I think really what this is saying is that these are complicated systems involving complicated chemistry, even without anything that interferes with that chemistry, and those things that interfere with the chemistry cause additional complications. And we're going to, in the end, rely upon a good modeling technique such as, I think and I hope, Serge Galant has presented, coupled with experimental data to really pin this down.

G. B. SKINNER: Getting back to Dr. Altman's comment, one thing that has impressed me from several of the talks is how rapidly hydrogen atoms diffuse in low pressure flames. I am sure this idea is involved in the answer to his comment - the hydrogen atoms are moving back and reacting with CF_3Br, and also with CH_4, in quite a low temperature region of the flame. So as I see it, the inhibitor in the Biordi flames and the ones Dr. Fristrom is working with was not reacting in the recombination area, but the reverse of that, rather well ahead of the flame. The inhibitor was reacting with various species in relatively cool gas. Perhaps the reason the inhibitors appeared to have only a moderate effect on the combustion process is that reactive species are continually being produced in the reaction zone to replace those that diffuse upstream and react.

R. M. FRISTROM: I have some general comments. We're an extremely diverse group, and I think our principal weakness is that do we do not all talk the same language. Practical workers are interested in putting out fires, and if a correlation exists which allows practical predictions, they're interested in this, not in the detailed chemistry. I would add one point: when you must worry about people and what happens to them in fires, chemistry is important. But I think that one of the lessons we can learn from this meeting is that we need more interchanges of this kind. We need a special effort from a chemist to summarize for the practical workers and the fluid dynamicists what we, as chemists, know that will help them. We need a fluid dynamicist to tell chemists what we need to known in terms we can understand. We need a practitioner to tell us what the problems are. Each one of us has his own special vocabulary, and it seems to me that a contribution that each one of you might think about is what questions did you come away from here about the other guy's baliwick. Based on a lot of the questions asked here, much of our problem is the language barrier.

E. R. LARSEN: I would like to make several comments at this point. During my talk I was trying to point out that from a practical standpoint we're trying to determine just what is involved in converting a flammable gas mixture into a non-flammable mixture.

If one examines a flammability limit diagram such as my figure 2, and imagines for the sake of this discussion that it is expressed in mass units, several things can be seen. In the pentane-oxygen-nitrogen system the amount of fuel present at all points along the lower limit curve, $\overline{FJ}$, is about 45 mg/l. This amount of fuel represents a heat producing capacity of about 0.3 - 0.4 calories per milligram of the total mass of the system. This is essentially true for all of the hydrocarbons, and appears to represent the minimum amount of heat required to off-set heat losses from all causes. Methane is exceptional in that it requires only 33 mg/l to produce the required 0.3 - 0.4 cal/mg of total mass.

At the point where the stoichiometric curve, $\overline{C_{st}C}$, intersects the "air" curve, $\overline{AD}$, we have about 75 mg/l of pentane present and have about the maximum energy production per milligram of total mass. This amount of fuel represents nearly 0.67 cal/mg of total mass of the system, or about twice the energy content of a limit mixture. At the experimental peak (P) produced by adding N_2 to the system the amount of fuel present has dropped to about 62 mg/l. The concentration of O_2 required to maintain flame propagation also drops drastically, from 267 mg/l to 152 mg/l, or from 8.0 moles of O_2/mole of fuel to 5.5 moles of O_2/mole of fuel.

However, if a heavy inert gas, for example SF_6, is added as the agent the amount of fuel needed at the peak in order to have flame propagation increases to about 88 mg/l. The oxygen demand is about 223 mg/l, or 5.7 moles of O_2/mole of fuel.

A "typical" halogenated agent, when added as the agent, causes the fuel demand at the peak to jump to ∿130 mg/l, and the oxygen demand of the system to increase to about 240 mg/l, or ∿6.0 moles of O_2/mole of fuel. The energy production here is still about

0.3 - 0.4 cal/mg of total mass.

Now, you might say that in a practical sense, "I don't really care what the chemistry is, since it would appear that if the energy producing ability of a gas mixture is greater than 0.3 - 0.4 cal/mg of total mixture the gas will allow flame propagation to occur, regardless of the chemical nature of the mixture." From a practical standpoint this is a relatively easy statement to make and is likely to be reasonably correct for gas mixtures at 1 atm and ambient temperatures. However, under other conditions and with other compounds, for example - ethers, such an approach is dangerous.

What I would like to see developed is a computer program, based on the chemistry of the system, where the composition of a mixture could be plugged in and predict whether or not the mixture is capable of flame propagation.

It is my understanding that in order to accomplish this it is necessary to know not just the initial composition of the mixture, but also the final products of the combustion processes. In this event the study of flame chemistry becomes extremely important.

Another problem area where flame chemistry appears to be extremely important is smoke generation. In lean flames smoke production is nearly non-existent, but when one adds a flame suppressant to the mixture the production of smoke becomes a problem. This problem is becoming increasingly important in flame-retarding plastics, and some information on the manner in which the flame suppressant modifies the flame chemistry would be of great value.

C. HUGGETT: I'd like to add a bit to Professor Skinner's remarks on the importance of the hydrogen atom. We've seen numerous examples of its role in the inhibition process in the last couple of days. This is a situation that occurs when you have strong concentration and thermal gradients in the flame, and it would seem this should account for some of the differences we observe when you add the inhibitor from the fuel side or from the air side. This may provide the link between the practical question of how you apply the inhibitor to the flame and the chemistry of the inhibiting agent. Another consequence is that the importance of hydrogen atom reactions is probably exaggerated because of the high H-atom diffusivity. Dick Gann pointed out this morning that there are a number of other reactions involving O atoms and other radicals that should be important in the inhibiting process. Going back yesterday to the paper from Berkeley on what was considered to be a well-stirred reactor, in this situation, these other reactions could conceivably be equally important; and I think that looking at inhibiting reactions in a well-stirred reactor might be very helpful in understanding the mechanism of the inhibition process.

R. G. GANN: I'd like to comment on the relation of the studies we're doing to the real situation. In using halon systems to suppress existing, fully-developed, hot, Class B fires, the halon is left virtually intact. Only small amounts are actually destroyed in the fire, yet the fire goes out. In the studies that are covered here, we have the opposite situation: total consumption of the suppressant with no flame extinction. I think this merits a significant amount of consideration, and obviously some of our studies could be oriented in that direction. This is more the type of problem that the fire protection person has to worry about.

R. M. FRISTROM: Basically we're modelling an unsuccessful attempt at putting out a fire.

J. C. BIORDI: One of the questions I wanted to ask when Charlie Ford was giving his talk was this. I have seen a film, (it may not have been a DuPont film, in fact I think it was a Fenwal film) where a torch was thrown into a model room. Then the fire detection system triggered and out spewed CF_3Br. At the rate at which it came out, it looked to me like it was blowing the fire out, as opposed to extinguishing it chemically. I think this is part of the question Dick Gann has raised. In the actual application, is the fire extinguished because of aerodynamics primarily?

C. L. FORD: It doesn't seem to be primarily due to aerodynamics, although it may be fair to say that aerodynamics plays a part in it. In some instances, a very rapid discharge, for example flooding maybe a 5000 cubic foot room in 1 second, will extinguish a fire which if you discharged the agent over 10 or 20 seconds might not be extinguished. I think the aerodynamics really plays a minor role rather than a major one in the extinguishment of fires.

J. C. BIORDI: Would you recommend the use of 1301 on a fully developed fire?

C. L. FORD: It depends on what you mean by "fully developed." If it's a pool fire thats been burning for a minute or two, so that the fuel has reached an equilibrium condition, yes, I would. If it's a house burning that's developed to a point where the roof has caved in, no, I wouldn't. I think we have to define the term "fully developed."

R. G. GANN: That's why I qualified my comment with "Class B."

E. R. LARSEN: Two years ago, when I presented a paper on this subject, I discussed an experiment that we had carried out. We found that when a flaming Nujol-soaked wick was inserted into a mixture of air and various halogenated agents, including CF_3Br, C_2F_5Cl, C_3F_8, and C_5F_{12}, the concentration of agent necessary to produce a 2 second extinction time was essentially independent of the nature of the agent. It was noted that the use of too little agent (<2-3 vol %) actually prolonged the extinction time considerably beyond that found when no agent was employed. The extinction time without agent was about 12 sec, as against extinction times of ∿22 sec when 2 vol % of CF_3Br was present and ∿36 seconds when 1.8 vol % C_5F_{12} was present.

At short extinction times little agent decomposition occurred while with long extinction times large amounts of agent are decomposed. If the brominated agent causes flame extinction by undergoing decomposition to liberate hydrogen bromide, or bromine, and these agents are the active flame suppressant that it is hard to understand why the combustion process is so prolonged under conditions of extensive agent decomposition.

This behavior seems to fit the pattern found in real systems and is difficult to explain in terms of chemical kinetics.

N. J. BROWN: I would like to comment about the so-called mechanism of flame inhibition. A real fire sitution is likely to be most similar to a turbulent diffusion flame. Premixed laminar flames represent another type of interaction between chemistry and fluid dynamics and a well-stireed reactor represents a third. All three combustion configurations will probably exhibit different mechanistic properties. Investigators who are attempting to elucidate the mechanism of flame inhibition should examine the results of inhibition experiments performed in different combustion configurations. Elucidating this complex mechanism will require the co-ordinated effort of chemists, fluid dynamicists and fire practitioners.

E. R. LARSEN: In this particular system, using a diffusion flame, the concentration of agent required to extinguish that flame is somewhat less than that required to inert a premixed system.

J. W. HASTIE: Continuing on with Nancy Brown's comment, I think one of the problems the purists have in this flame inhibition business is that we don't know enough about the chemistry of a real fire to be able to say what laboratory conditions we should set up for model studies that would be relevant to real fire conditions. For example, we need to have information on the character of these real fires from people with an idea of the distribution of chemical components and temperature. This would help us in setting up model experiments.

E. R. LARSEN: There is another point that I would like to make about premixed gas systems. One can take a fuel-oxidant mixture which is not flammable, in that it lies outside of the lower flammable limit, and by adding some halon 1301 convert it into a flammable mixture.

In much of the chemistry that I have been listening to, speakers have mentioned both promoters and inhibitors. In the case that I just cited I would argue that halon 1301 is a promoter. In other cases involving slightly different compositions halon 1301, it is argued that halon 1301 is an inhibitor. I am inclined to believe that much of our differences are semantic, rather than real.

W. G. MALLARD: I have a question indirectly addressed to this discussion. When the wick goes out, does it smoke at all?

E. R. LARSEN: When the halon is present at too low of a level the smoke becomes very thick. However, when enough halon is used to give a short (<2 sec) extinction time, little smoke is observed.

W. G. MALLARD: This is a question I don't know about

since the chemistry I've done is in very clean systems. Whenever we get anything dirty, you shy away from it. But something Joan Biordi showed earlier makes me ask: is it possible that one of the roles of the halon is simply to create smoke which has got to be a better radiator and therefore eliminate a lot more of the heat. It seems like this has to be taken into account if you're going from just a flame which can't radiate as much heat as smoke can. The smoke has got to be a lot blacker body than a flame.

E. R. LARSEN: One area of concern which no one has mentioned is what happens in upper limit, or rich, systems. In rich mixtures it appears that the flame is supported primarily by combustion of the hydrogen portion of the hydrocarbon and leaving the carbon portion to form tremendous amounts of soot. In a real fire situation the soot itself, once it has cooled to room temperature, should be a tremendous fire extinguisher, or flame suppressant, because of its high heat capacity and radiative properties.

W. G. MALLARD: It's got to do that at high temperatures. It's not going to radiate at low temperatures.

N. J. ALVARES: I've had an opportunity to work with a lot of smokey fuels and observed that when we add 1301 either in sub-extinguishing concentrations or in concentrations that will extinguish the fire, the smoke production is reduced. The character of the smoke is certainly changed. We will change the color and apparent density of the smoke, but we certainly don't visually increase the amount of smoke.

J. C. BIORDI: I'd like to get back to this business that was raised initially, that is, the toxic products generated from the decomposition of halogenated hydrocarbons in a real fire extinguishing situation. We have't talked about that at all, and it's something, at least in my one experience of this sort, the industry people have not wanted to hear about. Does the man in the DuPont ad hold his breath while he makes those comments?

C. L. FORD: No. He's a well-trained actor, but he doesn't hold his breath.

J. C. BIORDI: How come you don't get HF?

C. L. FORD: You do get HF, definitely. You get some HBr, possible some free bromine. The relative proportions of these things depend, to some extent, on the fuel, and the amount you get depends on things like the size of the fire and how quickly you put it out. In the ad, the fire is extinguished very quickly. There were quantities of HF and Hbr formed, but they were not analyzed. From other data we've obtained, I would guess the HF level after the fire was in the range of 25-50 ppm. This would be typical, I think, in an industrial fire-extinguishing application. We try to have a system designed in such a way that the post-extinguishment atmosphere will not contain more than, say, 50-100 ppm as a maximum. These are very irritating levels of HF, but for short exposures of a few minutes, they're not particularly harmful in a physiological sense. This is not to say, Joan, that the amount of HF that you would measure immediately above the flame is not considerably higher, maybe of the order of a percent or two, maybe higher than that. I don't know. But when that is diluted throughout the room in a total-flooding-type application, it's quite low. Hot HF is a fairly active species, and it can find its way throughout the room very quickly, and it does appear to be well distributed at a fairly uniform concentration and fairly low in situations when the fire is not too terribly big.

J. C. BIORDI: This is why I was asking about fully-developed fires earlier on. Except, the point is, I haven't seen any of these measurements.

C. L. FORD: Some of the data has been reported in the Academy of Sciences Symposium given in Washington in 1972.

W. H. MCLAIN: One's interest in the quantity and distribution of combustion products in a real fire often depends on the application. For industrial applications property damage is more important. For domestic or habitable space applications toxicity factors are more important.

It is difficult to accurately sample and measure HF, HCl and HBr in real fires. One reason is they react with water to form liquid particulates resulting in low gas phase concentrations. Although the toxicity of the liquid particulates is usually less than the gases, a low measurement for the concentration of HF based on gas phase sampling does not necessarily imply that there is no hazard to life.

From a practitioner's point of view, a fire protection engineer is more interested in the quantity of agent needed to extinguish a fire than in detailed gas kinetic mechanisms. Most real fires have non-ideal three dimensional geometries with a number of hot surfaces which act as reignition sites. The effectiveness of the agent often depends on physical processes associated with the delivery of agent to these fire sites. Thus a fire in a wall cavity and a fire in an open area of an industrial plant represent two quite different agent distribution problems.

R. F. KUBIN: I thought that I would mention that there is a difference even between inhibiting agents and quenching agents. Not much attention was given to that; in our talk, we didn't have time. We have the work of Rosser in which he puts small amounts of hydrocarbons in hydrogen flames and notices tremendous reductions in the flame speed. Yet on the other hand, we obviously know that you don't quench a hydrogen flame by adding a hydrocarbon to it. In our experiments, we found that, in a premixed flame where the hydrocarbon is added to the flame with enough air so the hydrocarbon can participate in the combustion, the amount of inhibitor needed to quench that flame is increased by the addition of hydrocarbons to the hydrogen flame.

F. A. WILLIAMS: I wish to amplify a little on what I think Bill McLain was heading toward. It seems to me that we haven't emphasized very much the logistical or fluid mechanical or heat transfer aspects of the problem, and these could well be very important. From what has been brought out so far at this meeting, it is conceivable to me at this stage that it might have been better for the Navy to install nitrogen rather than CF_3Br in engine rooms. It may be that in most practical applications, a fluid dynamical process is more important than the chemical methods we have been considering. In flooding the fire, you're mainly interested in cooling it quickly. The fluid dynamical contribution could be controlling. It may be that all of the chemistry is not very relevant to the practical situation. On the other hand, the kinds of tests that have been described here suggest that the chemistry is very important in suppressant evaluation. I wonder, though, how close to practical situations these "applied" tests are. Perhaps even the most "applied" tests that we've heard about are, in some respects, more idealized than they should be. I don't have any feeling for the answer to the question at this stage, but it seems to me that perhaps it's something we should consider further.

R. G. GANN: I think the reason we're talking about halons rather than nitrogen or CO_2 is that unless you're luckly enough to be able to blow out the fire with one of those two, the amount that you've realistically got to dump into the room is essentially going to preclude anybody in that room from getting out. If he's lucky, he might make it to the wall where there might happen to be an air supply, but barring that, the chance of getting out is going to be quite slim.

A. S. GORDON: I can't agree with you on that, Dick. We haven't scaled much, but the indication is that at about 15 mole percent of oxygen, a hydrocarbon flame will go out, and humans can survive in that environment.

R. G. GANN: That assumes rapid and uniform mixing. What size fire?

A. S. GORDON: So far we haven't put out very large ones.

R. F. KUBIN: I would like to amplify what Forman Williams had to say on the use of CF_3Br in ship engine rooms. Flooding the engine room with a halon agent has other consequences. The engine room is a highly ventilated place and supplies the oxidizer stream to the boilers. With CF_3Br in the atmosphere corrosion problems will be manifold. We have observed extensive corrosion in the hoods we use for testing due solely to the products of CF_3Br. With regard to liquid nitrogen, we have extinguished a 4 x 4 foot JP-5 pan fire, given a 1-1/2 minute preburn, in 8 seconds with very crude equipment and an application rate of 3/4 pound of liquid nitrogen per square foot.

R. G. GANN: There is a patent, by the way, on the use of liquid nitrogen to suppress fires (C. D. MacCracken, #3,438,445). We found this out in the course of our own extensive work on the technique of nitrogen pressurization as a technique for fire suppression in gastight spaces.

A. S. GORDON: With CF_3Br, the scale factor to big ones from little ones in the laboratory is not bad. We can say the same thing goes for nitrogen.

R. G. GANN: The reason I mention the nitrogen problem is, as you're well aware, we are scaling some nitrogen suppression, and it doesn't scale simply.

R. F. KUBIN: We find in our laboratory experiments that, on a weight basis, CF_3Br is twice as effective as nitrogen. But on the other hand, CF_3Br costs us $6/pound while the liquid nitrogen costs us 4 cents/ pound. You can't beat the economics.

C. L. FORD: Who are you buying your CF_3Br from?

R. F. KUBIN: It's in laboratory quantities in a little bottle, and we got it from Dow.

C. HUGGETT: I think this is a good point for me to get in my annual plug for CF_4.

R. G. GANN: For which Clay holds a patent.

C. HUGGETT: It's much more effective than nitrogen; it's much more stable than 1301 so that if you introduce it into a fire you get very little HF and, of course, no HBr, and it has a very low toxicity. I think we're discussing different types of situations here, and I think we should distinguish between closed systems and open systems. In an open system, you can use 1301 and then you can get out. We are now proposing to use these materials in closed systems - aircraft cabins, subway cars, submarines - where people are going to have to wait out the end of the fire incident.

A. S. GORDON: Our results say that CF_4 is, on a molar basis, about 2.5 times as effective as nitrogen. Then, of course, you can make the argument that nitrogen can be collected from the atmosphere if you're aboard a ship, while you have to have tanks of CF_4.

C. HUGGETT: I agree, there are a lot of logistic and engineering problems, but I would like to see someone make a thorough analysis as to the tradeoff in these systems. Another point that seems to be coming up recurringly here is that the way in which the agent is introduced into the fire is very important. Someone mentioned the relative inefficiency of halons in practical situations. If you have a fire in a room you almost have to fill the room to a critical concentration of the agent, yet its only the agent that gets into a very thin active flame zone thats effective. The rest of it might just as well not be there, except that we don't know how to introduce it only into the active flame zone. The other important point affecting efficiency is the rate of application and I think this is apparent from the simple flame propagation experiments that Charlie Ford described yesterday morning. If we have a long tube with a premixed gas with a little bit of inhibitor in it, below the critical range, the flame will propagate the full length, and we will generate toxic products from all the agent that is present in the sample. Now if we could concentrate that agent in the first 10% of the tube length to give 10 times the concentration, the flame would not propagate and we would have very little toxic products. It seems to me that the chemistry of all these agents we've been discussing is not all that different. The problem is, how do you get them to the active site in the most efficient way. This, of course, depends on such things as the boiling point, the density, and the design of the application system. As far as I can see, this is where we can make big advances in efficiency. I don't think we can do a heck of a lot with the chemistry of the halon extinguishants.

R. F. KUBIN: I would like to mention an observation from our work with CF_4. Before we went to all quartz systems, we found that CF_4 etched Pyrex chimneys just as rapidly as CF_3Br, indicating fairly substantial amounts of HF. This was true for both diffusion and premixed flames where we slowly approached the quenched point. Two possibilities for the HF formation are H abstraction from CF_4 or dissociation of the C-F bond in the hot post flame zone, the latter being less favorable energetically. In either case the CF_3 radical is generated and now the ensuing chemistry would be much like that from the use of CF_3Br.

As an aside I would mention that the white solid deposit on the walls of Pyrex chimneys, as noted by some authors when using CF_3Br, is sodium bromide, presumably from the attack of hydrogen bromide on the sodium borosilicate glass.

C. HUGGETT: CF_4 is not a very efficient extinguishing agent. It should be considered for inerting applications to prevent the occurrence of a flame. No flame - no HF.

W. H. MCLAIN: We have this continuing systematic problem here of what do you do about a fire, a real fire as opposed to what happens in a burner. If I'm going in after a real fire, I'm going to get in there, I'm going to have a high agent concentration, I'm going to put it out. I'm not going to run it for a long period of time. My temperatures are going to be relatively low, the reaction time is low.

R. F. KUBIN: I am going to deny part of that, Bill. Because of the way the Navy has set up its fire fighting system, as was described in the paper by Norman Alvares, when there is a compartment fire, one of the first things they are going to do to put the fire out is local application of an agent, for example, water, AFFF, $KHCO_3$, or CF_3Br. If this is not successful, they are going to get out. Now they have a hot fire; they are going to seal it off and flood with CF_3Br. Under these conditions you have a large amount of flame.

AUTHOR INDEX

SUBJECT INDEX

D

G

H